BECOMING A WILDLIFE PROFESSIONAL

Becoming a Wildlife Professional

EDITED BY **Scott E. Henke**
TEXAS A&M UNIVERSITY–KINGSVILLE

and

Paul R. Krausman
UNIVERSITY OF ARIZONA

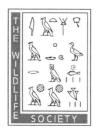

Published in affiliation with The Wildlife Society

JOHNS HOPKINS UNIVERSITY PRESS | BALTIMORE

Johns Hopkins University Press
2715 North Charles Street
Baltimore, Maryland 21218-4363
www.press.jhu.edu

Library of Congress Cataloging-in-Publication Data

Names: Henke, Scott E., 1960– editor. | Krausman, Paul R., 1946–
    editor.
Title: Becoming a wildlife professional / edited by Scott E. Henke and
    Paul R. Krausman.
Description: Baltimore : Johns Hopkins University Press, 2017. |
    Includes bibliographical references and index.
Identifiers: LCCN 2016046938 | ISBN 9781421423067 (hardcover :
    alk. paper) | ISBN 9781421423074 (electronic) | ISBN 1421423065
    (hardcover : alk. paper) | ISBN 1421423073 (electronic)
Subjects: LCSH: Wildlife managers—Vocational guidance. | Wildlife
    conservationists—Vocational guidance.
Classification: LCC SK355 .B43 2017 | DDC 639.9023—dc23
    LC record available at https://lccn.loc.gov/2016046938

A catalog record for this book is available from the British Library.

*Special discounts are available for bulk purchases of this book. For more
information, please contact Special Sales at 410-516-6936 or specialsales@
press.jhu.edu.*

Johns Hopkins University Press uses environmentally friendly book
materials, including recycled text paper that is composed of at least 30
percent post-consumer waste, whenever possible.

Title page photos: *left,* Dustin Angell, *right,* Scott Henke.

To our spouses, Crystal Henke and Carol Krausman, who gave us the love, confidence, and encouragement to be who we are. And to Bud and Gloria Henke and Ernst and Betty Krausman, our parents, who are our heroes of the Greatest Generation.

CONTENTS

CONTRIBUTORS

Craig Allen
US Geological Survey
Nebraska Cooperative Fish and Wildlife
    Research Unit
University of Nebraska–Lincoln
Lincoln, Nebraska, USA

M. David Allen
Rocky Mountain Elk Foundation
Missoula, Montana, USA

Ben Anderson
Rocky Mountain National Park
Estes Park, Colorado, USA

John M. Bates
Integrated Research Center, The Field
    Museum
Chicago, Illinois, USA

Rick Baydack
University of Manitoba
Winnipeg, Manitoba, Canada

Joe Beach
Texas Parks and Wildlife Department
Austin, Texas, USA

Becky Bennett
Rocky Mountain Elk Foundation
Missoula, Montana, USA

David L. Bergman
Arizona Wildlife Services
Phoenix, Arizona, USA

Vernon C. Bleich
California Department of Fish and
    Game (retired)
University of Nevada–Reno
Reno, Nevada, USA

Jessica L. Blickley
Occidental College
Los Angeles, California, USA

Ben Bobowski
Rocky Mountain National Park
Estes Park, Colorado, USA

Jessica Reeves Bogart
Tomball Independent School District
Magnolia, Texas, USA

Ryan A. Boyer
National Wild Turkey Federation
West Branch, Michigan, USA

Tyler Brown
South Carolina Department of Natural
    Resources
Columbia, South Carolina, USA

Monika Burchette
Caesar Kleberg Wildlife Research
    Institute
Texas A&M University–Kingsville
Kingsville, Texas, USA

Clint Carpenter
National Wild Turkey Federation
Edgefield, South Carolina, USA

David J. Case
D. J. Case & Associates
Mishawaka, Indiana, USA

Shawn Cleveland
State University of New York–
    Environmental Science and Forestry
    Ranger School
Wanakena, New York, USA

Apryle Craig
Rocky Mountain National Park
Estes Park, Colorado, USA

Kristy Deiner
University of Notre Dame
Notre Dame, Indiana, USA

Don A. Draeger
Friedkin Ranch Properties
Eagle Pass, Texas, USA

Rhys M. Evans
US Air Force
Vandenberg Air Force Base, California,
    USA

Joe Fargione
The Nature Conservancy
Minneapolis, Minnesota, USA

Kent Ferguson
Ferguson Consulting & Resource
    Management
Valley Mills, Texas, USA

Joseph Fontaine
US Geological Survey
Nebraska Cooperative Fish and Wildlife
    Research Unit
University of Nebraska–Lincoln
Lincoln, Nebraska, USA

Kelly Garbach
Loyola University Chicago
Chicago, Illinois, USA

Jillian Gericke
Rocky Mountain National Park
Estes Park, Colorado 80517, USA

Kristen Giger
National Wild Turkey Federation
Warren, Pennsylvania, USA

Jonathan H. Gilbert
Great Lakes Indian Fish and Wildlife
    Commission
Odanah, Wisconsin, USA

Ashley R. Gramza
Department of Fish, Wildlife, and
    Conservation Biology
Colorado State University
Fort Collins, Colorado, USA

Lisa K. Harris
Harris Environmental Group, Inc.
Tucson, Arizona, USA

Jim Heffelfinger
Arizona Game and Fish Department
Tucson, Arizona, USA

Scott E. Henke
Caesar Kleberg Wildlife Research
    Institute
Texas A&M University–Kingsville
Kingsville, Texas, USA

Cendy Hernandez
National Audubon Society
New York, New York, USA

Fidel Hernández
Caesar Kleberg Wildlife Research
    Institute
Texas A&M University–Kingsville
Kingsville, Texas, USA

Billy Higginbotham
Texas A&M AgriLife Extension Service
Overton, Texas, USA

Clayton D. Hilton
Veterinary Technology Program
Texas A&M University–Kingsville
Kingsville, Texas, USA

Serra Jeanette Hoagland
Northern Arizona University / US
    Forest Service
Flagstaff, Arizona, USA

Jessica A. Homyack
Weyerhaeuser Company
Vanceboro, North Carolina, USA

J. Dale James
Ducks Unlimited, Inc.
Ridgeland, Mississippi, USA

Winifred B. Kessler
The Wildlife Society
Prince George, British Columbia,
    Canada

Holley Kline
Caesar Kleberg Wildlife Research
    Institute
Texas A&M University–Kingsville
Kingsville, Texas, USA

Lianne Koczur
Caesar Kleberg Wildlife Research
    Institute
Texas A&M University–Kingsville
Kingsville, Texas, USA

Michel T. Kohl
Department of Wildland Resources
Utah State University
Logan, Utah, USA

John L. Koprowski
School of Natural Resources & the
    Environment
University of Arizona
Tucson, Arizona, USA

Blaise Korzekwa
Caesar Kleberg Wildlife Research
    Institute
Texas A&M University–Kingsville
Kingsville, Texas, USA

Paul R. Krausman
Professor Emeritus, University of
    Arizona
Tucson, Arizona, USA

Iara Lacher
Smithsonian Conservation Biology
    Institute
Front Royal, Virginia, USA

Emily W. Lankau
LandCow Consulting
Madison, Wisconsin, USA

Kevin McKinstry
The Westervelt Company
Tuscaloosa, Alabama, USA

David K. McNaughton
Pennsylvania Department of Military
    and Veterans Affairs
Annville, Pennsylvania, USA

Mariah H. Meek
Department of Natural Resources
Cornell University
Ithaca, New York, USA

Wyman Meinzer
Benjamin, Texas, USA

Kelly F. Millenbah
College of Agriculture and Natural
    Resources
Michigan State University
East Lansing, Michigan, USA

Darren A. Miller
Weyerhaeuser Company
Columbus, Mississippi, USA

Karen E. Munroe
Department of Biology and Geology
Baldwin Wallace University
Berea, Ohio, USA

Steve Nelle (retired)
Natural Resources Conservation Service
San Angelo, Texas, USA

Kerry L. Nicholson
kernicholson@yahoo.com

John P. O'Loughlin
HR Capital Partners LLC
Denver, Colorado, USA

Leonard L. Ordway (retired)
Arizona Game and Fish Department
Tucson, Arizona, USA

Lindsey Phillips
Caesar Kleberg Wildlife Research
    Institute
Texas A&M University–Kingsville
Kingsville, Texas, USA

Cindy Pinger
Birmingham Zoo
Birmingham, Alabama, USA

Lauren M. Porensky
US Department of Agriculture,
    Agricultural Research Service,
    Northern Plains Area
Fort Collins, Colorado, USA

William F. Porter
Department of Fisheries and Wildlife
Michigan State University
East Lansing, Michigan, USA

James M. Ramakka (retired)
US Department of the Interior, Bureau
    of Land Management
Aztec, New Mexico, USA

Terra Rentz
College of Environmental Science and
    Forestry
State University of New York

Syracuse, New York, USA

Heather Richardson
Birmingham Zoo
Birmingham, Alabama, USA

Michael J. Rochelle
Weyerhaeuser Company
Lebanon, Oregon, USA

Edgar Rudberg
Greener Point Consulting
St. Paul, Minnesota, USA

Greg Schildwachter
Watershed Results
www.watershedresults.com

Tony Schoenen
Boone and Crockett Club
Missoula, Montana, USA

Nova J. Silvy
Department of Wildlife and Fisheries
Texas A&M University
College Station, Texas, USA

Kelley M. Stewart
Natural Resources and Environmental
    Science
University of Nevada–Reno
Reno, Nevada, USA

Hilary M. Swain
Archbold Biological Station
Venus, Florida, USA

Jeffrey Tafoya
US Department of the Interior, Bureau
    of Land Management
Farmington, New Mexico, USA

Stephen M. Vantassel
University of Nebraska–Lincoln
Lincoln, Nebraska, USA

Kurt VerCauteren
National Wildlife Research Center
US Department of Agriculture, Animal
    and Plant Health Inspection Service,
    Wildlife Services
Fort Collins, Colorado, USA

Richard T. Watson
The Peregrine Fund
Boise, Idaho, USA

Marit L. Wilkerson
University of California–Davis
Davis, California, USA

Kai Williams
The International Wildlife
    Rehabilitation Council
Eugene, Oregon, USA

Ken Williams
The Wildlife Society
Bethesda, Maryland, USA

Stephen M. Williams
Arizona State Land Department
Phoenix, Arizona, USA

Steven Williams
Wildlife Management Institute
Gardners, Pennsylvania, USA

Eric Winford
Washington Department of Natural
    Resources
Ellensburg, Washington, USA

Sandra K. Wyman
US Department of the Interior, Bureau
    of Land Management
Prineville, Oregon, USA

John H. Young Jr.
Texas Department of Transportation
Austin, Texas, USA

FOREWORD

It's not about the money" is a phrase often used but rarely backed up by personal examples. But, as I have spent my entire 45-year career in the wildlife profession, I can say to all who choose to go down this path that "it's not about the money." There are far greater rewards at the end of a career in wildlife biology and management. Being at peace with the field you choose, one that is focused on a resource that cannot speak for itself, is gratifying and fulfilling way beyond material things.

The authors of *Becoming a Wildlife Professional* have covered, from A to Z, the whys, the hows, and what to expect, whether you choose to step into graduate school first or move right into a position after an undergraduate degree. The authors have provided aspiring wildlife biologists with step-by-step instructions and great advice on how to obtain their dream job in the wildlife profession. I have never seen a book of its equal. The authors should be especially commended for emphasizing involvement in professional societies, as well as how to act professionally, an attribute many students don't consider until they are immersed in their careers.

Some of the nuts and bolts chapters deal with required skill sets, such as writing resumes and the interview process, but one of the rich characteristics of the book is the collection of personal insights from wildlife professionals throughout North America who share their philosophy about and experiences with the numerous employment opportunities available within the profession. These will be of tremendous benefit to aspiring wildlife biologists and educators, for it is the intimacy of individual feelings and reflections when we see things as they are, not as they should be. This aspect of the book will help guide young biologists into the career path that is right for them.

Lastly, the authors should be commended for touching on diversity. For example, when I began my career, our profession had a very real challenge in addressing deficiencies of gender involvement, let alone ethnic diversity. Over the past 30 years, we have embraced this, and more women are joining our professional ranks. At the Caesar Kleberg Wildlife Research Institute, half of our graduate students currently are women, who are not afraid or intimidated by getting out "in the wilds" of the brush country of deep South Texas. This is not to say that the gender issue is resolved, because we also need to retain women in the workforce. In addition, there are many jobs that don't require an outdoor focus. But to me, our profession's future is indeed precarious if we do not correct the lack of ethnic diversity.

Having spent a cumulative 80 years in the education process, the authors are extremely qualified to present this book to young and gifted readers who have an interest in wildlife biology and management. There is not a glut of information on the subject, so this book has been greatly needed and has been a long time in coming.

Fred C. Bryant, PhD
Director, Caesar Kleberg Wildlife Research Institute
Texas A&M University–Kingsville

PREFACE

The wildlife profession is still in its infancy, having its beginning in the 1930s, with Aldo Leopold (1933) writing the first wildlife management book in North America, entitled *Game Management*. Since that time, the wildlife profession has made numerous advancements (e.g., from management focused on game species and hunters to more inclusive management designed for all wildlife and stakeholders). As a result, wildlife management is now based on science and governed by laws to serve the common good.

Unfortunately, one aspect where the wildlife profession has not been as successful is in making future generations aware of the multitude of employment opportunities one could enjoy as a wildlife professional. Many youths grow up with a love of the outdoors but have no idea (or guidance) that their passion for the outdoors could become a career. For example, I [SEH] was conducting a wildlife inventory on a property in Texas and came upon a group of high school students who were camping. In talking with them, they asked about my equipment. I explained that I was a field biologist and was surveying the local wildlife. Their puzzled looks were obvious, and I had to explain what I meant by *field biologist*. None of the campers had heard that term before, nor did they understand that not all biologists work inside a laboratory. They were amazed to hear that I got paid to be outdoors and conduct wildlife surveys. Not only did I open their eyes to the wildlife profession, but, equally, they opened mine. In addition, PRK went to grade school and high school in North Africa and Europe. He was always interested in wildlife, but did not know about the profession until he stumbled on an Ohio State University student chapter of The Wildlife Society when he was a college sophomore. Many incoming wildlife-major university students, including graduate students, also are unaware

of the variety of career options available to them within the discipline. In 2014, the University of Montana hosted a special graduate seminar to introduce wildlife graduate students to the numerous jobs available to them after graduation, because of their naïveté about the subject. Although our experience is limited to the regions where we have worked (Alabama, Arizona, Indiana, Montana, and Texas), similar stories are told by professors across the United States. Most wildlifers think the general public knows about our profession, but this assumption couldn't be more wrong. We conducted an informal survey of high school career counselors and asked them what advice they would give to high school students who said that they had a passion for animals, would rather be outdoors than inside, and loved outdoor activities, such as camping, hiking, hunting, and fishing. The responses we received from nearly 75 counselors were nearly identical: veterinarian and park ranger. Only a third of the counselors added game warden and, shockingly, none mentioned wildlife biologist—in any capacity (federal, state, or private)!

The majority of the general public is not aware of our profession. It's no wonder why today's youth play games in the virtual rather than the real outdoors. We, as biologists, need to make future generations aware of our profession. This book is intended as a step in that direction. Its purpose is to highlight and described the multitude of employment opportunities available within the wildlife profession, and to offer step-by-step guidance in becoming a wildlife professional.

This volume consists of 12 chapters, beginning with why someone might choose a career in wildlife in the first place (chapter 1). We then discuss the need for an undergraduate wildlife education and what to expect from it, as well as offer advice on how to be successful as a student and within the profession (chapters 2 and 3). Chapter 4 covers the skills needed for conservation careers. Chapter 5 is the central theme of the book and describes nearly 100 wildlife-related jobs in 35 agencies and organizations. Next, we discuss various practical aspects: the benefits of being a member of professional societies (chapter 6), the development of a quality resume (chapter 7), and the art of the interview process (chapter 8). Chapter 9 discusses the need to always act professionally. The following two chapters present the pros and cons of advanced degrees from the perspective of professors (chapter 10) and students (chapter 11). Lastly, chapter 12 examines cultural diversity within the wildlife profession.

It is never too late to plan your future and consider career options. It is our hope that this volume will aid future generations of wildlifers in finding the job of their dreams.

ACKNOWLEDGMENTS

We thank all the people that assisted with this book, as it could not have been completed without their help. Vince Burke, Julie McCarthy, Tiffany Gasbarrini, Kathryn Marguy, Meagan Szekely, and Catherine Goldstead of Johns Hopkins University Press were instrumental in shepherding approvals and production schedules and keeping us on time. Kathleen Capels was the copy editor and we appreciate her sharp eye and dedication to detail. Scott and Crystal Henke prepared the index.

We owe a debt of gratitude to the following individuals, who either served as peer reviewers, provided insights about what it means to be a wildlife professional, or helped us in a myriad of ways to make this the best book it could possibly become: Heidi Adams (Louisiana Tech University), James Anderson (West Virginia University), Walt Anderson (Prescott College), Jeff Archer (University of Arizona), Heather Bateman (Arizona State University), Jacob Bowman (University of Delaware), Scott Brainerd (Alaska Department of Fish and Game), Robert Brewer (Cleveland State Community College), David Brown (Eastern Kentucky University), Melanie Bucci (University of Arizona), David Buehler (University of Tennessee), James Cain III (New Mexico State University), Timothy Carter (Ball State University), Joe Caudell (Murray State University), Chris Comer (Stephen F. Austin University), Jeff Davis (Nukingsteet Pest and Wildlife Control, Enfield, CT), Daniella Dekelaita (Oregon State University), Victoria Dreitz (University of Montana), Susan Ellis-Felege (University of North Dakota), Jorie Favreau (Paul Smith College), Lena Fletcher (University of Massachusetts), Robert Garrott (Montana State University), William Giuliano (University of Florida), Dixon Herman (Blue Ridge Wildlife Control, Hickory, NC), Charles Holt (Advantage Wildlife Removal, New Richmond, OH), Reg Hoyt (Delaware Valley University), Chris Hunnicutt (City Wide Exterminating, Charlotte, NC), Danielle Jarkowsky (Oregon State University), Randy Larsen (Brigham Young University), Paul Lukacs (University of Montana), Patrick Magee (Western State Colorado University), Eric McCool (McCool's Wildlife Services, Franklin, PA), Kevin Monteith (University of Wyoming), Lara Pacifici (North Carolina State University), Eric Pelren (University of Tennessee–Martin), Dan Pletscher (University of Montana), Karen Powers (Radford University), Shannon Rabby (Haywood Community College), Janet Rachlow (University of Idaho), Jason Reger (Blue Ridge Wildlife Management, Roanoke, VA), Jim Schneider (Michigan State University), Mark W. Schwartz (University of California–Davis), Brandon Shore (Swat Services, Marietta, GA), Kelley Stewart (University of Nevada–Reno), Brandi Van Roo (Farmington State University), Jeff Voelker (Critter Control, Omaha, NE), Mark Wallace (Texas Tech University), and Rob Wingard (Washington Department of Fish and Game).

Finally, we thank the students who will use this book to initiate and advance their careers in the wildlife profession. We are grateful to Fred Bryant and James Burchfield, the administrators of our programs at Texas A&M University–Kingsville and the University of Montana, respectively, for supporting our efforts. We also thank The Wildlife Society Council for approving the publication of this book.

Financial support for the volume was provided by

BECOMING A WILDLIFE PROFESSIONAL

1

Why Choose a Career in the Wildlife Profession?

PAUL R. KRAUSMAN AND SCOTT E. HENKE

Humans dominate the earth, so wildlife depend on our ability to make decisions that do not compromise their existence. Mankind's early attempts to interact with wild animals over 10,000 years ago could be thought of as the birth of wildlife management. Aldo Leopold, considered the father of modern wildlife management, cited the Bible as the earliest written record of managing wildlife (Leopold 1933). Yet wildlife management is a young profession that only recently gained a foothold among the objective sciences in the mid-nineteenth century, when organized sport hunters worked to eliminate market hunting (e.g., hunting bison for their tongues only) and curtail pot hunting (hunting for food or profit, or taking part in contests solely to win prizes; Krausman 2002, Organ 2013).

The developing profession received heightened recognition when Theodore Roosevelt was president of the United States. As president, Roosevelt brought conservation into the homes of Americans and clearly stated that scientifically based wildlife management was necessary to enhance and maintain the existence of species. Leopold (1933) summarized Roosevelt's thoughts into what is called the Roosevelt Doctrine, which has three main components:

1. All natural resources, including wildlife, are one integral whole.
2. Conservation through wise use is a public responsibility, and ownership of wildlife is a public trust.
3. Science is the proper tool for discharging that responsibility.

The bold actions taken by the early pioneers in the profession served as the springboard to maintain habitats and ensure the existence of many animals in North America. The complex process of wildlife management also became a precursor for later professions, including conservation biol-

ogy, landscape ecology, restoration ecology, and the human-dimensions aspect of wildlife management. These new fields are important in the management and conservation of wild flora and fauna. Wise management of our natural resources may be the most important legacy that we can leave behind for future generations.

We are currently undergoing what some call the sixth extinction (Ceballos et al. 2015). The previous five occurred thousands to millions of years ago and were influenced by natural events, such as meteors hitting Earth, changes to the climate (e.g., the Ice Age), altered vegetation, and so forth. The sixth extinction is unique, because it is being caused by humans and is operating at an accelerated rate, compared with natural events.

Early naturalists realized that natural resources were essential to life on Earth. Without serious attention to the management of these resources, however, life as we know it will cease to exist, because human civilization can destroy them. All resources can be divided into either inexhaustible (e.g., sun, wind, tides) or exhaustible categories. Exhaustible resources can be further subdivided into nonrenewable (e.g., minerals, oil) and renewable categories (e.g., water, vegetation). Wildlife, if managed properly, is a renewable resource, but it is exhaustible if improperly managed. As such, to discharge this responsibility, the new field of wildlife management demanded individuals that believed in the need for and usefulness of science as a tool for conservation. They needed to be able to diagnose the landscape to discern and predict trends in the biotic community and, where necessary, to modify these trends in the interest of conservation. Further, the wildlife profession required individuals with knowledge of plants, animals, soils, water, and vegetation, as well as familiarity with other professions that had similar interests.

To be truly proficient within the wildlife profession, a

person should obtain formal training in the general sciences, such as biology (cellular and systematic, genetics), chemistry (inorganic, organic, environmental), physics, earth sciences (soil science, biogeography, meteorology, hydrology), traditional wildlife biology and management courses (principles of wildlife management, wildlife ecology and ecosystems, large mammal ecology, waterfowl, upland game birds, wildlife techniques, wildlife disease, parasitology), and mathematics (algebra, calculus, statistics). That training should also include social science disciplines, such as psychology, sociology, economics, environmental policy and law, conflict resolution, oral and written communications, and political science (Baydack 2009). In addition, wildlife students should take courses to update themselves to the latest technologies, such as geographic information systems (GIS), the global positioning system (GPS), and remote sensing, as well as gain hands-on experience in wildlife research and design, while also remaining current in basic courses, such as botany, plant identification, mammalogy, ornithology, and herpetology. Unfortunately, such a curriculum would require a wildlife biologist to take more than 250 credit hours of coursework to be successful within our profession.

Although this proposed curriculum is not considered feasible (it would take a student nearly seven years to complete all the courses for a wildlife degree listed above), it does illustrate that wildlife biology is a melting pot of many disciplines, and that today's wildlife biologists need to select their curricula wisely and obtain a diversity of wildlife-related experiences in order to be competitive within the profession. The profession needs bright minds that have the drive, dedication, and desire to work for the conservation and management of wildlife. Throughout our careers, students have given us numerous reasons why they want to earn a wildlife degree, which have included everything from using a bachelor's degree to advance toward additional degrees in environmental law and veterinary science, to doing so because they love hunting and fishing.

In spring 2014, we asked 179 students at the University of Montana and Texas A&M University–Kingsville who were taking an applied wildlife management course why they selected wildlife as their major. Most (47%) chose this career because of their commitment to and love of animals and nature, followed by their passion for hunting and fishing (15%), and a desire to conserve wildlife (13%). A host of other reasons followed, including their interest in science and desires to communicate about wildlife to the public, learn more about wildlife, work outside, film and photograph wildlife, protect wildlife, rehabilitate wildlife, restore habitat, and become a wildlife veterinarian. Wildlife management is a profession that encompasses many careers in ensuring the health and viability of wildlife populations. Thus it is difficult, even for those in the profession, to provide a blanket definition of what a wildlife biologist (*wildlifer*) is and list the background and experiences needed to become a wildlife biologist. This becomes apparent when new students come to a four-year college or university with their parents and the parents indicate that their

The majority of students who select a career in wildlife do so because they have a deep appreciation for animals and nature (photo: Scott Henke)

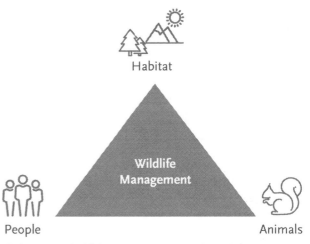

A holistic view of wildlife management covers the interactions among wild animals, habitat, and human needs and desires

child is not good at math and science but loves to hunt and fish and thus thinks he or she should become a wildlife biologist. In reality, a student needs a working knowledge of algebra (to use the formulas required in wildlife management) and calculus (to understand why the formulas work). A good definition of a wildlife biologist may help reduce the misperceptions that people have about the profession.

Let's look at the historical definition of wildlifers and see how their responsibilities have evolved since the beginning of the scientific profession of wildlife management and conservation. Leopold (1933) defined game management as the art of making the land produce sustained annual crops of wild game for recreational use. Today, *game* is a term that means whatever the legislature in the state in which the animal resides stipulates as game. The term can have a narrow, precise meaning, such as that which first appeared in British law in an act from the year 1389 (McKelvie 1985); in other words, game did not have a biological meaning, but was whatever the

The goals of wildlife management are to increase certain populations, such as endangered ocelots (photo: Mike Tewes); reduce populations, such as suburban coyotes; and maintain sustainable yields, such as for game species like white-tailed deer (photo: Tim Fulbright)

relevant legislative body decided that it was. The use of this word in British law migrated to North America and has led to many anomalies. For example, mourning doves in parts of the eastern United States are classified as songbirds and thus are nongame, but they are hunted as game in many western states. In addition, many species are not classified as game, but most states require a game license to either harvest or collect them. There are numerous definitions of wildlife, such as the one presented by the US Congress (1973), stating that *wildlife* refers to any member of the animal kingdom and includes parts (e.g., antlers), products, offspring, or any dead body parts. This definition is too descriptive, however, and does not take into account an animal's habitat and its relation to people. The Wildlife Society (the professional society for wildlifers) defines wildlife as "free-ranging animals of major significance to man." This is fairly simple, but it does incorporate wildlife, their habitats, and humans (Giles 1978). Until the 1970s, wildlife was synonymous with animals that were hunted, but with our interest in biodiversity over the past four to five decades, most free-living animals have become significant to humans. The term wildlife, however, is often restricted to terrestrial and aquatic vertebrates other than fish, because—due to political maneuvers and funding mechanisms—wildlife management and fisheries management have developed into separate disciplines, even though many of the basic principles between them are shared.

Wildlife does not have a universally defined meaning, and the word expands and contracts with the viewpoint of the user (Caughley and Sinclair 1994). In general, the wildlife profession considers wildlife to be native, free-living animals of major significance to man. The definition does not exclude plants or lower animals, because wildlife covers more than just the animal. When wildlife species are defined, the habitats that support them also have to be considered, because species and their habitat are interlocked and cannot properly be separated. Because wildlife in North America belongs to the public, humans also have to be considered with regard to wildlife issues. Giles (1978) presented a holistic view of wildlife that looked at the interactions among the animal, its habitat, and the humans who influence how that species is managed.

The goals of wildlife management can be broken down into four general categories under the umbrella of human dimensions. One goal is to increase a particular population, as is done for endangered and threatened species and others that have populations below management goals. The second general goal is to reduce populations that are above management goals or are detrimental to other populations. One example of the latter is the varied set of attempts to reduce coyote populations in urban and suburban settings, where they have dispersed and are not welcomed by society. In many such areas, deer populations have also increased, to the point where they are causing a significant number of undesirable human/wildlife interactions (colliding with vehicles, spreading disease, consuming ornamental landscaping, harassing children and pets). The third goal is to be able to harvest a population yet maintain a sustainable yield, as is the case for most game species. Thus hunting regulations are generally very conservative, because biologists want to ensure that game populations are not decimated. The fourth general goal is to do nothing for a population except monitor it over time. Our national parks provide a good example of this type of action, because parks act as areas where human interference is minimized. There are exceptions, such as the bison population in Yellowstone National Park. Their numbers have increased and thus are controlled when they exceed a social carrying capacity. These bison also exemplify the reality that management in lands outside of the parks influences management in national parks; everything is tied together.

Wildlife biologists are not restricted to only following a single goal. Management evolves over time, so they are usually involved with multiple animals in different jurisdictions that require varying goals. Moreover, all four of the above general goals are influenced by society. The public has been increasingly incorporated in wildlife management and will continue to be in the future, because citizens are the stakeholders that management has to satisfy. Biologists can have the most efficient management plans for wildlife, but if the public does not support these plans, they will not be initiated or will remain incomplete.

The classes you take to obtain an undergraduate wildlife degree are often centered on the coursework that is also required for certification as a wildlife biologist by The Wildlife Society. Specifics for this certification will be discussed in detail in chapter 2, but in brief, those of you who continue toward a wildlife profession will take classes that relate to an animal (wildlife management and biology, ecology, zoology), its habitat (botany, physical sciences), and human dimensions (humanities and social sciences, communications, policy, administration, law). There will also be other necessary courses to help you become a quantitative scientist (calculus, biometry, systems algebra, mathematical modeling, sampling, computer science, statistics). This suite of classes will prepare you for a future in the wildlife profession. It will take time, interest, and skill to complete them, but they will give you the background to work with and for the species you cherish.

In our experience, working independently, working with others, working outdoors, and working with wildlife and the habitats they depend on are some of the top reasons to join other wildlifers in enhancing our nation's natural capital. It may well be the most important thing anybody can do.

SUMMARY

Defining wildlife and its management is complex, because this involves animals, habitats, and the human dimensions that dictate what is managed and how that management will occur. As such, the profession is made up of diverse groups of individuals with different types and levels of expertise (e.g., managers, biologists, botanists, landscape ecologists, sociologists, information and education specialists, photographers, modelers). Students who want a career in wildlife should first obtain a bachelor of science degree from an accredited institution offering classes that can be used for certification by The Wildlife Society, and then seriously consider going to graduate school. A bachelor's degree will allow you to gain the basic credentials for a career in wildlife, and graduate school will allow you to specialize in a specific area of interest (e.g., wildlife populations, habitat, human dimensions, law enforcement, modeling). The profession needs individuals with solid academic credentials who are able to contribute toward the conservation and management of wildlife worldwide.

LITERATURE CITED

Baydack, R. 2009. The Wildlife Society ad hoc committee on collegiate wildlife programs: summary final report to The Wildlife Society Council, http://wildlife.org/wp-content/uploads/2015/03/Ad_Hoc_Collegiate_Wildlife.pdf.

Caughley, G., and A. R. E. Sinclair. 1994. Wildlife ecology and management. Blackwell Scientific, Boston, MA, USA.

Ceballos, G., A. H. Ehrlich, and P. R. Ehrlich. 2015. The annihilation of nature: human extinction of birds and mammals. Johns Hopkins University Press, Baltimore, MD, USA.

Giles, R. H., Jr. 1978. Wildlife management. W. H. Freeman, San Francisco, CA, USA.

Krausman, P. R. 2002. Introduction to wildlife management: the basics. Pearson Education, Upper Saddle River, NJ, USA.

Leopold, A. 1933. Game management. Charles Scribner's Sons, New York, NY, USA.

McKelvie, C. L. 1985. A future for game? George Allen and Unwin, London, UK.

Organ, J. F. 2013. The wildlife professional. Pp. 24–33 *in* P. R. Krausman and J. W. Cain III, eds. Wildlife management and conservation: contemporary principles and practices. Johns Hopkins University Press, Baltimore, MD, USA, and The Wildlife Society, Bethesda, MD, USA.

US Congress. 1973. A compilation of federal laws relating to conservation and development of our nation's fish and wildlife resources, environmental quality, and oceanography. US Government Printing Office No. 052-070-02871-4, Washington, DC, USA.

Wildlife Undergraduate Education and the University Curriculum

RICK BAYDACK

You are about to graduate from high school and want to work with wildlife, but your parents really want you to be a lawyer or a doctor. Your two brothers are successful engineers. And your sister is a renowned marketing specialist. Your best friends think that you're heading in a direction without good job prospects. So what do you do?

Working with wildlife generally takes a strong personal commitment. The field is exciting, dynamic, and filled with adventure. With proper preparation, students can enter the workforce ready to achieve their goals and move forward into successful jobs, as well as a career path that is most satisfying. Moreover, you have the opportunity to really make a difference in conserving a precious resource. This chapter explores the opportunities for wildlife education that exist across North America, examines the various academic requirements, provides a framework with which to compare educational institutions and programs, and identifies additional perspectives that students entering this field of endeavor should consider.

CONTEXT

What are the educational expectations that are associated with a wildlife career? In general, a bachelor of science degree from a recognized university is a prerequisite toward moving forward in the profession. Graduate education is often perceived as essential, and many agencies prefer to hire at that level. But in any case, the first step after graduating from high school is to find a university that meets your areas of interest. Throughout your university education, coursework will be essential, but additional considerations—such as volunteering; job shadowing; becoming a wildlife intern; attending conferences, seminars, and workshops; joining a professional society related to the wildlife resource that you would like to work with—are also important. Professional development

and life-long learning are critical features in becoming a wildlifer (Baydack 2009).

Students who want to enter the wildlife profession generally should have above-average skills and abilities in mathematics and science. They should ensure that their high school education and first year of university coursework emphasizes these disciplines. An ability to work with numbers and the use of problem-solving skills properly grounded in science are essential to a successful future in the wildlife field. Increasingly, knowledge of and competency in computer-based information technologies can provide significant benefits to wildlife professionals.

Another important consideration as you move forward with selecting wildlife as your career path is to think about the type of work you want to do. Perhaps most of the questions below will be difficult for you to answer when entering the field, but you will need to consider them as time goes on. You might try to answer them today, and then reflect on your responses as your perspectives about the field evolve:

- Are you an outdoors person, or do you prefer an office environment?
- Do you want to work with a particular species, a species group, a community, an ecosystem, or just wildlife in general?
- Are you more interested in wildlife habitats or wildlife populations?
- Do you have expectations of remaining in your hometown, or are you flexible in relocating to a potentially remote location?
- Is your passion for wildlife related to hunting, trapping, viewing, studying habitat, or some combination of these and other aspects of the profession?
- Are you more interested in research, administration,

fieldwork, public education, or something else? There is a lot to the profession and, as indicated in chapter 1, it involves not only animals, but also habitats and human dimensions.

• Do you have any other specific perspectives or viewpoints relating to wildlife work?

FIRST STEPS FIRST: FINDING THE RIGHT SCHOOL

Terminology is important here. In the United States, both colleges and universities offer four-year bachelor's degree programs, while community colleges offer two-year associate's degrees. In Canada, however, all four-year higher education institutions are known as *universities*, and two-year schools are referred to as *colleges*. This book uses *college* and *university* interchangeably to refer to all postsecondary institutions with four-year degree programs.

A large number of wildlife programs can be found at North American universities, estimated at about 400 as of 2008 (Wallace and Baydack 2009). Around 200 other higher education institutions across the United States and Canada also offer programs relating to wildlife. Universities generally offer four-year bachelor's degrees, which are the most popular and are appropriate for future wildlife work, although some three-year options do exist. Other postsecondary schools generally have two-year diploma programs, with some universities accepting transfer credits directly to their wildlife programs from designated institutions.

Wildlife programs at universities are housed in different academic locations. Some are in colleges or departments (called *faculties* in Canada) of agriculture, applied science, environment, forestry science, or some combination thereof. This may not seem important as you select a program, but the academic requirements for each will probably differ, depending on the program's home. Some may emphasize mathematical skills or fieldwork and the physical sciences more than others. Clearly understanding these requirements as you select a wildlife program could be the difference between a successful university experience and a difficult and unproductive one.

The location of a university in North America can influence its wildlife program (Wallace and Baydack 2009). In the northeastern and western regions of the United States, many wildlife courses are based in environmental programs, whereas in the southern and central areas, wildlife offerings tend to be located in wildlife and fisheries programs. Many Canadian universities have established their wildlife courses in environmental or environmental science programs, and these have seen increasing enrollments in recent years (Wallace and Baydack 2009). Coursework requirements for wildlife degrees vary greatly across North American universities. Some schools are affiliated with the National Association of University Fish and Wildlife Programs (NAUFWP), and these tend to offer more traditional coursework in the wildlife field,

although their relative presence is being reduced, due to the development of programs in other areas.

Wallace and Baydack (2009) reported that there are at least four times as many programs offering some kind of wildlife education than had been previously thought. Wildlife programs in biology/zoology or agriculture and forestry/fisheries have been diversifying their curricula, offering courses in natural resource management, conservation biology, toxicology, geographic information systems (GIS), and related topics. But other programs, most notably in biology and the environmental sciences, also have been expanding by adding wildlife, conservation, and natural resource courses to their curricula. This trend was particularly evident in Canadian universities, where environment-related programs have been identified as a growth area, particularly in times of economic downturn, when employees tend to return to universities for retraining or professional development (Wallace and Baydack 2009).

Most university wildlife programs have undergone a series of name changes that have sought to portray how they view their content and missions. Programs have uniformly moved away from terms like game management or wildlife management to those such as wildlife conservation, wildlife ecology, natural resource conservation, or natural resources and the environment. Some schools have modified the name of their wildlife concentration or degree program from an explicit management orientation to a less restrictive conservation or ecology title.

The rapid expansion and broadening of existing wildlife and biology programs and the growth of environmental curricula by schools competing for students has led many US and Canadian academic institutions to add courses and, in some cases, entire specializations in wildlife-related areas. New wildlife-focus fields are often placed in nontraditional departments, and therefore students selecting wildlife programs should be aware that these programs may exist in somewhat atypical academic locations.

A good place to begin your search for a school with a wildlife program is to visit The Wildlife Society's (TWS) web page for current student chapters (see appendix B). The majority of universities listed within this site have a direct Internet link to their respective student chapter websites, which contain information about recent activities of the student organization, the school's wildlife program, and useful people to contact (typically the organization's advisor or the department chair). Write or call these individuals to gain information about their particular wildlife program and university. Determine if the institution and its program are compatible with your desires and learning style. Ask questions. Is the university located within a large city or in a rural setting? What is the student population of the university (e.g., less than 10,000 students, more than 50,000 students)? What is the typical class size for the wildlife courses that are offered? Smaller classes (less than 30 students) are more conducive to hands-on learning opportunities. What is the proportion of courses that offer these types of opportunities versus typical lecture-only classes? How many students are in the wildlife program? Does the school offer graduate degrees in wildlife? You may not be interested in graduate school (now), but universities with graduate programs in wildlife typically offer opportunities for undergraduate involvement in wildlife research, which provides bachelor's degree students with invaluable experience. Does the curriculum of the wildlife program lead to professional certifications, such as those offered by TWS and the Society for Range Management? How many students graduate each year from the program? What percentages of graduates obtain wildlife employment after they finish their degree? Where have recent students been employed? Answers to these questions will help you select a university that is the best fit for you.

UNIVERSITY COURSEWORK

Universities are dynamic places, where changes to curricula always seem to be taking place. Administrators attempt to meet demands by the student body, seek to ensure that employers are content with the product their program produces, and strive to satisfy parents and other caregivers that their children are receiving a quality education. Therefore, students often find themselves in a fluctuating academic environment, where change seems to be the order of the day. How do those who want to excel in the wildlife profession ensure that they take the right courses in the proper sequence so their goals can be achieved?

One guideline is for students to get involved with a professional society or organization that is related to the wildlife profession, such as TWS, the Society for Range Management, or the Environmental Careers Organization of Canada (Eco-Canada). These agencies offer certification standards and guidelines (often based on university coursework) through which prospective wildlife professionals can plan and select appropriate wildlife-related courses and emphasis areas. It is critically important for students to consider this option as early as possible in their university career, so the proper prerequisite courses and timing of requirements can be achieved.

CERTIFICATION REQUIREMENTS OF THE WILDLIFE SOCIETY

Let's investigate how these certification requirements can be used to assist a student who decides to register in a wildlife program by examining those of TWS (see appendix B). Download a copy of TWS's application for certification, which should help as you read through the remainder of this chapter.

The Wildlife Society was founded in 1937. Its mission is "to inspire, empower, and enable wildlife professionals to sustain wildlife populations and habitats through science-based management and conservation." The organization enhances its members' networking and learning opportunities, as well as their professional and career development, and provides numerous ways for them to get more involved in creating a

better future for wildlife and wildlife habitats. And an important component of its programming is certification.

Certification is a process based on meeting minimal educational, experiential, and ethical standards. By completing this process, a certified wildlife biologist will have demonstrated expertise in the art and science of applying principles of ecology to the sound stewardship and management of a wildlife resource and its associated environment. The objective of TWS's certification program is to provide public and private clients and employers with accountability standards regarding professional advice in matters concerning wildlife resources. There are several benefits to the public from certification:

- It defines minimum standards for professional wildlife biologists.
- It ensures that all practicing wildlife biologists meet such standards.
- It creates and maintains public confidence in the advice and opinions of certified wildlife biologists.
- It assists the public in evaluating the qualifications of wildlife biologists by establishing a procedure for critical peer evaluation.

For students entering the wildlife profession, there are many benefits to becoming certified. Certification offers a recognized credential and increasingly is being required or preferred as an employment criterion. It provides students with a recognized affiliation with a professional organization and thereby ensures that a person adheres to stated ethical standards. Certification enhances an individual's credibility when speaking or testifying. And, in many cases, an increased starting salary can be attained or negotiated. In short, certification as a wildlife biologist is the preferred track for future employment, so students should plan accordingly as they complete their wildlife degrees.

Certification provides students with a recognized affiliation with a professional organization (photo: Scott Henke)

Certification needs to encompass the full breadth of the wildlife field. Initially, it is based on satisfactory completion of specified educational requirements, thereby leading to the designation of Associate Wildlife Biologist® (AWB). With the completion of five years of relevant experience after graduation, professionals can seek a designation as a Certified Wildlife Biologist® (CWB). All applicants for certification must formally attest to upholding the Code of Ethics of TWS (see appendix B). Students requesting certification will have their applications reviewed by a five-member Certification Review Board.

The minimum academic requirements for certification include completion of 36 semester hours of coursework in the broad category of biological sciences; 9 semester hours in the physical sciences; 9 semester hours in the quantitative sciences; 9 semester hours in the humanities and social sciences; 12 semester hours in communications; and 6 semester hours in policy, administration, and law. Within each of these categories, descriptions of the requirements for qualifying courses are provided below. For some categories, specific subject concentrations may be needed.

Course Requirements in the Biological Sciences

Biology and ecology form the foundation on which the wildlife profession is based, and they should be a dominant component of any wildlife curriculum. In the biological sciences, coursework must meet or exceed the minimum requirement in each of five different subcategories: (1) wildlife management (six semester hours), (2) wildlife biology (six semester hours), (3) ecology (three semester hours), (4) zoology (nine semester hours), and (5) botany (nine semester hours). Let's examine each of these components further, as they can be useful to students in designing their undergraduate programs. Note that many academic institutions are unfortunately not familiar with the requisites for certification by TWS or similar professional organizations, so students often are responsible for ensuring that they meet the appropriate criteria through their degree programs.

Wildlife management courses are extremely important. They must emphasize the principles and practices of wildlife management and demonstrate training in understanding and manipulating habitat relationships and population dynamics, as established by human concerns and activities. Wildlife biology courses should focus on the biology and behavior of birds, mammals, reptiles, or amphibians. Eligible courses should ensure familiarity with the biology of wildlife species and their habitat relationships. They must include at least one class dealing solely with the science of mammalogy, ornithology, or herpetology, or a class that combines these three fields. The ecology requirement can be satisfied by the successful completion of any course in general plant or animal ecology (human ecology is excluded). Zoology classes should focus on the taxonomy, biology, behavior, physiology, anatomy, and natural history of vertebrates and invertebrates. Courses in genetics, nutrition, physiology, disease, general zoology, or fisheries biology are also accepted. The botany requirement can be met by classes in general botany, plant genetics, plant morphology, plant physiology, or plant taxonomy. At least one botany course must be primarily concerned with plant taxonomy or plant identification.

Additional Course Requirements

Certification requirements extend well beyond the biological sciences, because wildlife professionals must have a broad-based educational background. The physical sciences component consists of classes in chemistry, physics, geology, or soils, with at least two of these disciplines represented. All students must have an understanding of the basic quantitative aspects of wildlife science. Courses in quantitative sciences must include at least one class in statistics and two additional classes in calculus, biometry, advanced algebra, systems analysis, mathematical modeling, sampling, computer science, or other quantitative science. A solid background in the quantitative sciences is becoming increasingly important in the wildlife profession.

The humanities and social sciences requirement can be satisfied with courses in economics, sociology, psychology, political science, government, history, literature, or a foreign language. Wildlife management often is people management. Learning and understanding differing viewpoints within society can be an essential skill for a wildlife biologist. Therefore, do not readily dismiss the importance of the humanities in your education.

The communications category can be met with courses designed to improve communication skills, such as English composition, technical writing, journalism, public speaking, or the use of mass media. Every wildlife-related employer lists good oral and written communication skills as one of the top credentials for new hires.

Policy, administration, and law includes courses that demonstrate a significant content or focus on natural resource policy or administration, wildlife law or environmental law, or natural resource / land use planning, as well as classes that document contributions to the understanding of social, political, and ethical decisions for wildlife or natural resource management. As students prepare to enter the workforce, it is critical that they have exposure to and an understanding of natural resource policy and the issues that impact the implementation of resource management in the real world.

Equivalency Considerations

Professional experience may be used to satisfy these educational requirements in instances where specific deficiencies exist. Examples may include published papers or a completed thesis in place of courses in English composition or technical writing. Also, documentation of demonstrated professional experience may satisfy the criteria for classes such as botany, resource policy, administration, land use planning, or public speaking. When using professional experience to substitute for a course, students must make sure that at least one university-level class has been completed in that category and any other specific requirements are met.

Certification Summary

Although there is probably no single perfect wildlife program, TWS certification requirements for coursework represent the core areas that should be present in any high-quality wildlife program. Students should be certain that the wildlife program they are considering provides the necessary complement of classes that will enable them, if they wish, to achieve certification once they graduate.

THE IDEAL WILDLIFE PROGRAM

This section provides ideas for what the components of the ideal bachelor's degree curriculum might be. At the undergraduate level, we cannot expect students to master everything a potential employer might require; thus we focus on core competencies that should prepare students either for employment or for continuation on to graduate study. There is great diversity among programs that offer bachelor's degrees in wildlife, and departments vary in terms of faculty expertise. Accordingly, in addition to the core competencies identified here, we suggest some potential options or concentration areas that might build on that core, taking advantage of faculty strengths.

Core Competencies of the Ideal Program and Other Areas of Focus

In general, TWS certification requirements for coursework represent the core areas of competency that should be present in any high-quality wildlife program purporting to prepare individuals for a wildlife career. Additional important attributes, however, go beyond these classroom skills.

Teamwork and stakeholders: Employers have noted that gaps in the abilities of wildlife graduates to be productive members of a team and to work with stakeholders are important ones to rectify. Universities with courses that emphasize interdisciplinary work involving various stakeholders are preferred.

Field experience: Wildlife students now often suffer from nature-deficit disorder and lack the outdoors experience that historically was common. Thus a core component of any wildlife curriculum should be to include as much field time as is feasible for the students. Field skills can only be developed in the outdoors. Seasonal internships with wildlife agencies can be an effective way to gain this type of hands-on knowledge.

Critical thinking: Every wildlife graduate should be able to think critically. Instruction in wildlife programs should not focus on telling students *what* to think, but rather helping them learn *how* to think.

Building on the Core

The basic educational competencies for a wildlife degree usually can be acquired on university campuses. In most wildlife programs, the faculty members are diverse and can

A core component of any wildlife curriculum is hands-on field experience (photo: Jim Heffelfinger)

represent a variety of specializations. Thus there often are opportunities to build on these strengths in various programs by developing options or concentration areas that provide students with additional instruction and training that will enhance their core understanding. Some potential (but not exhaustive) options include geospatial analysis, math and quantitative sciences, behavior, ecology, and conservation biology.

Geospatial analysis: GIS is used in most field studies, so it will benefit nearly all students to have exposure to and an understanding of GIS and remote sensing. It could be reasonably argued that in today's world, GIS should be one of the core requirements for a wildlife graduate. Various courses or practical field experience in GIS, remote sensing, geography, mapping, and related areas could be used to develop a student's expertise in this area.

Mathematics and quantitative sciences: There is an increasing demand for wildlifers with strong mathematical and quantitative skills. Courses in modeling, regression, experimental design, survey design, programming, multivariate analyses, and quantitative population ecology are preferred.

Behavior: In programs with faculty whose strengths are in this sector, it might be reasonable to develop a program option in animal behavior. This would include courses in behavior, psychology, evolution, and supporting areas.

Ecology / evolutionary biology: For students who may not have a strong interest in management and conservation, a program option in basic ecology / evolution might be suitable. Additional courses to supplement the core offerings might include community ecology, plant and population ecology, and evolution.

Conservation biology: Many programs already include conservation biology in their names or have that option within their curriculum. Coursework in this area should focus on the application of ecological principles to the conservation of biological diversity. These probably would differ somewhat from wildlife management classes, in that they would concentrate on wild species that are rare or endangered and would lack an emphasis on topics such as harvest management.

ADDITIONAL PERSPECTIVES ON UNDERGRADUATE WILDLIFE PROGRAM REQUIREMENTS

At this point you should have identified a university and a wildlife program that appear to be right for you. And you may have already been admitted to the program of your choice. But you should not stop and simply wait for knowledge to be offered to you. Now is the time to become proactive and consider several different initiatives that can be used to your advantage as you pursue your chosen goals. In short, it is time to stand out from the crowd so others will take notice of you and your skills and abilities. The sections that follow provide advice from a variety of wildlife professionals on how you can make that happen.

Theory versus Practice

To best prepare students for working in an increasingly complex world, universities need to provide a strong foundation in basic science and ecological theory and ensure that training in applied skills also is included in their curricula. Critical needs include giving students opportunities to appreciate and understand the principles of the North American Model of Wildlife Conservation, the role of wildlife and habitat management as an important part of the profession, the tenet that there is more to conservation than wildlife research, and the political realities of managing wildlife populations. Students must be taught to be flexible, in order to meet new challenges presented from shifting stakeholders and to understand that, for many agencies, some stakeholders (hunters) are customers. Students are therefore responsible for choosing a program that best fits their interests yet is also broadly based. Employers must be realistic about the work readiness of new graduates and collaborate with universities to help identify and develop the skill sets that they need. Universities should allow as much flexibility in coursework as possible, while still providing instruction in basic theory and practice. Life-long learning is essential for professional growth and development, and the use of continuing education to maintain and enhance the skills of wildlife professionals should be encouraged.

Coursework versus Experience-Based Learning

Early in the history of humankind, all learning was the result of experience. Later, humans managing to live long enough to produce offspring began, at some point, to educate their

A component of wildlife education is involvement with lay groups, such as local business organizations (e.g., Rotary Club, Lions Club) and schools. Here, PhD candidate Serra Hoagland demonstrates how to use a trail camera to local youths at a science fair in Flagstaff, Arizona (photo: Serra Hoagland).

Wildlife curricula contain considerable time away from traditional classroom settings and involve training outdoors (photo: Dustin Angell)

children about survival. As societies became larger and more complex, teaching styles evolved, eventually becoming the successful classroom style that is so prevalent today. Much knowledge has been imparted to students within the structure of the classroom, but this type of education goes only so far. Learning also arises from many things experienced outside a classroom. In few disciplines is this more evident than in wildlife science, where much of what we study is, by definition, wild—and outdoors.

In the past decade or so, more emphasis has been placed on experiential learning techniques, even as early as middle school. Evidence of the effectiveness of this type of learning exists in classes on wildlife techniques. Examples include courses at the University of Missouri–Columbia and Michigan State University, where students are presented with telemetry projects and allowed to discover the best techniques to use through trial and error. Another tool for experiential learning is on-campus research. Texas A&M University–Kingsville has successfully incorporated on-campus research on squirrels, feral cats, and bats. In all of these examples, projects are first introduced in the classroom, prior to actual work in the field.

Certain themes develop when we move from a discussion of *how* learning takes place to *what is taught* in the first place. Until recently, most undergraduate courses dealing with wildlife statistics were taught in mathematics or statistics departments. These departments, while sound on the quantitative aspects, often overlooked the nonmathematical elements of statistics, such as planning and managing scientific studies, defining the scope of the problem, explaining data-collection protocols, and developing sampling schemes. In the quantitative sciences and in other fields involving techniques and research, a problem-solving approach has been found to work best. The Department of Fisheries and Wildlife at Michigan

State University, for example, uses a problem-solving model to teach quantitative analysis skills.

In an attempt to obtain an up-to-date look at state wildlife agency requirements and expectations for entry-level candidates for wildlife biologist positions, the National Wild Turkey Federation conducted an informal email survey of their technical committee members. This committee is composed of biologists in each state wildlife agency who are responsible for wild turkey management in their state. The survey was simple, with only three questions, but the results provide a good picture of what state wildlife agencies are looking for in entry-level biologists. From the 17 agencies that responded, 59% required a bachelor's degree as the minimum amount of education for an entry-level applicant. One to two years of experience sometimes was a hiring criterion, but in several states a master of science (MS) or higher degree was an acceptable substitute. Over 82% of the agency biologists stated that basic biology, population management, and data-analysis skills were important for their agency's jobs. The top response in terms of people skills was the ability to communicate effectively with the public.

Current research has shown the effectiveness of experiential learning in promoting interest and aiding retention, from grade school all the way through university programs. Many university programs are unable to take full advantage of available techniques, however, due to time and budget limitations. In addition, state wildlife agencies are looking for applicants possessing both advanced degrees and on-the-job experience, and they complain that qualified applicants are hard to find. Tools for experiential learning are in place, and some schools have taken advantage of these techniques. Others, due to administrative constraints, are seeing a trend away from experience-based learning within their programs and toward a combination of coursework and either internships or summer work experience.

Employer-Desired Competencies

Curriculum development should be based on an inventory of what wildlife professionals have expressed as the knowledge and competencies they prefer in students. In January 2009, all TWS members with email addresses ($N = 7,381$) were sent a personalized electronic invitation to participate in a survey that identified what students needed in terms of university training. Responses were received from 1,750 TWS members (418 state agency personnel, 342 federal agency personnel, 111 nongovernmental organization [NGO] personnel, 235 members from the private sector, 218 university personnel, 27 local governmental personnel, and 8 Native Americans).

Respondents from all nonacademic sectors of the profession identified nontechnical aspects of the job as the most important areas of competency (effectively communicating orally and in writing, working well in teams, and interacting with stakeholders). Although nonacademic respondents tended to rate the proficiency of recent entry-level hires highest in these areas of competency, their mean ratings of importance and proficiency differed more for nontechnical than technical areas of competency. (Chapter 4 addresses other skills needed for nonacademic wildlifers.)

Nearly all universities emphasize the importance of writing and speaking skills in their curricula. Yet the results of this survey suggest that despite that emphasis, students may not be leaving school with the desired levels of competency in communication skills. One potential strategy for improvement is to incorporate more writing assignments in university courses, although this approach is time-consuming and potentially costly, in terms of additional faculty needed. The perceived problem with written communication skills may be related more to the types of written communication that students learn versus what they need on the job. Most writing assignments in college, whether they are term papers or theses, focus on research. Many wildlife professionals, especially those who work for governmental agencies, spend more time writing environmental assessments or management plans than research publications or reports.

Improving students' ability to interact with stakeholders or in team settings was identified as an increasingly important area for university instruction. Although many classes incorporate team assignments and some universities offer courses emphasizing human dimensions—which are skills that have been added to the certification requirements for both TWS and the American Fisheries Society—human dimensions are still an underdeveloped area in the curricula of many universities. Students seem to readily understand the need for stakeholder involvement in making decisions about wildlife conservation. But they have a more difficult time in understanding the pros and cons of alternative approaches to public involvement in decision making. This suggests that although

students should be exposed to the various aspects of human dimensions in their college courses, the most effective way of improving their ability to interact with stakeholders may be through on-the-job training, as well as continuing education for wildlife professionals who have gained some real-world experience.

The overall picture that develops from TWS's survey is that gaps remain between what is taught by professors and what is desired by potential employers of graduates from wildlife programs. In particular, faculty and students in this area see proficiency in oral and written communications and in teamwork as ancillary rather than necessary skills. It seems reasonable—and important—for agencies and academic institutions to continue to strive to bridge this divide, shaping well-trained professionals who are ready to enter the workforce.

Identification of Customers and Customer Needs

Another important aspect related to the preparation of graduates for careers in wildlife conservation is in interactions with stakeholders. There is a widespread perception among employers that their new wildlife hires often lack a background in and experience (or even familiarity) with hunting and trapping; increasingly come from urban backgrounds; and arrive at their jobs without practical field knowledge and hands-on basic training in the use of field equipment.

Employers' needs have become more varied, especially as a result of the emergence of NGOs (e.g., The Nature Conservancy, The Peregrine Fund) as significant employers of professional wildlife staff. In addition, the duties in agencies and NGOs are more complex, due to an increased emphasis on collaboration, demands for government to be more transparent and responsive to all stakeholders, and systems-based approaches to management. It is impossible to define a single set of skills, knowledge, and training that agencies need in their professional wildlife managers, and no one person has all of these diverse attributes. Rather, employers need to hire and develop employees with a range of abilities and interests to ensure that those varied needs can be met over time as these individuals develop throughout their careers.

Employers can help to address this gap in a number of ways. They can broaden their recruitment efforts; review their minimum requirements in terms of education and experience for new hires; participate in university advisory councils and curricula reviews; support university efforts to provide experiential learning opportunities; engage TWS in discussions related to certification, to ensure that this is relevant to the work of their organizations; support university graduate programs with projects relevant to the employer's work; and develop and fund internship programs to give students real-world experiences. Employers can also support lifelong learning and the further development of their employees by instituting workforce planning processes that define both the future work and the employee skills and experiences needed to do that work; creating employee orientation programs; requiring employees to formulate development plans; providing mentoring opportunities; encouraging employee participation in leadership programs; implementing programs geared toward developing the skills and abilities needed at all levels in the organization; supporting employee participation in professional meetings and conferences; and providing financial support for employees interested in pursuing continuing education opportunities. Generally speaking, universities that encourage the inclusion of employer advisory committees are probably more closely connected to the real-world job market than those that do not. Hence students may wish to find out if their chosen university offers employers the opportunity to become involved in curricular development, and then make their selection based on this criterion.

Dialogue and debate around this topic are healthy and necessary ways to encourage discussions and collaborative approaches between educators and employers in addressing the challenge together. Educational institutions and employers need to be in constant communication. Universities cannot define their programs without input from employers; and employers cannot place unreasonable demands on universities or their graduates. In short, professional education does not end once you earn a degree. Instead, graduation is only the beginning of the second phase of a long and continuous educational road.

CONSTRAINTS AND CHALLENGES

Too much to know, too little time. The field of wildlife management / biology / ecology / science has changed substantially over the past several decades. There is much more for students to grasp, they often come into their undergraduate programs less prepared than previous students, and there is precious little time to provide them with everything they need to know (or what we think they need) in four years. Additionally, as budgets become tighter, some higher-level administrators are pushing for shorter semesters (certain universities now have 13-week semesters, compared with a traditional 15-week term) and fewer credits necessary to graduate. Faculty members often feel that, even after four years, they are not able to fully prepare a student to be gainfully employed, and many employers agree.

SUMMARY

By following TWS certification requirements for coursework, universities should be able to provide wildlife students with all of the necessary classes in the biological, physical, and quantitative sciences; the humanities; communications; and policy, administration, and law. Students' preferred universities should be those that offer additional core areas, including participation in teamwork, an understanding of stakeholders, field experience, and critical thinking skills. Some potential (but not exhaustive) options for universities to build on these core competencies include geospatial analysis, additional quantitative skills, behavior, evolutionary ecology, and conservation biology. Students must recognize, however, that everything they need for job success cannot be provided in a four-year program. Further study and lifelong learning are essential for any student seeking a career in the wildlife field. Therefore, professionals in this area should be sought out and their advice taken before, during, and after a student becomes involved in a university wildlife program.

LITERATURE CITED

Baydack, R. K. 2009. The Wildlife Society ad hoc committee report on collegiate wildlife programs: summary report to The Wildlife Society Council, http://wildlife.org/wp-content/uploads/2015/03/Ad_Hoc_Collegiate_Wildlife.pdf.

Wallace, M. C., and R. K. Baydack. 2009. The diversity of options for wildlife education. Transactions of the North American Wildlife and Natural Resources Conference 74:54–61.

Advice for a Student Entering the Wildlife Profession

A Professor's Perspective

SCOTT E. HENKE AND PAUL R. KRAUSMAN

As the heads of range and wildlife programs at different universities, we often get asked to talk to incoming students and their parents. Part of what we say is informational, explaining the workings of a university (e.g., required coursework, scholarship opportunities), but much of it is motivational. Our goal is to let new students know how they can succeed in a university and subsequently position themselves to enter the wildlife profession. Students often ask about potential social life and joining student organizations, while parents most often inquire about employment opportunities after graduation. Our speeches appear useful to both audiences, because sometimes we see the students nod their heads and smile, while, during other aspects of the talk, it is parents who appear to like what is being said. We believe, however, that what we have to say contains pertinent information for many groups. Much of our advice in this chapter comes from anecdotal stories that have occurred over the course of our careers as professors. There is no particular order to the following thoughts, because we have found that people and their needs differ greatly, but we have categorized our comments under three main groupings: general advice, input about academics, and pointers about gaining work experience. Our hope is that by reading this chapter, you will find value in the suggestions that are offered, which can help guide you in your wildlife career. Because many of our anecdotes are personalized, they were written in the first person, but similar stories could be espoused by either of us. Also, while these are real examples, names were changed to protect people's identities.

GENERAL ADVICE
Luck Has Little to Do with Success— Hard Work Pays Off

I have heard some say that I am lucky, because I have a good job, with flexibility. If one of my children is sick or has a ball game, I can take time off to take care of them or attend their functions. But was it luck that I made good grades, was accepted into graduate school, and earned a PhD? Or that I studied some evenings, rather than going out with friends? Or that I missed holidays with my family while I was out in the field conducting research for that PhD? Or luck that I often worked late into the evenings, rather than sleep? Were all these things luck—or my choices? I would argue that they were conscious decisions on my part. I deliberately gave up some of the fun stuff early in my life to reap the benefits later. Sometimes the best option is not "live for the moment." Life is full of choices. Every decision can have great ramifications in your life. Always be aware of the options available to you and carefully weigh them. If you do not, you could find yourself on a less desirable path.

Attending a University Is a Privilege, Not a Right

Unfortunately, I have met students who seem to believe that, if they pay tuition, they are entitled to a diploma. But college does not work this way. Instead, think of college as being similar to the lottery. You pay your money (tuition) for an opportunity to win (receive a diploma). The good news is that your chances of succeeding in college are much greater, because you are in control of the choices that can dictate whether you graduate or not. It is important to understand that what your tuition buys you is the precious opportunity to be educated. However, the ability to turn that opportunity into academic and professional success cannot be bought. It must be earned.

School Is an Investment in You—Who Better to Invest In?

It seems as if education is one of the few enterprises in which the client (the student) is usually very happy to get less than what he paid for. Students too often choose to do the minimum to get by, or take classes just for an easy A. They appear to be more interested in graduating quickly, rather than receiving a good education. The problem with this kind of thinking is that the majority of employers desire quality, both in a person's past experience and in his or her current work. Take the time to always perform at your very best and produce top-notch work. If several professors teach the same course, sign up for the faculty member who has a reputation for demanding the higher standards. Such efforts get noticed, and they will pay off in the end. For example, Josh was a student with a get-by attitude. He passed his courses and received a bachelor of science degree in wildlife management. After working for a few years, Josh discovered that most of the wildlife jobs that interested him required a minimum of a master of science (MS) degree. Josh contacted me to inquire about our graduate program, so he could earn an MS degree and make himself marketable for these jobs. Unfortunately, Josh graduated with a 2.52 grade point average (GPA) for his undergraduate degree. The majority of wildlife graduate programs in the United States require a minimum GPA of 3.0 for admission. Josh soon learned that merely getting by early in life limited him to a wildlife career without much opportunity for advancement. Therefore, use your time in school as an investment in yourself. If you don't take your education seriously, then why should a future employer do so?

Life Has Consequences—Always Be Aware of Them

Although it is very easy to live for the moment, you should always pay attention to the potential consequences of your actions. You need to consider the short- and long-term outcomes of each choice. Weighing them out should help you make better ones. For example, if you drink alcohol and then drive your vehicle, potential negative results could arise. For instance, you could cause a serious accident. One short-term consequence could be that you get arrested for driving while intoxicated (DWI). You plead guilty, receive a fine, and perhaps are required to perform some community service. You've paid your debt to society, so that's the last you'll ever hear about this, right? Wrong. Years later, you've just graduated with your wildlife degree and have applied to a state wildlife agency for a biologist position. As part of the job requirements, you will be driving a state-owned vehicle. Unfortunately, your record shows that you once were arrested for DWI, and the person doing the hiring does not want to create a potential risk for the agency by offering you the job. Hence, a long-term consequence! It is one that is very realistic and needs to be con-

sidered before you get behind the wheel of a vehicle if you've been drinking alcohol. Likewise, experimenting with drugs in your college career can preclude a career in wildlife law enforcement, as well as other options. Many state game and fish agencies have restrictions on employing a person convicted of any level of unauthorized drug use. With such a history, your wildlife career can be over before it even begins.

Develop Your Ethics

Ethics are the principles that guide us, especially when no one is looking. Employers want people on whom they can rely. They need people who can get the job done, both well and on time. A solid work ethic will be noticed by professors and employers and will result in excellent reference letters. For example, Lance was an undergraduate research assistant who helped with numerous graduate student projects. He did good quality work and was an extremely honest person. If he made an error during data collection or analysis, he would acknowledge it and rectify the problem. He would let others know if a job was completed or not. He would show up for work on time, or he would call to say he would be late, even if only by a few minutes. After a while, supervisors knew they could depend on the quality and extent of Lance's work; they did not have to inquire if he showed up for a project or check on his progress. Whatever Lance said was always true. This is the type of employee that everyone should strive to become—someone who is competent, dependable, and, most of all, completely trustworthy. Needless to say, when Lance graduated, he received excellent letters of recommendation that praised his high level of integrity to future employers.

Attitude Is Everything

There is a story about a bystander who gets shot in a convenience store robbery. He's taken to the hospital and, while lying on the emergency room table, is asked if he is allergic to anything. The man opens his eyes and says "Yes." The doctors and nurses stop what they are doing, knowing that giving him certain medicines could have dire consequences, and ask, "What are you allergic to?" The man again opens his eyes and replies, "Bullets!" The moral here is that he kept a positive attitude and joked, even when his situation was critical. A positive attitude is always appreciated by everyone nearby.

I have lived in southern Texas for nearly 25 years. It is fair to say that this region is quite hot during the summer months. In fact, it has been referred to as *Hell's furnace*, with temperatures exceeding 110° F and humidity greater than 90%. It doesn't help to complain about how hot it is. Trust me, everyone already knows that. Life is 10% about what happens to you and 90% about how you react to that 10%. So if work conditions become stressful, think of the story about the man on the emergency room table. Stop complaining about workloads or assignments, which gives the impression that you are a whiner. Instead, lighten your coworkers' stress by laughing about the situation. You will find that most people prefer to work with a person who is upbeat.

You Are Always in the Public Eye

You are being evaluated every day by the people around you, including how you dress, how you drive, and what you say. You only have one chance to make a good first impression, so use it wisely. For example, I had a student working on a research project in a nearby town who used a university truck to travel to and from his research site. One day, he thought he was running late and decided to drive recklessly (speeding and cutting in and out of traffic to pass other vehicles). One person he passed wrote down the individual vehicle number, which appears on the back of every university truck. That person's complaint was tracked to my student, through the vehicle number. He was reprimanded for his erratic driving while on university-related business.

Another example of being careful about your public appearance involves social media, such as Facebook and similar sites. Keep in mind that potential employers are likely to investigate your social media presence as they consider your candidacy. It can be challenging to keep privacy settings locked down for every account you have, so be thoughtful about what you post. If you post 10 different photos of you drinking alcohol, it may give someone who doesn't know you the impression that you have a substance abuse issue. Is that really what you want to convey to a potential employer? In one example, Nate was completing his final semester as an undergraduate wildlife major. He had a 3.6 GPA, a variety of experiences with animals and habitat manipulation, and solid references, all strong qualifications for a particular job. A state wildlife agency was considering hiring him as a biologist. As part of their background check, the agency pulled up his Facebook page and saw that he had posted numerous photographs of himself drinking alcohol with friends. Because the position required the use of the agency's trucks and boats, he was perceived to be a liability as a potential employee. When Nate learned that he lost the job due to his Facebook page, he removed the photographs, but the damage had already been done. Like it or not, public forums such as Facebook are fair game for employers, so be sure to project a professional image if you want to be taken seriously.

Act and Dress Like a Professional at All Times

I am not saying that you cannot have fun, but remember, you're always in the public eye. Make sure your behavior (and language!) seems appropriate to the people around you, especially if you are in a uniform or a vehicle that can help identify you. Speaking as an employer, I would not want to hire a person whose language in the grocery check-out line would be more appropriate in a Samuel L. Jackson movie, because much of wildlife management is people management. Nor am I alone in this belief. You must keep in mind that the average age of a person doing the hiring for wildlife jobs is about 50 years old and may not come from the same cultural background as you do, and may not share the same attitudes and beliefs, so it is important to be respectful of others, in the same way you want them to respect you. For instance,

Remember that while you are in public, especially in a company uniform or vehicle, someone is always watching (photo: Scott Henke)

a previous generation only wore hats outside and removed them when entering a building. Such behavior was exemplified by Andrew "Bum" Phillips, the head football coach for the Houston Oilers (now the Tennessee Titans). Bum Phillips was known for wearing a 10-gallon cowboy hat during games, but he never wore his hat during home games, which were held in the Astrodome. When asked about this, Coach Phillips replied he was taught to never wear a hat indoors—and the Astrodome has a roof. Remember this story as you enter a building. Remove your hat, and I guarantee that you'll make any nearby older folks smile.

Facial tattoos and piercings can be an important part of some people's cultural heritage, but if you consider these simply for the sake of embellishment, you should weigh the impact they can have on the first impression of a potential employer. When I was 20 years old, I worked in Kenya and had many friends who were Maasai. Members of the Maasai tribe are known for fashioning large loops in their earlobes when they reach the age to be considered adults. One night my friends decided that I was old enough and to have my earlobes cut. Although such minor surgery would make me fit in well with my Maasai friends, I knew that one day I would return to the United States, where it would be difficult to explain these loops. Needless to say, I talked my way out of having this done, and I'm confident I would have had difficulty obtaining employment if I had not.

When addressing people, either verbally or in writing (via letters, emails, and texts), make sure to use a proper title. If you are not sure what it is, find out in advance or ask people directly what they prefer. For example, someone elected to the House of Representatives or the Senate should be addressed as Congressman (or Congresswoman) X or Senator Y, respectively. A person with a PhD degree should be called Doctor Z. Taking the extra effort to learn people's titles and refer to them appropriately shows respect, which will be reciprocated. Also, never use a person's first name until you are asked to do so. Again, it's all about respect.

To end this section with a positive story, after graduating

with my bachelor's degree, I applied for a temporary position with the Indiana Department of Natural Resources. As I was leaving the interview, an older gentleman in the department entered the room, stating that he was having car troubles and needed to get across town quickly. I said I was heading that way and could take him to his destination. While driving there, we came across a funeral procession heading in the opposite direction. Even though our paths would not cross, I was taught that it was polite to stop your vehicle and allow the procession to pass. After these vehicles had gone by, I dropped the gentleman off at his destination and went home. The next day I received a phone call saying that I got the job, in part because I stopped for the funeral cortege. The caller explained that much of the work I was being hired for involved dealing with the public, and stopping to let the funeral cars go by displayed sympathy and concern for people, which is what they wanted in an employee. Thus even what you may think are minor actions or behaviors are noticed, so always be mindful of what you do.

Own Up to Your Mistakes

Take responsibility for your actions. No one is infallible. If you make a mistake, do what you can to correct it. Do not shift the blame, lie, or otherwise attempt to cover up your error. That only makes matters worse. In my experience, all mistakes eventually surface, and it's just a matter of time until they become known. Learn from your errors and don't repeat them. Look on them as a way of bettering yourself. People, especially employers, will respect you more if you admit to a mistake as soon as you learn about it and offer ways to rectify it. Several years into my career as a professor, I had a student named Lee, a member of the college football team, who did not pass my Introduction to Wildlife Management course. In the past, I had several other players fail that class, and they all provided excuses as to why that happened: football required too much of their time; the coaches wouldn't let them study past 2200, because they needed to sleep; they got hurt during practice and the trainer told them to rest, so they couldn't come to class; and so on. Many others earned an F in the course without offering any explanation. Lee was different, however. He came to me and apologized. He stated that he loved the class and enjoyed the material, and not passing was entirely his own fault. After our conversation, I decided to hire Lee as a research assistant, because of his ethics in owning up to his mistake. This was a good decision, because Lee was an excellent employee and maintained his impeccable standards throughout his job.

Life Is About Balance

You may have heard the old adage, "Do things in moderation." You need to find a good balance between your school life (i.e., your job) and your social life. Part of the university experience is to have fun, but do not let this become your main goal while you are in school. If you do nothing but work, you could burn out early in life. At the same time, you can't always play and entirely forget about schoolwork. Unfortunately, I see the latter occurring all too often. I work at a university that is located near the Gulf of Mexico. If the fishing is good, students frequently neglect to attend class. Develop and maintain priorities. School should be your top priority, but on occasion, do have fun. Keep a healthy balance between the two while you are achieving your ultimate goal of becoming a wildlifer.

ACADEMIC ADVICE
You Must Attend Class

The transition from high school to college can be difficult. Typically, this is the first time students are on their own, making daily decisions for themselves. This new freedom, however, comes with additional responsibility. You no longer can rely on parents making sure you wake up on time and go to school, or asking if you did your homework. Currently, the nationwide retention rate is lowest for university students between their freshmen and sophomore years, meaning that if students drop out of college, this typically occurs before they enter their second year. Academic suspension for poor grades is the biggest reason why students leave college, which is linked to lack of attendance. Students need to develop good habits, both in going to class and consistently arriving on time. Such priorities will make your life easier when you enter the workforce. An employer expects employees to arrive on time and do their work. Why should professors expect less from their students? From a student's perspective, would you buy concert tickets and then not show up for the event? Tuition is paid at the beginning of each semester. Since you've already invested in your education, why not attend class and reap the benefits? You'll be surprised at how your grades will improve.

Most Students Are Capable of Receiving an A in Every Course

I have had this debate with students nearly every year that I have been a professor, but I adamantly believe the above statement is true. I have had students tell me that math or chemistry is just too difficult for them, and that it is impossible to get an A in such subjects. They typically name a person in their class who does very well, with seemingly no effort. Sure, we all know someone who could read a chapter once and remember everything, and could ace an exam with little to no effort. We disliked those people when I was in school, and I'm sure the same still applies for students today. Most of us have to put in a lot of effort to obtain that result. Some individuals are better in math, others excel in English, and still others think science is easy. Courses in which we are weaker require more on our part to achieve top results. Good grades depend on how much work you are willing to put in. Remember, life is about choices. Greater effort today can pay off later in life.

Each Class Requires Study Time

This concept goes hand in hand with the one above. In general, about 10 hours of studying per class per week should

As a student, treat school like a job. Attendance is expected (photo: Scott Henke).

Find a healthy balance between your work life and your social life (photos: Scott Henke)

be expected. For example, if you are taking four courses during a semester, then you should be spending about 40 hours per week studying. Students should consider school to be similar to a job. A typical workweek is 40 hours, and the same amount of time and effort should be given to school-work. Subjects in which you excel may not require as much time, but subjects that you find difficult will require more. Know your strengths and weaknesses. Many universities offer free tutoring services, so use them when needed.

Get to Know Your Fellow Students

Although your classmates can become potential competitors for future jobs, they also can become future employers and reference sources. Get to know them and begin networking while you are a student. You never know what the future may bring. For example, Jennifer and Erin met in college and became friends. Jennifer graduated the year before Erin and got a job with a wildlife consulting company. The company did well and wanted to increase the number of biologists on staff, so Jennifer recommended Erin. Today, the two friends still work together.

Grades and Your GPA Are Important

As a student, you should know what your GPA is at all times. As a professor, it is disheartening if I ask students about their current GPA and they tell me they don't know what it is. This tells me one of two things: either they are embarrassed to mention their GPA, because it is low, or they don't care about school. If you, as a student, aren't concerned about your career or your future, then why should anyone else be interested, especially potential employers?

Your GPA gives employers an insight into the quality of your work and your work ethic. Employers rightly believe that if you excel as a student, the same level of performance should continue when you are an employee. If employers consider GPA to be an important hiring criterion, then you should, too. One aspect of my job as department chair is to keep track of our students' employment. Undergraduates have been shocked by the trends I've discovered. Career opportunities are significantly correlated with graduating GPAs. Also, as GPAs drop below 3.0, certain job opportunities appear to be no longer available. To illustrate this point, I tracked the immediate employment of more than 300 students who had earned their bachelor's degree in range and wildlife management from Texas A&M University–Kingsville during 2005–2015. Of those students who had a GPA greater than 3.0, more than 90% had accepted wildlife employment or graduate school positions by the day of their graduation. The percentage of students who were employed in the wildlife profession right after graduation decreased exponentially as their GPAs declined from 3.0 to 2.0 (a C average), and it was extremely rare for a student whose GPA was less than 3.0 to be accepted into a graduate program.

In addition, students whose GPAs were less than 2.75 typically had wildlife ranching enterprises as the only employment option in their field of study. Therefore, it is important to realize that the wildlife profession is very competitive, and you must maintain your edge. A GPA of more than 3.0 will keep you in the running for most jobs. Also, bear in mind that most wildlife graduate programs in the United States have a minimum admission standard of a 3.0 undergraduate GPA.

You may not think graduate school is right for you at the moment, but what about in your future? As a case in point, Jacob was an average student who graduated with a 2.4 GPA and began working on a wildlife ranch in Texas. Initially, Jacob was content with his job, but after several years, he felt that he had reached the pinnacle of advancement on the ranch. Jacob desired more challenges from his career and inquired about graduate school, in order to increase his marketability for wildlife employment. Unfortunately, he quickly learned that his GPA, although adequate for graduation with a bachelor's degree, was substandard for admission to graduate school. Although his academic advisor had told him this could be a future problem, Jacob did not listen to this counsel and destined himself to a wildlife career with limited opportunities.

Lastly, maintaining a high GPA can pay off later. For example, I knew two recent graduates who both were hired by the same US Fish and Wildlife Service wildlife refuge. Both young men worked side by side, doing the same jobs each day. Because one student graduated with a higher GPA (more than 3.5) than the other (ca. 3.0), the former was hired at a bigger annual salary. Moreover, the discrepancy between their rates of pay became larger with each cost-of-living percentage increase. I've told students this story, and many do not think it is fair, because the two men were doing the same job. Whether fair or not, it still is a reality. Wouldn't you rather be the employee who receives the larger annual salary? You can, by keeping your grades as high as possible.

Don't Overthink Simple Problems and Questions

For each class, I provide students with a review sheet for exams. Unfortunately, some students believe I will trick them on tests and ask questions concerning topics that were not on the review list. Most professors, however, want to assess a student's knowledge, and this cannot be accomplished through

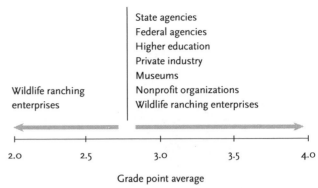

A comparison of wildlife-related opportunities for students whose final overall GPA was 2.75+ (based on a 4.0 scale) with students whose final overall GPAs were less

tricks. It can be done by asking straightforward questions and wanting relevant answers. Do not think about possible exceptions to the rule. To make my point, I have asked wildlife students, when they see ducks and geese flying in a V pattern, why one side of the V is always longer than the other. They might contemplate animal behavior and the fluid dynamics associated with wind resistance, but the answer is simple—there are more birds on that side! People tend to overlook obvious solutions. Try, instead, not to overthink the situation.

Don't Rush Through Exams

There is no additional credit for being the first person to turn in a test. I've had students complete what I considered to be a three-hour final exam in 30 minutes, although it is rare for a student who leaves an exam so quickly to have performed well. Therefore, use all the time allotted to you. If possible, recheck your answers. Students who walk out of an exam thinking it was easy typically are not smarter than other students, they were just better prepared. Remember, each class requires study time. Make sure you have studied properly before you enter the exam room. Do not cram for tests, since the information gained in this manner is rarely retained. Instead, study the material that was taught during the week. At the end of each week, review what was covered then and in the previous weeks. You will find that your confidence in mastering the material for an exam will increase substantially, because you will be better prepared.

Stick to the Questions Asked

Carefully read the actual question on an exam, and make sure you don't mentally insert additional words into the question. Do precisely what is requested. If your response to an exam question does not stick to was asked, it can appear as if you are rambling and perhaps don't know the actual answer. Professors refer to this approach as the *shotgun method*, with students hoping that at least one portion of what they write is on target. Sometimes what a student sets down may be correct,

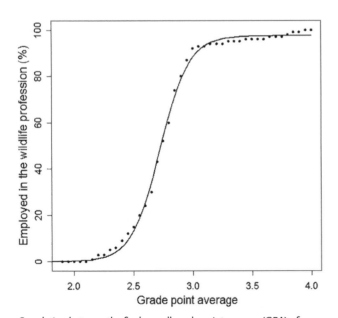

Correlation between the final overall grade point average (GPA) of a student graduating from the wildlife program at Texas A&M University–Kingsville (*N* > 300) and the percentage of those students who had accepted employment within the wildlife field or been admitted to graduate school by graduation day

but it does not answer the question. For example, it might be 1000 on Tuesday. Suppose I were to ask you, "What time is it?" and you reply, "It's Tuesday." Well, yes, it is Tuesday, but that's not what I asked. Even if the correct answer is included within a student's response, I typically deduct points, because it is not obvious that the student knew which portion was right. Therefore, do not expand on your answer *unless* the additional information is relevant to the question asked. Also, do not rewrite the exam question and then tailor your reply to the change that you made. For example, Charlie was a senior in my population dynamics class. He altered the wording for the only question on the test and did an excellent job of answering his revision. Unfortunately, it did not apply to the question being asked, and his final grade went from an A to a C. Such attempts fool no one. They give the impression that, as a student, you are as unprepared as Charlie.

Do Your Work Promptly

Professors, as well as employers, expect assignments to be completed correctly, thoroughly, and on time. Teachers may not have the ability to fire a student for a poor performance, but they can assign a low grade for inadequate or tardy work. Such students should expect poor grades, especially if they habitually procrastinate. Deadlines for assignments generally can be met, unless you put off getting the work done. For example, I often require students to complete a semester-long project and, as such, I expect a thoroughly prepared assignment, given the length of time that is allotted. Recently I had a cohort of students complain that my 16-week deadline was unreasonable. Because I had never given such an assignment in the past, I was unsure of the actual time required for completion; therefore, I granted a four-month extension. When the work was turned in, I asked students how many hours it took to complete it. All of them stated that it required 20–25 hours, which averaged out to 1.5 hours per week during a 16-week semester—an achievable task. The lesson here is that procrastination can make a doable task seem impossible. Promptly beginning assignments and properly managing your time, however, can make it possible to finish even the largest tasks on schedule. Remember the story of the hare and tortoise: steady and slow will win the race.

Develop Self-Discipline and Time Management Skills

Humans are creatures of habit, and psychologists say that it requires at least 28 days to develop a habit. So start today to form good ones. Self-discipline when in school requires time management. Wasted hours and days are wasted forever. You cannot turn back the clock. I knew of a nontraditional student who had been an oilfield worker but decided to come back to school for a wildlife degree. He used to leave for work at 0700 each weekday morning and return home to to his wife and two young children by 1800. His evenings and weekends were reserved for his family. He treated school just like his former job. Once again, he left each morning at 0700, regardless of when his first class of the day was scheduled. When he was not in class, he was in the library, completing assignments, reading textbooks, or studying. At 1800, he packed up his books and went home to be with his family. He graduated with a bachelor's degree in wildlife without ever working after 1800 or during the weekends (except for field trips). Proper time management can make this scenario possible for any student. Get into the habit of not wasting time, and you will find that you have enough for both school and fun.

Do Not Sit Back and Stay Quiet—Participate!

Professors have a tendency to remember two groups of students: those who perform exceptionally well, are intelligent, and are go-getters; and those who spend most of their time trying to figure out how not to do the work. Those who sit quietly, who never ask questions, or who do not offer their opinions—in other words, those who do not visibly participate—tend to be among the forgotten masses. Do not let this happen to you. Professors are good sources of employment recommendation letters, but they will have to know you well to provide you with a reference that can help you obtain a job. Get to know your professors!

EXPERIENTIAL ADVICE
Gain Experience Early

Get involved in student organizations, and with professors and graduate students. Employers demand employees with a diversity of experiences. Therefore, start participating in wildlife activities as a freshman. Do not wait until you are a senior before beginning to make yourself marketable to potential employers. A major advantage of gaining practical expertise early is that it can help you decide if your desired career path is right for you. For example, in Texas we have developed a shadowing program in which a number of wildlife professionals have agreed to allow students to shadow (follow along with) them during a typical day's work. I encourage all my students to participate in this program. More often than not, students return from a shadow day stating that the job was not what they thought and that they are no longer interested in it. Although this reaction may seem negative on the surface, it actually is positive. Wouldn't you rather determine that a particular job is not right for you early in your academic career, while you still have time to make a change? The average student requires about 3–4 shadows before they find a wildlife career path that they truly enjoy.

Be open to any wildlife experience that is offered, even if it is not your main interest, because it can pay off in the future. For example, when Michele was an undergraduate, she wanted to work solely with white-tailed deer. If an animal didn't have antlers, she wasn't interested. Michele did well in her ornithology class, however, and became well versed in the birds of southern Texas. This skill was advantageous when she landed an internship on a ranch, where she led wildlife tours that included spotting birds. Michele did such a good job that she was hired as a permanent employee on the ranch, where she was able to demonstrate her knowledge of and desire to work with deer. Today, Michele is the head biologist of a more than 120,000 ha wildlife enterprise, where she is involved with white-tailed deer daily. She owes her dream job to birds. Therefore, do not be too selective. Acquire experience with as many species, habitats, and techniques as possible. Who knows which ones will be valuable in the future.

Get to Know Your Professors—You'll Need Job References

Every job announcement and application asks for a minimum of three references from individuals who know the quality of your work and can attest to your ethics and attitude. Begin developing a solid relationship with them now. Employers often expect one or two of these people to be or have been your professors. If not, this lack can serve as a red flag to an employer. Were you not a good student? Did you cause trouble in class? Professors can be helpful references, but they need to know you well in order to write good recommendations on your behalf. Keep in mind that professors are not obligated to provide reference letters for students. It is their choice to do so. If you were just another student in class and did not assist with research under a faculty member's supervision or work with graduate students, and if you were not actively involved in the school's student professional chapters, then what attributes could a professor describe in a letter on your behalf? Do you remember the story of Lance, discussed in the "Develop Your Ethics" section? I knew Lance well: how he reacted when things went wrong, his attitude under extreme weather conditions, his work ethic, and the quality and quantity of his work. I wrote a three-page letter of recommendation for

Lance, praising these characteristics. This is the type of letter to earn from your references.

Networking Is Important for Your Career

Making professional contacts (networking) is extremely important in the wildlife profession. So crucial, in fact, that networking is covered in two other chapters (7 and 9). You need to attend wildlife conferences and meet professionals early on in your academic career. This is not a time to be shy or introverted. Wildlife professionals can assist you with internships, jobs, and scholarships. For instance, Brent was a wildlife graduate student who did not fit into the stereotype: he was from an urban environment, he had never been hunting, and he questioned the system of managing game as opposed to bettering conditions for all species. Up to this point, Brent had applied for various wildlife scholarships, but he had never received one. He then attended a state wildlife conference, where he met a man associated with a local

Professors can give you excellent references for jobs, but you must take the time to get to know them (photo: Scott Henke)

It's never too early to begin gaining experience (photo: Scott Henke)

Attendance at professional conferences is an excellent way to make job contacts (networking) (photo: Wini Kessler)

chapter of the Safari Club. During their conversation, the man asked Brent about his views concerning hunting. Brent confided that he had never been hunting, so the man invited him on an elk hunt, which Brent accepted. Later, Brent was notified that he had received a scholarship from the Safari Club. The gentleman Brent met from the club argued on behalf of Brent's application. As this example illustrates, the benefits of networking are numerous and should not be ignored.

Your GPA Is Important

We can't stress this enough, so we bring up this topic one last time. Do not kid yourself while you are a student and downplay the importance of your GPA. It demonstrates your job quality to employers, and a B average or greater helps keep all wildlife career paths available to you. Do not shut doors to potential opportunities in your future.

SUMMARY

Congratulations, you made it to the end of this chapter. We hope you have found some value in what we have to offer. Please keep in mind that much of our advice is based on 75 years of combined experience within the wildlife profession. We have helped literally hundreds of students obtain their dream jobs, and hope our stories can be of benefit to you, too.

The chapter has offered guidance on how best to become a wildlife professional, from general suggestions to advice concerning academics and how to acquire experience within the profession. Take what we have to say to heart. It really can be useful as you develop your wildlife career.

Skills for Nonacademic Wildlife and Conservation Careers

ERIC WINFORD, MARIT L. WILKERSON, LAUREN M. PORENSKY,
IARA LACHER, KELLY GARBACH, KRISTY DEINER, JESSICA L. BLICKLEY,
AND MARIAH H. MEEK

One of the challenges you will face when entering the job market is determining what skills employers are looking for in a prospective candidate and how these align with your own skill set. Equally daunting can be figuring out how to successfully communicate to a prospective employer, through your application materials, that you have indeed developed those desirable skills (chapter 7).

A wildlife or conservation degree can lead to a wide variety of different career paths (chapter 5). In this chapter, we present information about the skills needed for nonacademic conservation careers and share insights from a survey of 19 individuals who work in the field of wildlife biology or conservation and have experience in hiring undergraduates. They are employed by a mix of nonprofit, for-profit, and federal and state governmental organizations. In aggregate, the group has hired people for field positions, lab openings, seasonal jobs, and volunteer work. Throughout this chapter, we present useful solutions to help guide you in developing your general career goals and direct you toward highly desirable skill sets. We present general skills that apply to a broad suite of potential employers, as well as the expertise needed for a particular job in a specific sector. In addition, our hope that is you will be able to better define your career goals by ascertaining which jobs use the proficiencies that best align with your interests. The right approach is the one that works for you.

In this chapter, job capabilities have been separated into two categories: disciplinary and nondisciplinary (Blickley et al. 2013). Disciplinary skills are specific to the wildlife and conservation field, such as the ability to identify a particular species from its morphological traits, use a statistical analysis program, or operate and maintain the equipment and vehicles needed for fieldwork. Nondisciplinary skills are those that are associated more with interpersonal interactions and a good work ethic. Examples include being able to work as part of a team, write a clear and concise technical report, proactively identify and solve a complex problem, and complete a task on time and on budget.

WHAT ARE THE TOP SKILLS EMPLOYERS WANT?

The field of wildlife biology offers a wide range of possible careers and positions, each with varying expectations of experience required by potential employees. As a student interested in pursuing a job in wildlife biology, you must strategically plan a curriculum that includes appropriate academic courses and targeted extracurricular activities that will give you the skills you need to get the job you want (chapters 7, 8, and 9).

Expertise in the job-specific field of study (disciplinary skills) is commonly cited as an important qualification in publications that evaluated skills and competencies sought by employers in their entry-level wildlife and conservation employees (Adelman et al. 1994, Brown and Nielsen 2000, Kubik 2009, Sample et al. 1999, Saunders and Zuzel 2010). A strong foundation in wildlife and conservation science is of primary importance for many employers. For example, employers surveyed by DeLany (2004) generated an exhaustive list of 384 areas of proficiency required for entry-level positions in wildlife management, 57% of which were disciplinary.

Through our surveys, we also found disciplinary skills to be of primary importance to employers in multiple aspects of wildlife biology. When asked "What are three of the top skills undergraduates need in order to get a job in the field of

---

Contributors to this chapter share first authorship, with the exception of Mariah H. Meek.

wildlife biology (or conservation more generally) after graduation?," interviewees named disciplinary capabilities most frequently (14 out of 17 respondents), citing formal training, knowledge of the study organisms and systems, and hands-on experience with specific tools and programs. Respondents described formal training as a "strong coursework record" (grades or grade point average) and certification, such as a "biology-related degree." They also included "know[ledge of] wildlife issues (species and ecology)" and the ability to use specific tools, such as geographic information and positioning systems (GIS and GPS).

Even though disciplinary skills are decidedly important, surveyed employers also emphasized several crucial nondisciplinary ones. The most important appear to be communication and interpersonal skills (Brown and Nielsen 2000, Kubik 2009, Sample et al. 1999). In at least two studies (Sample et al. 1999, Saunders and Zuzel 2010), employers ranked expertise in these areas as more important than any disciplinary skill. DeLany (2004) reported that among the 73 criteria ranked as "highly important" by wildlife professionals, 34% were related to communication or interpersonal skills. Respondents in our survey mentioned these attributes in 47% of their replies, and work experience in 41%. Other key nondisciplinary skills included ethics (DeLany 2004, Sample et al. 1999), basic computer modeling (Brown and Nielsen 2000, Saunders and Zuzel 2010), critical thinking, innovation, and problem solving (Brown and Nielsen 2000, DeLany 2004, Kubik 2009). Although some studies mentioned project management or program leadership (Adelman et al. 1994, Sample et al. 1999), most did not identify them as critical for undergraduate-level applicants. Similarly, interdisciplinary training was often mentioned (Allison and McBride 2003, Dietz et al. 2004), but it rarely ranked near the top of the list. In our survey, 53% of the replies cited nondisciplinary skills, such as the ability to work both independently and in a team and to "maintain self-motivation." Twelve percent focused on other traits, which included "interest in" and "passion for" a particular field of work, "cultural sensitivity," and a diversity of references who can attest to accomplishments beyond the classroom. Many of these nondisciplinary skills can be gained through previous work experience or on-the-job training (Blickley et al. 2013, Bonine et al. 2003).

To better understand the balance between the two, we asked the question, "When hiring undergraduates, how important are disciplinary skills (such as identifying birds or using a GPS unit) relative to nondisciplinary skills (such as project management or interpersonal skills)?" Of the 19 responses to this question, 8 stressed disciplinary skills, 6 emphasized interpersonal skills, and 5 viewed both as equally important. The take-home message we find here is that you should spend an equal amount of time and effort in gaining competency in these two areas and be able to indicate that when applying for positions. Once you have your career goals more clearly defined (if they are not already), you should have a better idea of where that balance might lie.

COMPARISON OF UNDERGRADUATE- AND GRADUATE-LEVEL POSITIONS

Unsurprisingly, jobs requiring an undergraduate degree have different expectations about an applicant's skills and experience than those needing a master's or a PhD degree. That variability may be found more in the depth of your skill set than its breadth. Previously, we sifted through job advertisements that stipulated graduate degrees in conservation science (Blickley et al. 2013). We quantified how often different competencies were mentioned in the advertisements for three nonacademic job sectors: nonprofit, governmental, and private. Disciplinary skills made up 37% of an average position announcement and were clearly critical for graduate-level applicants, but several nondisciplinary attributes also stood out. These included project management (11% of an average job advertisement), interpersonal skills (8%), written communication (6%), program leadership (6%), and networking (6%). Different forms of communication, including written, oral, and outreach, made up about 14% of an average advertisement.

Our more recent survey specifically asked, "How do [the previously listed top three undergraduate] skills differ from the skills that graduate students need?" Fifty-six percent of the respondents felt that graduate students should have more-advanced disciplinary skills (e.g., data analysis) and 44% expected graduate-level applicants to have more expertise in nondisciplinary areas, such as scientific writing, independent work, leadership, and management. A third of those who replied felt that experience is the main difference between the two education levels. One person captured the general tone by noting that "graduate students are typically hired for higher-level positions, and therefore would additionally need more-refined interpersonal skills, language proficiency, and higher-level disciplinary proficiency." Only 2 out of 18 replies indicated that there is not a marked difference between an undergraduate- and a graduate-level education in terms of required competencies.

From our past job advertisement and interview results (Blickley et al. 2013) and our more recent survey, we conclude that disciplinary skills, communication, and interpersonal skills continue to hold their importance as students transition from undergraduate-level to graduate-level jobs. Employers hiring applicants with master's or PhD degrees also want evidence of increased disciplinary ability and greater management and leadership capacities. Most of our survey respondents suggested that graduate-level applicants should have more-advanced skills and greater experience. In other words, employers expect graduate school to transform students from active contributors into managers and leaders of scientific endeavors.

Despite these clear differences in expectations, the competitiveness of today's wildlife job market causes applicants with graduate degrees to apply for jobs that technically only require an undergraduate one. This makes it difficult for people with less experience to get a permanent wildlife position without an advanced degree. Applicants without at least an MS degree may need to build up their experience and skills through temporary, seasonal, or term positions, in order to be competitive for higher-level permanent positions. Linking up with a governmental agency, nongovernmental agency, or private company during your undergraduate career can be a way to fast-track this process. Alternatively, you could view your time as an undergraduate student as a way to build up the training, experiences, networks, and good grades that will make you competitive to enter a graduate program.

THE IMPORTANCE OF JOB SECTORS

Outside of academia, wildlife biologists are typically employed in the private (15%), governmental (65%), and nonprofit (5%) sectors (Doyle et al. 1999). Employers across the three different areas value many of the same skills and experiences in job candidates, particularly when seeking to fill entry-level wildlife biology positions. Employers may prioritize different skills and competencies, however, due to variances in their mission, their culture, and the requirements of the specific job they are seeking to fill.

As a jobseeker, what strategies can you use to determine the breadth and range of skills that an employer is looking for when filling a specific job? For early-career conservation professionals with a graduate degree, these three sectors emphasize different abilities in their hiring announcements (Blickley et al. 2013). Private-sector job postings asked for more-specific analytical and disciplinary competencies than either governmental or nonprofit organization ads. Project management expertise came up more often in the nonprofit listings. Communication and networking skills were commonly mentioned in governmental and nonprofit ads.

It is important to understand precisely what an employer is looking for. A good place to start is with a careful reading of the job advertisement. Most of them explicitly list the skills that are required or desired in a successful applicant. You also can glean valuable information by researching the mission and the approach of the hiring organization and by investigating the background of employees in similar positions. As you make decisions on where to focus your energy and effort, do so with an understanding of your skills and abilities, as well as what types of work you want to pursue. As you apply for positions, craft your letter of interest and resume with those specific jobs in mind (chapter 7).

SIGNALING YOUR SKILL SET TO POTENTIAL EMPLOYERS

Knowing when and how to convince a future employer that you have what it takes to excel in a wildlife biology career can be difficult. Once you have identified and targeted the skills and experience needed for your chosen career, the next step is to determine how your own experience and interests have prepared you, either directly or indirectly, to take on the duties

for specific jobs. Your avenues to communicate this information will probably be traditional job application materials—a cover letter, resume, and writing samples (chapter 7)—and an interview (chapter 8).

Effectively communicating your abilities using these written and in-person tools takes effort and practice. In fact, self-promotion may be a skill in and of itself. Being able to list your various capabilities on a resume and provide evidence that you have used them showcases you as someone who is well qualified for a career in wildlife biology and conservation (Martinich et al. 2006). By tailoring your job application materials to specifically address the requirements of a particular job, you indicate to an employer that you understand the duties of the job and are sufficiently prepared to work in that position. Because communication is an important skill that employers are seeking in prospective employees, you should think of the job application process as your first opportunity to demonstrate your ability to effectively communicate to others.

Listing relevant coursework demonstrates a solid foundation in a particular field of study (Nelson et al. 2008), and citing concrete experiences shows how you put that knowledge to use (Colón-Rivera et al. 2013). Undergraduate research activities also present a good opportunity to highlight your disciplinary, organizational, and interpersonal skills. When employers were asked directly how an applicant could communicate competence in several specific areas, Blickley et al. (2013) reported that demonstrating experience, above and beyond coursework, is the best way to do so. Employers mostly obtained this information from a resume, by looking at an individual's volunteer positions or internships. Additional confirmation can come from a personal interview or, in some cases, talking with a candidate's references (Blickley et al. 2013). Many professional organizations have certification programs that serve as an effective means of showing you have various basic or advanced skills related to wildlife careers (chapter 2). For example, The Wildlife Society has a certification program for wildlife biologists, with different designations that are based on educational and work experiences (see appendix B). Certification from a respected professional organization indicates to a prospective employer that you meet the professional standards for a career in that field and are an active participant in the relevant professional society (chapter 6).

When respondents in our recent survey were asked, "How can undergraduates demonstrate to potential employers that they have competency in the desired skills?," they corroborated the findings from Blickley et al. (2013). Their answers most frequently stated that evidence came from an applicant's activities outside of a university setting, either as a volunteer or in another work environment. One respondent described a positive association between a diversity of skills and extracurricular activity, stating that "volunteering at an agency [as] an undergraduate can show [that individuals have] gathered a skill set around: (1) leadership, (2) project management, (3) fundraising, or (4) partnership development." This could also indicate "proof of initiative . . . to improve their skill set and experience level."

Cover letters and interviews give undergraduates the chance to detail relevant linkages between their extracurricular activities and their other capabilities, which resumes may not. For example, one of our respondents noted that "if they haven't had a field job before, but they write about spending a lot of time outdoors, or an experience volunteering in a lab, and [show] how that can apply to what they will be doing, that is helpful." You should explicitly think about making such links. Another respondent stated: "While it is useful to mention that one has studied anatomy and conducted a necropsy in a lab course, that point gains more relevancy if it is accompanied with a description of how one has butchered their [sic] own meat each hunting season or assisted with field necropsies with professionals. The latter experiences indicate (1) likely a solid understanding of wildlife anatomy, (2) skill in the tools used for the job, and (3) comfort in doing the job." Thinking outside the box and focusing on holistic experiences that encompass several different abilities could greatly benefit undergraduates applying for positions. We recommend outlining that information in a cover letter, especially if what you have done is well outside of your job or academic experiences, and make sure to follow up with more-detailed descriptions during an interview.

WHAT ELSE MIGHT YOU CONSIDER AS YOU EMBARK ON YOUR WILDLIFE BIOLOGY CAREER?

Eighteen of the 19 respondents who answered the question "What other advice would you give to undergraduates who want to become wildlife biologists?" provided a wide range of useful information for current undergraduates. We identified six common themes within their open-ended replies. First, 83% of the respondents said that undergraduates should build up their disciplinary skills, and most of them emphasized the value of varied experience. Second, 28% said that relevant extracurricular work experiences are so important that students should try to volunteer if paid positions are not available in their field. Third, 39% discussed the value of building communication and interpersonal skills, and several mentioned that experiences outside of the wildlife field can help a student build up these nondisciplinary talents. Fourth, 33% discussed the importance of attitude, advising students to be enthusiastic, optimistic, determined, and persistent. Fifth, 17% noted the value of building a good network of contacts and references. Sixth, 22% advised students to take some time to explore their world, with suggestions ranging from living abroad to just going outdoors. This included continued education, as two respondents urged undergraduates not to rush into graduate school. They counseled taking the time to know what you want out of graduate school before you start, and to choose a graduate program, advisor, and funding situation with care. Possibly most importantly, heed what one of our respondents said, "Follow your passion, gain a broad array of experiences, get involved, and dedicate your time to something you and others believe in."

SUMMARY

To be competitive in the wildlife biology job market, you must take control of your own educational and early professional experiences. Viewing your undergraduate education as the opportune time to gain valuable workplace skills—such as communication, leadership, project management, and networking—in addition to the more traditional disciplinary knowledge will make you competitive for a range of possible career options after graduation. In addition to developing the relevant skills for a job, you have to effectively signal to a prospective employer that you possess those skills. Lacking direct experience in a particular area should not necessarily dissuade you from applying for a job, particularly if you can communicate to an employer that capabilities you developed in a different context can be applied to it, or that you have the ability and enthusiasm to acquire skills quickly once you are on the job. By being strategic, you can find a good fit for a career in the rapidly evolving field of wildlife biology.

The following are nine points to help guide you through the process of gaining experience and getting the job you want (adapted from Blickley et al. 2013):

1. Focus on key skills. Gain competency in transferable skills. Broaden your potential for future opportunities by focusing on areas of overlap. We recommend that students take extra courses beyond their degree curriculum, if financially possible, to gain additional training related to their wildlife discipline. Undergraduates should review their degree curriculum and compare it with coursework needed for certification by various professional societies, as well as consider taking extra courses where their curriculum is deficient in meeting professional society requirements. In addition, develop your oral communication skills while you are an undergraduate. This can be done by creating ecological and animal-related presentations that can be given to students in elementary and secondary schools, or to business organizations, such as local Rotary or Lion's Clubs. Gain leadership skills by becoming an officer in a student organization.

2. Decide on a career track as soon as possible and tailor your coursework and experiences accordingly. Students need to read chapter 5 (describing a plethora of wildlife careers), decide which wildlife jobs excite them the most, note the educational and technical skills required for such positions, and, while still in college, actively acquire the skills listed within those job descriptions. If a job position of interest to you requires a graduate degree, then make sure your academic performance as an undergraduate is sufficient to allow you to be admitted into a graduate program.

3. Be creative, and go beyond minimum requirements. Don't graduate with only having done coursework. While in college, get to know your professors and the types of research they conduct. Volunteer to assist their graduate students, or determine if opportunities exist for undergraduate research. College is not a time to be introverted. Remember, you're in college to gain the necessary skills to become marketable for employment. Don't wait for these to be handed to you—go after them.

4. Start collecting job information early. Make it a practice to scan and evaluate job advertisements long before you are ready to apply, so you can develop the skills needed for the positions that appeal to you.

5. Be strategic. Identify the job skills that make you stand out, match these with positions that require these abilities, and then augment your profile with complementary ones. For example, if you are a better field biologist than a people person, build your resume to that strength by gaining quantitative analytical skills.

6. Be proactive. If your program does not offer what you need, make your own opportunities and build your own network. Contact potential future employers and volunteer. One way of encountering potential employers is to attend meetings and actively seek out professionals who have the jobs you desire.

7. Do not undervalue your experiences. Potential employers may give weight to what you have done with regard to group leadership, event planning, or volunteer positions, even if these activities were not directly linked to your undergraduate degree or gained during that period.

8. Talk to wildlife professionals. Take the time to learn about the career paths and training of people with jobs that you would be excited to have.

9. Recognize time constraints. Budget your time among the conflicting interests of finishing your degree, fulfilling work obligations, and gaining a breadth of additional skills.

LITERATURE CITED

Adelman, I. R., D. J. Schmidely, and Y. Cohen. 1994. Educational needs of fisheries and wildlife professionals: results of a survey. Fisheries 19:17–25.

Allison, E. H., and R. J. McBride. 2003. Educational reform for improved natural resource management: fisheries and aquaculture in Bangladeshi universities. Society and Natural Resources 16:249–264.

Blickley, J. L., K. Deiner, K. Garbach, I. Lacher, M. H. Meek, L. M. Porensky, M. L. Wilkerson, E. M. Winford, and M. W. Schwartz. 2013. Graduate student's guide to necessary skills for nonacademic conservation careers. Conservation Biology 27:24–34.

Bonine, K., J. Reid, and R. Dalzen. 2003. Training and education for tropical conservation. Conservation Biology 17:1209–1218.

Brown, R. D., and L. A. Nielsen. 2000. Leading wildlife academic programs into the new millennium. Wildlife Society Bulletin 28:495–502.

Colón-Rivera, R. J., K. Marshall, F. J. Soto-Santiago, D. Ortiz-Torres, and C. E. Flower. 2013. Moving forward: fostering the next generation of Earth stewards in the STEM disciplines. Frontiers in Ecology and the Environment 11:383–391.

DeLany, B. W., Jr. 2004. Entry-level job skills needed by wildlife management professionals. PhD dissertation, Louisiana State University, Baton Rouge, LA, USA.

Dietz, J. M., R. Aviram, S. Bickford, K. Douthwaite, A. Goodstine, J. L. Izursa, S. Kavanaugh, K. MacCarthy, M. O'Herron, and K. Parker. 2004. Defining leadership in conservation: a view from the top. Conservation Biology 18:274–278.

Doyle, K., S. Heizmann, and T. Stubbs. 1999. The complete guide to environmental careers in the 21st century. Island Press, Washington, DC, USA.

Kubik, G. H. 2009. Projected futures in competency development and applications: a Delphi study of the future of the wildlife biology profession. PhD dissertation, University of Minnesota, Minneapolis, MN, USA.

Martinich, J. A., S. L. Solarz, and J. R. Lyons. 2006. Preparing students for conservation careers through project-based learning. Conservation Biology 20:1579–1583.

Nelson, C. R., T. Schoennagel, and E. R. Gregory. 2008. Opportunities for academic training in the science and practice of restoration within the United States and Canada. Restoration Ecology 16:225–230.

Sample, V. A., P. C. Ringgold, N. E. Block, and J. W. Giltmier. 1999. Forestry education: adapting to the changing demands on professionals. Journal of Forestry 97:4–10.

Saunders, V., and K. Zuzel. 2010. Evaluating employability skills: employer and student perceptions. Bioscience Education 15:Article 2, www.bioscience.heacademy.ac.uk/journal/vol15/beej-15-2.pdf.

Wildlife Careers

From A to Z

SCOTT E. HENKE

This chapter describes the job opportunities available to a person with a wildlife degree. I ask incoming college freshmen two questions. The first is, "What do you do for fun?" I'm often told that they camp, fish, hunt, and hike. I explain that I asked this question because, if pleasurable activities became their job, and they could get paid for doing them, then that would be an ideal line of work. I continue by saying that since working people spend nearly 50% of their life at a job, they should do something they truly enjoy. It's better to be a person who wakes up in the morning and is excited about going to work, rather than one who dreads the prospect. My second question is, "What kind of career do you want after you graduate?" Inevitably, about 95% of the students provide the same answers—either being a game warden or working within a private wildlife ranching enterprise. I've come to realize that the reason I get the same responses each year is because these are the jobs to which youth in Texas are exposed. As wildlife professionals, we have not made the breadth of potential career opportunities that are available within our field widely known. Also, I have found that even if beginning wildlifers know about some of these other possibilities, they do not have a realistic view of what the job duties truly are.

This chapter is intended to expose future wildlife professionals to the wide array of potential careers available to them and help them decide which job best fits their interests. We encourage you to read each position description carefully, select the ones that suit you best, and then locate and contact individuals in your region who hold such positions. Ask these people if they would be willing to be a mentor. You will be surprised how many professionals want to help young aspiring wildlifers break into our field. For example, at Texas A&M University–Kingsville, we have developed a shadow program—a list of individuals who are willing to allow students to tag along during a typical day at work. We have more than 50 people on the current list, ranging from employees in federal and state agencies (e.g., US Fish and Wildlife Service, Texas Parks and Wildlife Department) and private corporations (e.g., wildlife ranching industry) to researchers (e.g., university professors), all of whom are willing to serve as mentors. Another way to decide if a particular area of interest is right for you is to become an intern with the relevant agency or organization. Do not be discouraged, however, if the job is not what you thought it would be. A negative experience as an intern is never a waste of your time. It's better to spend three months at something you decide is a poor fit for you than to spend your entire academic career making yourself marketable for a position that you later find you do not like. Therefore, intern with several organizations while you are still a student. Participating in such programs, or acquiring a mentor in a sector where you think you might like to be employed, is a great way to verify that you would enjoy the work involved with that job. It is up to you to make the initial call, however. Doing so is making an investment in your future.

Spend time finding the career path that is best for you. Once you determine which one that is, use the job descriptions in this chapter to make yourself a good prospect for work in that sector. Read what the agency or organization looks for in a new employee, what type of background is useful, and what the specific educational requirements are. Then go out and fulfill those criteria. Use this chapter as a guide to obtain the position and career of your dreams.

We have identified 35 agencies and organizations that offer nearly 100 wildlife-based positions where you can use your wildlife degree. Each job description is written by an expert in that field; each author is either currently employed by or is recently retired from the relevant agency or organization. The following job descriptions are divided into one of three sectors: governmental, nonprofit, and private. Each listing

Table 5.1. A summary of the federal pay grade system (in dollars), based on 2016 salaries

Grade	Step 1	Step 2	Step 3	Step 4	Step 5	Step 6	Step 7	Step 8	Step 9	Step 10	Within-grade amounts
1	18,343	18,956	19,566	20,173	20,783	21,140	21,743	22,351	22,375	22,941	varies
2	20,623	21,114	21,797	22,375	22,629	23,295	23,961	24,627	25,293	25,959	varies
3	22,502	23,252	24,002	24,752	25,502	26,252	27,002	27,752	28,502	29,252	750
4	25,261	26,103	26,945	27,787	28,629	29,471	30,313	31,155	31,997	32,839	842
5	28,262	29,204	30,146	31,088	32,030	32,972	33,914	34,856	35,798	36,740	942
6	31,504	32,554	33,604	34,654	35,704	36,754	37,804	38,854	39,904	40,954	1,050
7	35,009	36,176	37,343	38,510	39,677	40,844	42,011	43,178	44,345	45,512	1,167
8	38,771	40,063	41,355	42,647	43,939	45,231	46,523	47,815	49,107	50,399	1,292
9	42,823	44,250	45,677	47,104	48,531	49,958	51,385	52,812	54,239	55,666	1,427
10	47,158	48,730	50,302	51,874	53,446	55,018	56,590	58,162	59,734	61,306	1,572
11	51,811	53,538	55,265	56,992	58,719	60,446	62,173	63,900	65,627	67,354	1,727
12	62,101	64,171	66,241	68,311	70,381	72,451	74,521	76,591	78,661	80,731	2,070
13	73,846	76,308	78,770	81,232	83,694	86,156	88,618	91,080	93,542	96,004	2,462
14	87,263	90,172	93,081	95,990	98,899	101,808	104,717	107,626	110,535	113,444	2,909
15	102,646	106,068	109,490	112,912	116,334	119,756	123,178	126,600	130,022	133,444	3,422

contains a job description, including duties and the likely percentages of time spent in the field and in an office; the necessary background required; information about useful experiences needed to acquire the position; education required, including helpful coursework, certifications, and advanced degrees; and a pay scale, to provide a realistic idea of compensation and benefits. Because the federal government uses the same pay scale across all agencies, we describe the federal pay grade system, called the General Schedule, in the beginning of this chapter and then refer to the pay schedules for each federal job that is described.

Read this chapter carefully and highlight the positions that initially pique your interests. Then, during your academic career, reread this chapter. Often you will find that your interests change with new experiences. Finding the right job—and a fulfilling career—are important choices, so make them wisely. See how each position aligns with your beliefs, ethics, and career goals. Remember, much of your adult life is spent working, so select a job that is enjoyable and intellectually satisfying to you.

FEDERAL PAY GRADE SYSTEM

Pay scales for most federal employees are summarized by what is known as the General Schedule (GS). The GS system is organized by grades and steps, with salary amounts that can be changed annually by Congress, and pay ranges that vary by location within the United States (table 5.1). Each grade represents a level of complexity in job duties and responsibilities, with 1 being the lowest and 15 being the highest pay for that grade. In general, individuals with a high school diploma and no additional experience typically qualify for GS-2 positions; with a bachelor of science (BS) degree, GS-5; with a master of science (MS), GS-9; and a doctorate (PhD), GS-11. A step reflects the level of pay as determined by years of service and

performance during those years. Steps 1 through 10 have a formula for advancement. This formula, and more information regarding pay, can be found on the US Office of Personnel Management website (appendix B).

Entry-level positions are often advertised as multigraded positions, such as GS-5/7/9/11. This means that, if an individual's job performance is satisfactory for one year, he or she normally is promoted to the next higher grade level automatically, without having to meet additional educational requirements. A person starting at the GS-5 level would be promoted to GS-7 after one year, and to GS-9 after two years. After three years of satisfactory performance, the final promotion to GS-11 would result. After reaching the highest grade level for a position, automatic step increases would go into effect. Salary step increases occur at one-year intervals for steps 1–3, at two-year intervals for steps 4–6, and three-year intervals for steps 7–9.

In addition, the federal government adjusts salaries according to the geographic location of the position, and salaries may also change with general inflation patterns. The locality pay depends on a variety of factors but is adjusted to account for the average living expenses in that particular area. Think of the cost differences between living in a remote location in the Midwest versus living in downtown New York City. The locality pay schedules try to account for this disparity. For example, someone who has just completed a bachelor's degree and has no other applicable work experience could be hired as a GS-5 step 1. If the position is based in Denver, Colorado, the annual salary for 2016 is $34,742. A person with the same background, hired at the same step level but working in Phoenix, Arizona, receives slightly less ($33,100).

The federal government also provides general awards, student loan repayment incentives, retirement benefits, access to premier health care benefits, sick and annual leave, and travel opportunities. If you have received positive reviews on your annual job performance, then you may be eligible for a cash performance award. Often an agency provides the option of paid time off instead of a financial award for exceptional performance. Cash awards vary, depending on your grade and step level. You also can receive spot awards, extra effort awards, quality step increases, keepsake items (e.g., coffee mugs, t-shirts, briefcases, badges) and many other awards that are based on exceptional performance. It is important to note, however, that federal budgets are constantly in flux, so these awards may or may not always be available. In terms of student loan repayments, a federal employer can apply up to a maximum of $10,000 per year, for a total of not more than $60,000, for any one employee. This incentive comes at the discretion of the agency, and employees receiving it are required to sign a service agreement. Service agreements, more commonly known as continued service agreements, are relatively common in federal agencies. They are a form of contract between the employee and the agency, where a person commits to work for the government for a preestablished length of time in exchange for federally sponsored training or education. You can find more information about continued service agreements online (appendix B). This program is used to attract and retain highly qualified individuals.

Lastly, the federal government offers excellent retirement benefits. Federal agencies match an employee's retirement contributions in an amount equaling up to 5% of that person's annual salary. Full-time employees have access to exceptional health care. They also earn eight hours of sick leave each month, and the equivalent of 13 days of paid annual leave during their first three years of service. Earned leave then increases to 20 days per year and, after 15 years of service, rises to 26 days per year.

GOVERNMENT AGENCIES: FEDERAL ORGANIZATIONS

BUREAU OF LAND MANAGEMENT
James M. Ramakka, Jeffrey Tafoya, and Sandra K. Wyman

Agency: US Department of the Interior, Bureau of Land Management

The Bureau of Land Management (BLM) is an agency tasked with performing a perpetual balancing act as it strives for sustained multiple uses for a vast area of public land. The BLM administers more land than any other agency in the United States. Approximately 16% of the surface acreage (99.2 million ha) and 45% of the subsurface mineral acreage (283.3 million ha) of the United States is managed by the BLM. Most of these publicly owned lands are in the 12 western states and are managed for livestock grazing, with nearly 18,000 permits and leases, 221 wilderness areas, and the federal government's subsurface mineral property.

The Federal Land Policy and Management Act of 1976 directs the BLM to manage lands under its jurisdiction for mul-

tiple—and often competing—uses, such as recreation, range, timber, oil and gas, coal, uranium, watershed, wildlife, and fisheries, as well as sites with scientific and historical values. This act, however, did not repeal BLM mandates under earlier legislation, including the Mining Law of 1872, the Mineral Leasing Act of 1920, and the Taylor Grazing Act of 1936. Nor does it exempt the BLM from more-recent legislation, such as the National Environmental Policy Act (NEPA), the Endangered Species Act, the Wild Horse and Burro Act, the Clean Water Act, and the Clean Air Act.

BLM-managed lands contain important habitat for species as diverse as mule deer, desert bighorn sheep, moose, northern spotted owls, desert tortoises, rainbow trout, and brook trout. Over 30% of the breeding habitat for greater sage grouse and more than half of the breeding distribution for tundra-nesting species (e.g., snow geese, Steller's eiders, spectacled eiders) occur on BLM land (North American Bird Conservation Initiative, US Committee 2011).

To accomplish its mission, the BLM employs approximately 10,000 people in 107 field offices, 50 district offices, 20 national monuments, 16 national conservation areas, a National Operations Center, and a National Training Center. All of these offices are overseen by the BLM's national office in Washington, DC. At the end of fiscal year 2014, 217 full-time and 11 term-appointment (limited duration position) wildlife biologists were employed by the BLM (G. Walsh, BLM, personal communication).

Rangeland Management Specialist
Job Description

Rangeland management specialists oversee public rangelands to establish and maintain rangeland health, ensure their orderly use to support the western livestock industry, and tend the land for the communities that rely on them. Rangelands and livestock grazing are what often come to mind when you hear the title "rangeland management specialist," so you may be wondering what a person in this position has to do with wildlife. Both the BLM and the US Forest Service allow livestock grazing on public lands, and this program is usually directed by rangeland management specialists, also known as range specialists. The primary responsibilities of range specialists are to manage livestock grazing permits and ensure that these lands are meeting rangeland health standards and functions. Yet there are also many indirect aspects of the job that weigh heavily on the benefits provided to wildlife and wildlife habitat.

Typical job duties for range specialists include the administration of assigned grazing permits and grazing allotments. This involves monitoring range condition and productivity, establishing and adjusting livestock stocking rates (the number of animals and the duration of their grazing), keeping accurate records, billing, and implementing range improvement projects. Other duties include customer service and work with grazing permittees to ensure that they understand the agency's mission regarding range management and the proper use of public rangelands. Rangeland specialists work closely with all the other members on an interdisciplinary team—biologists, botanists, hydrologists, archeologists, soil scientists, and planners—to work collectively to meet the agency's objectives for land management. Good rangeland specialists recognize the balance between providing rangeland forage for livestock and other uses of public lands. Rangeland specialists find themselves spending much of their time (30%–40%) with grazing permittees, discussing livestock stocking rates, pasture rotations, and forage utilization, as well as looking for ways to improve range conditions. Approximately 20%–40% consists of monitoring the range condition, vegetation communities, precipitation, and drought. Depending on funding, 10%–30% covers developing and implementing range improvements. Another 10%–30% might be used to write environmental analyses or National Environmental Policy Act analyses for the issuance of grazing permits or implementation of projects. All of these percentages vary by state, by office, and by funding priority. Rangeland specialists may find themselves involved in reviewing mining or oil and gas extraction proposals, recreational plans or events, or transportation plans. A review of proposed transportation plans, events, mineral leasing, and other resource projects is done to ensure that, under the BLM's multiple-use mission, there are no impacts, or that any impacts to rangelands can be mitigated. Whatever issues an agency's office deals with are addressed in an interdisciplinary environment. Rangeland specialists might also find themselves working with wild horses or burros, instances of trespass by feral livestock, state livestock inspectors, and, occasionally, state and county animal damage control personnel. A fair amount of time also is spent doing paperwork. Yes, paperwork! Almost every action performed in an agency by any employee results in paperwork. Documentation of what you do is very important when you work as a public servant.

So how does all of this relate to wildlife? Wildlife needs habitat that consists of four primary elements: food, water, cover, and space. Rangeland specialists manage and provide all of these items. Ultimately, healthy rangelands equate to viable wildlife habitat, watersheds, vegetation communities, and water. Range improvement projects vary by individual offices, geographic location, and funding availability. These projects include practices such as range seeding, prescribed burning, vegetation manipulation, weed control, fence construction, and water developments (e.g., spring improvements, water wells, pipelines). Vegetation treatments typically try to improve plant communities by attempting to reach early or mid-seral (intermediate) stages in ecological succession. Water and fencing projects usually are designed to facilitate livestock management and ensure proper grazing use of the lands. These types of projects benefit wildlife directly and indirectly. Improved vegetation communities often provide increased forage and habitat for wildlife as well as livestock. Vegetation communities can be manipulated to supply an array of positive characteristics, such as seasonal types of forage that benefit livestock and wildlife. For example, thinning woodlands and initiating prescribed fires can often increase browse

Rangeland specialists establish and adjust livestock stocking rates by monitoring range conditions and productivity (photo: Jim Ramakka)

species that provide better nutrition for ungulates. Grasses and forbs can be increased or planted to furnish early spring and summer forage. Vegetation can be manipulated to produce edges (areas where wildlife have access to forage close to cover, for escape if needed). Water sources that are developed to aid in the distribution of livestock, allowing these animals to move to newer or more-distant areas to graze, are frequently taken advantage of by wildlife. When water sources are created, range specialists and wildlife biologists often work together to coordinate the storage and availability of water for wildlife, even when livestock are not present. This ensures that all kinds of fauna, from the smallest birds to rodents and big game, have water throughout the year. For many rangeland specialists, the most rewarding part of the job is implementing and seeing the completion of successful range improvement projects that make a difference on the landscape.

Rangeland specialists often attend on-the-job training programs to learn monitoring techniques; acquiring skills such as how to become pesticide applicators, ride and pack domestic horses, fly in helicopters, work in fire management, and possibly handle and manage wild horses; and become more knowledgeable about litigation and how to best integrate laws with land management. In many offices there is usually an opportunity to cross-train, working with other disciplines to gain a variety of experiences. Rangeland specialists and wildlife biologists often work closely together to combine efforts and achieve desired habitat conditions. There is occasionally time and funding for research projects. In my current assignment, I [JT] have assisted and partnered with university personnel to study (1) the effects of vegetation manipulation on erosion control and sediment movement, and (2) possible solutions for managing invasive cheatgrass by inoculating plants with fungal endophytes that limit these plants' ability to produce seed; participated in a study using water produced from natural gas wells to facilitate vegetation reclamation; and initiated a project using a recently approved herbicide to manage cheatgrass. I have been fortunate to meet and work with various professionals in multiple fields that are all tied to the natural environment.

Background Needed

In addition to fulfilling the educational requirements, rangeland specialists benefit by and are more successful if they possess and hone additional skills. An ability to work with diverse groups and individuals, and to communicate effectively with a variety of customers and the general public, are both very important. Handling controversial and emotional situations with professionalism is critical. Whether you're making decisions that affect permittees or dealing with a public that may be passionate about an issue, how you communicate with the people involved and manage difficult circumstances frequently determines the outcome, as well as the amount of grief it takes to resolve issues. One difference between range jobs and many other resource management positions is that rangeland specialists must make decisions that may impact permittees' abilities to generate revenue from their grazing operations. Writing skills are extremely valuable. Agencies appreciate employees who are able to carefully and tactfully convey messages and decisions—both of which may later be used in litigation—to customers and the public. The interpretation of laws, policies, and regulations is another important skill, as you make important decisions that must align and be consistent with many of these requirements. Natural resource management specialists should be prepared to perform tasks (often alone) in the outdoors, handle a four-wheel-drive vehicle, be able to hike in uneven terrain, tolerate hot and cold weather, and endure long hours on a job, not to mention being able to work in an office environment for extended periods.

Education Required

To qualify as a rangeland management specialist, a bachelor's degree, with very specific coursework, is required. Rangeland specialists have one of the most stringent sets of course requirements in the natural resource fields in governmental agencies. The following is taken from "Individual Occupational Requirements" in the US Office of Personnel Management's Rangeland Management Series, 0454.

A bachelor's degree in range management or a related discipline is required, which must include at least 42 semester hours in a combination of classes in plant, animal, and soil sciences, as well as natural resource management. Of that 42 hours, at least 18 semester hours of coursework must be completed in range management, including such areas as basic principles of range management, range plants, range ecology, range inventories and studies, range improvements, and ranch or rangeland planning; a minimum of 15 semester hours in directly related courses in plant, animal, and soil sciences, including at least one course in each of these three areas, such as plant taxonomy, plant physiology, plant ecology, animal nutrition, livestock production, and soil morphology or soil classification; and at least 9 semester hours in related resource management subjects, including courses in such areas as wildlife management, watershed management, natural resource or agricultural economics, forestry, agronomy, forages, and outdoor recreation management.

I [JT] pursued BS and MS degrees in wildlife science. One requirement as an undergraduate was to take a small number of courses or credit hours outside of my major field. I chose range, because I had an interest in range science. After enjoying my first few classes, I continued taking more senior-level range courses on my own, in hopes of receiving a minor in range. I worked as a wildlife technician throughout college and found the work rewarding. My choice of a location in which to work, however, led me to apply for a rangeland specialist position in Utah. Shortly after starting that job, I knew I was still in a career that enabled me to serve wildlife through good land management. After moving to New Mexico, I realized even more of the potential I had to influence wildlife habitat through good range management. When the state director of the New Mexico BLM initiated an effort to secure more funding for landscape-type programs, I was able to implement $100,000 worth of projects every year for approximately five years. The position, in my opinion, was the perfect combination of range and wildlife management.

Pay Scale

Entry-level rangeland specialists are usually hired at the GS-7 level (table 5.1). Most rangeland specialist jobs are offered with growth potential, where the employee starts as a GS-7 and may work toward a GS-11 level, depending on the length of time spent in that position, experience, and performance. Students graduating with an MS degree in range science or similarly acceptable coursework may qualify to start at the GS-9 level. With more time spent in a grade and satisfactory performance, an individual may reach the GS-11 pay scale. There are also opportunities to become supervisory rangeland specialists, which entail overseeing range staff members in an office. A supervisory position could potentially be at the GS-12 level. In addition to the base salary, oftentimes employees can supplement their income by participating in the fire program. With their manager's approval, employees can train and be certified to assist with fire activities and can work for additional time periods.

Wildlife Biologist
Job Description

Wildlife biologists with the BLM have a great deal of responsibility, but also tremendous opportunities to have a long-term influence on wildlife habitats over large areas. The majority of positions are located in district or field offices. It is rare for a field office to have more than two wildlife biologists (and often only one) on staff, even though a field office may manage more than 404,000 ha of land.

The primary role of wildlife biologists is to advise BLM decision makers, such as field office managers or district managers, on the effects of proposed actions on wildlife and wildlife habitat, and to design and implement wildlife habitat improvement projects. In many offices, the endangered species and fisheries programs are included as part of the wildlife biologist's duties.

BLM biologists are expected to be aware of current habitat conditions and trends, as well as the status of wildlife populations (both locally and regionally), and to have knowledge and understanding of the latest techniques for habitat improvement and mitigation. They also are expected to be the office expert on laws, regulations, and recent court decisions that may have an influence on wildlife. Wildlife biologists regularly interact with other BLM professional staff, including geologists, engineers, range managers, foresters, soil scientists, hydrologists, geographers, land use planners, archeologists, and real estate specialists.

Depending on the office, 50% of a wildlife biologist's time may be spent in the field, reviewing other entities' proposed projects, such as oil and gas wells or new roads; working with range managers on grazing allotment reviews; monitoring habitat conditions; surveying for high-priority species; building and maintaining existing habitat improvement projects; and other duties. In my [JR] own field office experiences in four states, I monitored mule deer, elk, pronghorn, and desert bighorn sheep habitat; mapped and monitored the nests of golden eagles, ferruginous hawks, and prairie falcons; surveyed habitats for the presence of endangered butterflies and threatened tiger beetles, inventoried breeding bird atlas blocks (data collection units); served on a riparian habitat evaluation team; established monitoring studies for endangered plants; and many other similar duties. Working with volunteers, I also built and maintained habitat improvement projects, such as water catchments, riparian enclosures, and shrub plantings.

While fieldwork is an essential part of the job, time spent in the office is equally important. Managers expect their wildlife biologists to be able to strike a balance between field and office work, in order to provide timely and professional input for the management team. Office duties can include preparing reports or briefing papers that summarize the results of monitoring studies, attending interdisciplinary team meetings to discuss proposed projects, writing or reviewing pertinent sections of National Environmental Policy Act documents and land use plans, and responding to public comments and concerns. Coordinating with other federal or state agency counterparts, writing accomplishment reports to justify budget expenditures, and preparing budget requests for future fiscal year projects are also important parts of the job. Wildlife biologists who are responsible for threatened or endangered species programs spend time preparing biological assessments and conducting

BLM wildlife biologist Jim Ramakka returns from monitoring ferruginous hawk nests in the Bisti Wilderness, New Mexico (photo: Jim Ramakka)

Section 7 (interagency cooperation) consultations for endangered species with the US Fish and Wildlife Service.

With a few exceptions, time in the field becomes rare for wildlife biologists positions at the GS-12 level or above. These jobs are usually located in district or state offices and focus on program management, budget development and fiscal tracking, accomplishment reporting, and internal and external program review. Only 5% of BLM wildlife biologists are in GS-13, 14, or 15 positions, and half of those are in the national office in Washington, DC. Staff members there are primarily responsible for coordinating major budget and policy issues, responding to congressional inquiries, and cooperating with other agencies and nongovernmental organizations on national-level issues.

Background Needed

The educational requirements listed below are just the first step in qualifying for a BLM wildlife biologist position. Time spent conducting fieldwork in remote situations, familiarity with four-wheel-drive vehicles, and farm or ranch experience are all helpful, but not essential. Candidates with prior work experience, such as a summer job with BLM as a biological technician, have a better chance of making the list of candidates referred to the selecting official. This person will call previous supervisors to ask about a candidate's performance. Managers I [JR] have talked with stress that, in addition to preferring candidates with a knowledge of ecology, population biology, and conservation biology, they want to hire individuals who can communicate effectively, are honest, are willing to put in long hours when necessary, can work as part of a team, and know how to disagree without being unpleasant. They also look for individuals willing to leave their desks and computers and actually go out in the field. Personal characteristics shared by effective BLM wildlife biologists include adaptability, good negotiation skills, and an ability to work effectively with others who may have different viewpoints.

Good oral and written communication skills are essential, as is an ability to function independently (often with minimal supervision) while still providing timely, accurate information for use in management decisions. A desire to learn from counterparts in other specialties (e.g., foresters, range managers, fire management specialists, geologists) is extremely helpful.

Recommended reading for anyone considering a job with the BLM includes Skillen (2009), for an excellent history of the agency, and, although they may seem outmoded, Cutler (1982) and Thomas (1985, 1986). The latter three publications provide excellent insight into what is expected of wildlife biologists in multiple-use agencies.

Education Required

To qualify as a GS-5 wildlife biologist, applicants must have a bachelor's degree in wildlife biology or biology, including 9 semester hours in wildlife-related subjects, 12 in zoology, and 9 in botany.

Candidates can apply to start at the GS-7 level if they have at least one year of experience at the GS-5 level, have completed one full year of directly related, full-time graduate studies, or can meet specific requirements for superior academic achievement (upper one-third of your graduating class for a bachelor's degree, a GPA of at least 3.0, and induction in a recognized national academic honor society).

At the GS-9 level, applicants must demonstrate either one year of experience equivalent to GS-7 requirements or the completion of at least two years of study toward an MS degree in wildlife biology or a related field. To start as a GS-11, applicants must demonstrate at least one year of GS-9 level work or at least three years of wildlife-related graduate study at the PhD level.

Pay Scale

Entry-level wildlife biologist positions are often advertised as multigraded positions, such as GS5/7/9/11 (table 5.1). Current federal salary tables showing pay grades and step increases can be downloaded from the US Office of Personnel Management website (appendix B). In addition, flexible work schedules are common in most BLM offices.

Range Technician
Job Description

Range technician positions may serve to train you for technical range management work at a higher grade level or as support staff for higher-level technicians and rangeland management specialists. Range technicians provide technical support and assistance in range management and rangeland improvements, with duties varying throughout the different field offices, districts, and states. Range technician positions may be seasonal in nature, depending on the location and budget constraints of the agency and a particular field office. Unlike many rangeland management specialists, range technicians are often in long-term positions at one location for many years and provide continuity in their knowledge of and services on public land allotments. They spend the majority of their time in the field and, during the off-season, in office duties. With this on-the-ground knowledge of allotment pastures, range technicians are integral to grazing management on public lands. They generally spend most of their time conducting upland and riparian monitoring by assessing the condition of fences, water developments, weed infestations, and use by livestock and other animals (e.g., wildlife, wild horses, trespass livestock), as well as noting other uses within each allotment. Range technician positions may serve as a stepping stone to getting established in a federal agency or provide opportunities for employment while finishing required classes for a professional-series position (e.g., rangeland management specialist or wildlife biologist).

Range technicians are often the eyes and ears of the agency on rangeland activities, reporting to their supervisory rangeland management specialist or to other specialists as necessary, depending on the situation and the information acquired. Normal interactions for range technicians include personal contact with coworkers in the office, permittees and other users of public land, contractors, and the general public, in order to provide information, receive and clarify instructions, and report on work results or any problems encountered.

Range technicians can expect long days in the field, where exposure to extremes of heat and cold, dry and wet conditions, rough terrain, insects, snakes, and other hazards may be common. There are often lengthy travel distances to pasture or allotment sites. Range technicians may ride through allotments to ensure the proper distribution and movement of livestock, identifying paint marks, brands, and ear tags; observe, record, and report any signs of unauthorized grazing use; check the placement of supplements and water, according to allotment management plans; and assist in the roundup of trespass livestock. Range technicians also may collect and record data for range studies, such as the land's condition and overall ecological trend, usage maps, utilization, actual use, and climate information. They also may serve as team members for range inventory efforts, including mapping, collecting, compiling, and recording vegetation data for ecological site descriptions; update, compile, and record water-source data (e.g., wells, pipelines, reservoirs) and boundary lines (e.g., fences, allotments, natural barriers) for range inventory mapping efforts; provide assistance in identifying needed improvements (e.g., wells, reservoirs, pipelines, fences, vegetation projects), as well as inspecting or inventorying the condition of existing projects; observe wildlife populations and activities; conduct wild horse and burro inventories; or possibly serve on fire crews (if qualified and needed).

Time spent in the office by range technicians usually constitutes the less desired part of the job. Most of this work is devoted to inputting data collected throughout the field season; updating allotment maps; documenting situations observed in the field; contacting permittees or other federal land users; and keeping current on specific rules, regulations, and procedures that are grounded in policies and guidelines. The accuracy, reliability, and acceptability of the work of rangeland management specialists, as well as the operations of the range program as a whole in the area administered by a field office, are greatly affected by the range technicians' work and varied duties.

The GS 455 classification series also includes range aid jobs at the GS-2 and GS-3 levels. These are trainee positions and include basic work in accumulating information.

Background Needed

What can you do to improve your skills and abilities in order to increase your chances of becoming a range technician? Previous experience working on a ranch and with livestock is very beneficial, as is experience backpacking and horseback riding in rough, steep terrain. Other helpful criteria are an ability to use two- and four-wheel-drive vehicles and all-terrain vehicles in both dry and inclement weather, and to know how and when to avoid dangerous situations. Good skills in verbal and written communications, which can be understood by a broad spectrum of people with various backgrounds in grazing and natural resource management, are imperative. Range technicians work with diverse groups and must do so in a respectful manner at all times. Therefore, it is important to maintain an open mind about options for achieving natural resource management goals and objectives. There is often more than one way to accomplish rangeland improvements. Range technicians must have good organizational skills, not need direct supervision, and be able to work in interdisciplinary team settings while conducting inventories and assessments. They should be knowledgeable about computers and software such as Word, Excel, PowerPoint, and mapping programs. To be successful in this position, you must also understand how to use topographic, aerial, and other tools, both to help keep agency maps up to date and to avoid losing your way in the field.

Seasonal work experience—a period of at least three months of continuous employment on a seasonal basis—in government agencies is applicable toward acceptable amounts of previous experience. Shorter periods may be combined to equal a season, provided that a total of at least three months' experience is demonstrated. For any period where the work took place over more than three months, that excess time is credited toward an additional season of experience. One season of experience qualifies you for a GS-2 level position; two seasons, for GS-3; and four seasons, for GS-4. Examples of qualifying specialized experience include work in the conservation, regulation, and use of federally controlled lands for grazing; range research activities; work in the operation of a livestock ranch or assistance with the management or preservation of lands comparable to the public range (e.g., Natural Resources Conservation Service, state wildlife agencies); or range or forest fire control, prevention, or suppression work.

Education Required

For GS-4: Successful completion of two years of classwork, including at least 12 semester hours in any combination of courses such as forestry, agriculture, crop or plant science, range management or conservation, wildlife management, watershed management, soil science, natural resources (except marine fisheries and oceanography), outdoor recreation management, civil or forest engineering, or wildland fire science. No more than 3 semester hours in mathematics can be credited.

For GS-5: Successful completion of a full four-year course of study leading to a bachelor's degree, with either a major emphasis in forestry, range management, agriculture, or a subject matter field directly related to the position, or coursework that included at least 24 semester hours in any combination of classes, such as those for GS-4 requirements. No more than 6 semester hours in mathematics can be credited.

For GS-6: Successful completion of at least one year at the GS-5 level, or further coursework in resource-related topics (graduate-level courses or an MS degree).

For GS-7: Successful completion of at least one year at the GS-6 level, or further coursework (MS degree or higher).

Any range-related classes, including plant identification, rangeland and riparian ecology, wildlife management, soils, and any courses that help you read the landscape or watershed, such as those found in the range management, wildlife biologist, or other similar professional-series class requirements, are beneficial.

Pay Scale

Range technician positions begin at the GS-4 level and may advance to GS-7 if your designated within-grade work is adequate or above expectations (table 5.1).

DEPARTMENT OF DEFENSE LANDS
Rhys M. Evans and David K. McNaughton

Agency: US Department of Defense

Though the US Department of Defense (DoD) may not be the first federal department that students think of when considering a job in wildlife management, it is a challenging and rewarding place to work. The DoD manages 12 million ha of land in 49 states and two US territories (Puerto Rico and Guam) and at dozens of foreign installations. While some of the facilities simply do not amount to much more than an airfield or a few warehouses, others exceed 404,686 ha in size.

Installations can be found in all ecosystems, from tundra to desert to tropical rainforest, and some even include coral reefs. DoD lands encompass outstanding biodiversity! In fact, the DoD manages more threatened and endangered species per hectare than any other federal or state agency and maintains active hunting, fishing, and outdoor recreation programs. The DoD also deals with urban wildlife / pest management issues, invasive/exotic species issues, forestry programs, airport/air-field wildlife management, wildfires, wetlands management, and many other focus areas. The DoD secures the boundaries of its lands, another factor resulting in a high degree of bio-diversity. This keeps the public out (and safe from munitions) and leaves some lands in the United States free from develop-ment.

The DoD partners with local, regional, national, and even international organizations (e.g., Partners in Flight coalition, Partners in Amphibian and Reptile Conservation, Bat Con-servation International, universities) to share resources, ideas, and management techniques. The management of military land is guided by the Sikes Act Improvement Act of 1997 (16 USC 670a et seq.), which requires (1) conservation and reha-bilitation of natural resources on military installations, (2) no net loss in military capability, (3) sustained, multipurpose use, including outdoor recreation, and (4) natural resource law enforcement.

Each of these items can be broken down into several sub-categories. The DoD uses proactive management measures to maintain healthy ecosystems in order to prevent wildlife and fauna and flora from becoming at-risk species. Depart-ment of Defense lands are a great place to conduct research on land management, resource recovery and rehabilitation, erosion prevention, and the effects of and need for prescribed burning. Personnel in such positions in the DoD can easily explain to an installation commander that healthy trees and diverse native vegetation are a great way to camouflage ar-tillery from a simulated opposing army. While the use of armored vehicles, artillery, and air assets can cause environ-mental damage, it is possible to minimize widespread harm by avoiding the repeated use of the same impact areas for the same activities. Rest, restoration, and rehabilitation of training areas are important and effective management tools, similar to rotational grazing.

Congress and the citizens of the United States have deter-mined that the US Armed Forces require areas in which to train, and that on those lands, military readiness and training take precedence. The public would be surprised to learn just how much conservation can occur on active training land. For example, while artillery training is taking place at Fort Hood in Texas, juniper habitat is also being actively managed to pro-tect two endangered bird species, golden-cheeked warblers and black-capped vireos. Bird/Animal Aircraft Strike Hazard activity (BASH; another thing about DoD is that we love our abbreviations) is another widely recognized area of main-taining a balance between wildlife protection and military readiness, as demonstrated in the commercial arena by the near-disastrous air incident in New York City in 2009 involving Canada geese. Managing airfields, including airway approach and departure corridors, to reduce the risk of wildlife conflicts is challenging, rewarding, and fun.

Outdoor recreation, including hiking, birdwatching, and other nonconsumptive uses, is common on military lands. Many installations include fishing and hunting as resource management tools (and benefits). These opportunities can be open to the general public or limited to active duty personnel, depending on many factors, such as base security, seasons, and ongoing military operations. Managing deer, waterfowl, up-land game, and other wildlife resources for sustained harvest is an exacting task anywhere, but when you incorporate military operations, unexploded ordnance, and other issues that are specific to DoD, it is even more challenging.

Management of the people who appreciate and use wildlife resources is also important, so installations with hunting and fishing programs usually have game warden or conservation law enforcement programs (though some of these duties may be conducted by employees of cooperating agencies, such as a state agency game warden or a federal US Fish and Wildlife Service officer).

Wildlife Biologist

Job titles in this area can be highly variable. Entry-level posi-tions could be classified as wildlife biologist, wildlife techni-cian, park ranger, natural resource specialist, or many other possible titles. To make it even more confusing, each job title has several numbers attached to it, such as GS-0401-5/7/9. The four-digit series (e.g., 0401, for a natural resource special-ist) is important. Focus your search on series that start with 00 (e.g., 0025, park ranger) or 04 (e.g., 0408, ecologist). The 5/7/9 part is also very important, as that's the grade level directly related to qualifications and salary.

Job Description

Managing DoD lands involves several thousand dedicated employees, from wildlife technicians to postdoctoral research fellows. Many biologists who work on military lands are em-ployed by the federal government; in some cases, however, they might be employed by state governments, consulting firms, cooperating agencies, or universities. Entry-level jobs can start as part-time or seasonal work and focus on one par-ticular species or issue, but as you progress in your career, responsibilities and types of authority can become much more challenging and complicated as you assume responsibility for new programs or different faunal groups. You also need to become skilled at legal compliance, budget management, in-teragency communications, and long-term planning as your career continues.

Sometimes DoD needs specialists who work exclusively with certain species, such as reptiles or riparian birds. At other times it needs generalists who can switch their focus from snakes to swallows at the drop of a hat. Entry-level biologists also must attend meetings, analyze data, and draw conclusions from these analyses (in a report or as paperwork). They might write scientific papers, contribute to formal research, or give input regarding a species or an installation management plan. As you progress through higher-level jobs, more of your time is spent indoors than out—management authority and higher salaries beget more meetings and additional hours in the office. Take advantage of the great opportunities for field-work early in your career!

A majority of military job openings are announced on the USA Jobs website (appendix B). Electronic resumes that typically are much longer and more detailed than those expected by private industry must be submitted. It frequently takes more than six months from the date of an announcement to the actual start date for an employee. The USA Jobs website also allows and encourages a user to create specific and recurring vacancy searches (e.g., for GS-0401-05 positions in the state of Washington). Another route into federal service with the DoD is the USA Jobs Pathways Program, which is usually directed at college sophomores and juniors (see appendix B).

Positions with consulting firms and cooperating agencies and organizations are announced and filled using an entirely different system, one more easily understood by applicants— the typical requirement to submit a resume and a cover letter. Many entry-level jobs are filled through contracting organizations, such as universities, research firms, and state government agencies. These are typically the easiest means of recruiting personnel for salaried positions, because they allow candidates to demonstrate their skills and adapt to the environment of an installation without a long-term commitment.

Background Needed

It can be difficult to qualify for your first job with the DoD. Military service, Peace Corps / AmeriCorps service, and seasonal/term work (the latter being for a single fixed period of time) can lead to special (expedited) hiring authorities. Abilities not immediately associated with wildlife (e.g., photography, off-road driving) can be helpful. Verbal communication skills at all levels, from kindergarten to postgraduate students and from one-on-one conversations to speaking to hundreds of people at a time, will serve you well. Written communication is also essential. As a DoD biologist, you may be asked to write a multipage position paper or a 50-page environmental assessment, while other circumstances may dictate that you prepare a bottom line up front (BLUF), in which you have to present most of your important information in two sentences or less. When applying for a DoD position, always mention any background you have with the military, whether it's growing up in a military family, serving in the US Armed Forces, or running a field study on a military installation in college. The vast majority of federal positions are only open to US citizens, and all male applicants must register for the Selective Service. Positions filled through contractors may be more open to foreign residents.

The Sikes Act mandates that cooperative plans for conservation and rehabilitation programs be conducted on military reservations. Here, fledgling great horned owls are roosting on a light amphibious vehicle (photo: Rhys Evans).

Education Required

The majority of DoD jobs require at least a bachelor's degree, typically in biology, wildlife sciences, ecology, environmental sciences, or a closely related field. Advanced degrees can qualify an applicant to be hired at a higher level. Additional formal and informal training (e.g., computer skills, geographic information systems, mapping, defensive driving, specific software training), and wildlife-specific skills (e.g., telemetry, animal capture and handling) are helpful.

Pay Scale

A bachelor's degree should qualify an applicant for the GS-5 level, an MS degree for the GS-7 level, and a PhD degree (although very few entry-level jobs are at this grade) for the GS-9 or higher level (table 5.1). Pay scales vary by geographic location, and nearly all positions include full benefits, retirement, and a Thrift Savings Plan, similar to private-sector 401(k) plans. Salary increases and promotions are almost always guaranteed, as long as competence is demonstrated.

Seasonal and contract jobs may start at an hourly wage and last from three to nine months for temporary positions. Some federal and state positions may be capped at three years, due to funding mechanisms, but often these jobs evolve into more-stable positions with career tracks. Some installations even have jobs with educational benefits, paying for higher-level degrees in return for research on the installation. All of these opportunities can be very diverse, and you should consult your nearest installation through their public affairs office to find out the mechanisms their natural resource office uses for hiring.

FOREST SERVICE
Serra Hoagland and Anonymous (adapted by Scott E. Henke and Serra Hoagland from the USFS website, www.fs.fed.us)

Agency: US Department of Agriculture, US Forest Service

The US Forest Service (USFS) is under the US Department of Agriculture and is the primary federal agency responsible for sustaining the nation's forests. Established in 1905, the USFS is the country's oldest and largest land management agency and is commonly known for its professionalism, effectiveness, pride, and loyalty (Apple 1997). The USFS is composed of three main branches: research and development, the national forest system, and state and private forestry. Each branch has a distinct mission. For example, its research and development section is the world's largest forest research organization, while the national forest system administers 78 million ha of national forests and grasslands. These lands range from mixed conifer forests and wilderness areas in Alaska, to tropical forests in Puerto Rico, and deciduous forests in the Southeast. National forests provide a wide range of ecosystem services,

including habitat for threatened and endangered species, such as coho salmon and marbled murrelets in the Pacific Northwest, and Indiana bats and red cockaded woodpeckers in the southeastern United States. The USFS also partners regularly with nongovernmental organizations, private landowners, Native American tribes, and state wildlife agencies on habitat management for countless game and nongame wildlife species. Its mission can be summarized as "caring for the land and serving people." As an agency, the USFS works to sustain the health, diversity, and productivity of the nation's forests and grasslands by applying scientifically based, sustainable, multiple-use management concepts to meet the needs of current and future generations. In every forest, wetland, grassland, or rangeland across the country, the USFS has a presence.

The USFS has more than 30,000 permanent employees in hundreds of locations across the country. Agency personnel focus their skills on managing and improving the nation's forests and grasslands. Many USFS employees work on various national forests, several produce primary research related to forest and range management, some develop the skills of others at Job Corps centers, and others provide expertise in state and private forestry partnerships across the country.

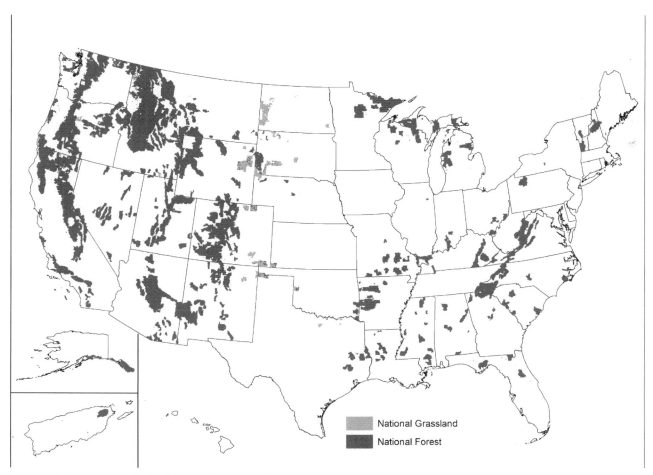

National Grassland
National Forest

Location of US Forest Service land holdings within the United States (adapted by Humberto Perotto)

Biological Scientist
Job Description

If biological scientists are not in remote forests conducting fieldwork, they are generally in their offices, analyzing data, writing reports, or preparing presentations. On average, biological scientists spend about 30% of their time on fieldwork, depending on the season. Because much of this activity can occur in areas that may be inaccessible during the winter, biological scientists tend to do most of their fieldwork during the spring and summer. For example, my [SH] official duty station is in Asheville, North Carolina, but my fieldwork is located in mixed conifer forests on the Mescalero Apache Indian Reservation in southcentral New Mexico. I also attend graduate school at Northern Arizona University, in Flagstaff, Arizona. Luckily, with the flexibility of the USFS, I can accomplish many things while being in remote work locations, which is one of many benefits in working for this agency.

The primary duties of biological scientists are to design and implement research studies and contribute to research projects that are focused on environmental threats and biological conservation. These scientists gather, compile, and provide data for natural resource databases, in order to create recommendations for forest and wildlife management. Much time is spent collecting and analyzing data to help protect wildlife species. For example, my [SH] current species of interest is the federally threatened Mexican spotted owl. I lead field crews of two to five individuals, depending on the task. I regularly coordinate and work directly with professionals from federal, state, and local agencies; private entities; tribal governments; and university faculty and students.

Since the mission of the USFS focuses on using sustainable, multiple-use management techniques, the agency has a broad array of opportunities for entry-level scientists. For instance, although my [SH] academic training and experience is primarily as a wildlife biologist, I have the opportunity to work on forest fire–related projects and formal national-level tribal consultations. Therefore, if you are interested in invasive species or pest management, soil conservation, recreation, or hydrology, you most likely will have an opportunity to branch out into those disciplines while working with the USFS. Whether it's by applying your existing knowledge and skills to pursue various interests or augmenting your training and engaging with a variety of projects and issues, the opportunities are available to you!

Fieldwork: In a typical field day, biological scientists could either be working in the forest, conducting Mexican spotted owl occupancy and reproduction surveys; measuring trees and downed woody debris in vegetation plots; handling woodrats, deer mice, and chipmunks for a mark and recapture study; training or working with student interns on various survey protocols; or helping with a prescribed burn operation in the forest. Some fieldwork also may involve partnering with other state and federal agencies on various wildlife projects, such as elk and mule deer captures, water resource improvement projects, and bat blitzes (helping to estimate bat populations by surveying many sites during a short time frame). As a forest service scientist, your research specialty or species of interest typically guides your fieldwork, but there are always opportunities to collaborate on new projects with the USFS or assist with ongoing research by other agencies. An advantage of being in a scientist position with the USFS is that you can bridge the gap between science and on-the-ground management. You also have a fair amount of flexibility in determining your research projects. So if you are interested in butterflies, bears, or beavers, you get to decide the approach, methodology, and types of analyses, and you are responsible for publishing the results.

Office work: When in the office, biological scientists are either planning and coordinating fieldwork, writing up results and reports, responding to emails, attending training sessions or meetings, reading new scientific publications, running analyses, working with other scientists, organizing presentations, or speaking on the phone with various collaborators and partners. As a federal agency, the USFS is obligated to maintain its trust responsibility to federally recognized Indian tribes, which means that the agency has to uphold and honor Indian treaty rights. Treaty rights include various aspects, such as the Indians' grazing, hunting, fishing, and water rights. When USFS actions may affect Indian treaty rights, then the agency must consult with tribes to minimize any negative impacts on those treaty rights. One component of this trust responsibility is to support tribes by organizing natural resource–related training programs, as well as providing new tools and techniques that allow tribes to manage their resources. For example, in my position, I [SH] have hosted, organized, and implemented training workshops that are targeted to tribal audiences and tribal natural resource management professionals.

Biological scientists also regularly spend time with youth and with academic partners at a nearby college or university. They frequently give presentations in local classrooms about the benefits of wildlife and how to track various creatures, and even work on projects with local youth to encourage them to pursue natural resource–related professions. Also, entry-level scientists are constantly keeping up to date on the latest methods and scientific findings related to wildlife and forest management. Biological scientists have the freedom and flexibility to think critically about new methods of and approaches to conserving wildlife.

There are countless benefits to working for the federal government, which include job security, favorable health and retirement benefits, and other advantages, such as teleworking (working remotely), flexible work schedules, and wellness programs to help employees maintain their personal health and well being. Like any job, however, there are some drawbacks. The USFS is a large federal agency, with over 30,000 permanent employees. As such, it suffers from many of the bureaucratic and administrative ills common to large organizations. Furthermore, conflicting public perspectives on the appropriate use of our national forests can place USFS

A female Mexican spotted owl grabs a mouse in front of a Forest Service biological scientist on the Mescalero Apache Indian Reservation in southcentral New Mexico (photo: Serra Hoagland)

employees in an uncomfortable cross fire of public debate. Forest Service employees generally are highly motivated and want to be effective, yet they can be frustrated by long bureaucratic processes and strict organizational regulations that seem overly time consuming. Most USFS employees, however, have an overwhelmingly positive response regarding job satisfaction and a sense of pride toward working for the agency.

Background Needed

Some of the hands-on skills that are encouraged (but not necessarily required) for wildlife-related fieldwork include an ability to operate a four-wheel-drive vehicle in various conditions (snow, mud, sand), knowledge of how to use a handheld GPS and compass, and the physical and mental capabilities to work in remote areas under various weather conditions. For fieldwork, it is also critical that you be skilled in orienteering and other forms of navigation, as well as trained in animal handling and identification, basic botany, and plant identification. Safety is paramount within the USFS. Therefore, your ability to follow proper procedures, accept various job hazards, and participate in regular safety-training events is critical. Some of the technical skills that are helpful for office

work are a knowledge of GIS, Microsoft Office programs, and statistical software, such as SAS or R. Strong social skills and facility with oral and written communications are essential for all positions in the USFS.

Education Required

According to those who responded to a 2014 federal survey, approximately 10% of the employees within the federal government have a PhD, 24% have an MS, 34% have a bachelor's degree, 26% have some college education (through a certificate, an associate's degree, or a trade program), and 6% have a high school diploma or the equivalent (Office of Personnel Management 2014). To be hired as a biological scientist requires a basic foundation in the concepts, principles, and methods applied in biology and ecology. Most new hires in professional-series biological scientist positions within the USFS have an MS degree in a field related to natural resources (biology, ecology, wildlife management, forestry). It is important to know, however, that you can work in nonprofessional-series jobs in the USFS without an advanced degree. Such positions form the USFS's technical workforce, which is the backbone of the agency. Many technicians help collect and process data on wildlife species, recreation, fuels, and timber management. Other technicians include those who are associated with fire crews or who work in law enforcement and administer permits for special use. An MS and a PhD qualify you for the higher-paying professional scientist positions. A PhD is required for a research scientist position, where the focus is on creating novel, innovative research methods and publishing the findings. Research scientists undergo a rigorous, periodic review process, where a panel of their peers assesses their work programs and accomplishments.

Pay Scale

Similar to other federal positions, USFS employees receive salaries based on the GS pay scale. Biological scientists are usually hired at the GS-11 level (table 5.1). You can also look up the exact pay tables for a variety of locations throughout the United States at the US Office of Personnel Management website (appendix B).

In working as a biological scientist, your travel to various conferences, meetings, and training programs may be paid by your employer, which offers an opportunity to present your research at conferences and meetings. For example, I [SH] have been fortunate and attended 20 different ones (e.g., annual and regional meetings of The Wildlife Society, Society of American Foresters conferences, Intertribal Timber Council meetings) that were fully or partially sponsored by the agency over the past four years. This resulted in over 30 different presentations: posters, symposium panels, and invited talks. I was able to travel to New York, Alaska, Hawaii, and several places in between to present my research and network with professionals across the nation. The USFS sponsors employee travel, in order to encourage interactions between agency scientists and natural resource management professionals. Face-to-face contacts help support the agency's professional development mission. It also ensures that USFS scientists pursue research topics that have management applications and allows research results to be discussed and implemented on a broad scale, to promote science-based environmental stewardship.

Forester
Job Description

As a forester, you may work at a variety of sites, from glaciers to laboratories, tropical rainforests to grasslands, or offices to mountainsides on national forest lands. Your responsibilities may span the full spectrum of natural resources—waters, soil, air, range, fish, wildlife, minerals, recreation, and wilderness—but your primary focus as a forester is on trees and vegetation management. With nearly 81 million ha of National Forest System lands to manage, USFS foresters directly affect the environment through a multitude of activities that range from designing fuel reduction or habitat improvement projects on forest lands to fostering partnerships with nearby owners of private forest lands and state foresters. Still others, such as silviculturalists, are engaged in monitoring various aspects of forest management.

Foresters are called on to manage natural resources for the benefit of different segments of the public. They may work closely with specialists in various fields, including archeology, botany, chemistry, computer science, electronics, engineering, entomology, geology, hydrology, soil science, wildlife, and fisheries biology. Foresters tend to work as part of an interdisciplinary team, sharing ideas and developing implementation plans with other agencies and with citizens' action groups. As a forester, the scope of your work might span a variety of duties or be more specialized. You could be involved in planning for, maintaining, using, or preserving the forest and its resources to meet the needs of the twenty-first century. You could develop, implement, or administer plans that span a wide range of mission responsibilities, including wilderness protection, timberland improvement, forest habitat analyses and enhancement, timber sales, tree nursery operations, recreation, and prescribed fire management and wildfire suppression.

Background Needed

Experience in silvicultural practices, forestry techniques, and forest management is important, as is work in plant identification, dendrology, and forest health. A strong, demonstrated ability in oral and written communications is required, because foresters must deal with a variety of people, from citizens' action groups to politicians, and from national forest visitors to the scientific community. An ability to maintain your composure, even when confronted by angry user groups, is required. A USFS forester is always in the public eye.

Education Required

Most research forester positions require advanced degrees. A substantial number of forester positions in the USFS are at the GS-11 level, with opportunities to move up to higher grade levels, particularly for those who are willing to relocate among different USFS duty stations across the country. To meet entry-level requirements as a forester, you need a four-year degree in forestry or in a related field that includes at least 24 semester hours in forestry coursework, and at least 6 semester hours in any combination of the biological, physical, and mathematical sciences and engineering. Your curriculum must be sufficiently diversified to include classes in each of the areas below.

Management of renewable resources: Eligible courses include silviculture, forest management operations, timber management, wildland fire science or management, utilization of forest resources, forest regulation, recreational land management, watershed management, and wildlife or range habitat management.

Forest biology: Eligible courses include dendrology, forest ecology, forest genetics, wood structure and properties, forest soils, forest entomology, and forest pathology.

Forest resource measurements and inventory: Eligible courses include forest biometrics, forest mensuration (measuring the contents of standing or felled timber and estimating growth and yields), forest valuation, statistical analyses of resource data, renewable natural resource inventories and analyses, and photogrammetry (taking measurements from photographs) or other forms of remote sensing.

Pay Scale

Foresters are hired at many different grade levels. College graduates may start at the GS-5 or GS-7 level. They spend their first year or two in training and developmental positions, and then may be promoted to the GS-9 level. Others with advanced degrees and experience may be hired at higher grade levels (table 5.1). Visit the US Forest Service website (appendix B) to determine current salary ranges for each grade level and step within a position.

Forestry Technician
Job Description

Forestry technicians perform a variety of duties. They suppress wildfires, maintain facilities in campgrounds, provide visitors with information (e.g., explaining fire, safety, and sanitation regulations), clear or repair hiking trails, help protect a wilderness area by monitoring uses and enforcing regulations, and plant trees or shrubs to rehabilitate a damaged site or stabilize a slope. Forestry technicians also analyze tree stands for growth conditions, disease, and insect infestations; improve timber stands by either planting or thinning trees; and visit prospective timber sale areas to determine species, types, and qualities of timber, as well as to select and mark trees to be cut and those to be preserved for natural reseeding or as wildlife habitat. They even interpret aerial photos to identify types of timber and habitat, inventory sites for prescribed fire management activities, and inspect recreation and timber sale sites or reforestation activities for compliance with special permits.

Background Needed

Past experience working in remote areas is beneficial. Proficiency in using all-terrain vehicles and mechanical expertise is highly desirable. Forestry technicians need to be physically fit in order to perform required physical activities in the field. Technicians must be able to work both in groups and alone. As with jobs in most natural resource fields, good oral and written communication skills are necessary. Often it is the technician that visitors to the national forests first encounter.

Education Required

To qualify for entry-level positions, you need either a high school diploma or the equivalent and three months of general work experience for a GS-2 job. A GS-3 position requires either six months of general work experience and one year of study beyond high school, including at least 6 semester hours in courses such as forestry, agriculture, range, wildlife conservation, watershed management, soil science, outdoor recreation management, civil or forest engineering, and fire ecology. A person with either 12 months of specialized work experience, such as firefighting positions, and two years of education above the high school level, including at least 12 semester hours in the courses previously described, can qualify for a GS-4 position.

Pay Scale

The pay scale typically begins at the GS-2 level and rises to the GS-4 level (table 5.1). Visit the US Office of Personnel Management website (appendix B) to determine current salary ranges for each grade level and step within a position.

Rangeland Management Specialist
Job Description

Rangelands constitute about 47% of the world's land area and almost two-thirds of the lands administered by the USFS. Rangelands serve a variety of uses that are necessary to maintain and enhance the nation's quality of life. Among them are habitat for many species of plants and animals; forage for wildlife species and permitted livestock; water for wildlife, agricultural, human, and other uses; a broad spectrum of outdoor recreational activities; open space; and natural beauty.

The duties of rangeland management specialists include conducting vegetation, soil, hydrologic condition, and trend surveys and analyses; developing coordinated resource management plans, including grazing management; issuing and administering grazing permits; identifying and developing rangeland improvement projects, including rangeland soil and vegetation restoration; conducting a successful invasive and noxious weed abatement and eradication program; planning and implementing habitat management programs for sensitive riparian and wetland habitats; conducting botanical surveys for sensitive plant species; and assisting other specialists in the development and evaluation of diverse multiple-use public land management activities. Rangeland management specialists typically work with a variety of natural resource management personnel, including ecologists, botanists, soil scientists, foresters, hydrologists, and wildlife biologists. In addition, they partner with ranchers, grazing associations, conservation groups, other agencies, and members of the community on rangeland management and conservation issues.

Background Needed

Sound management of the nation's rangelands, based on ecological principles, is required to gain the full measure of benefits and values that these resources offer. Rangeland ecology and conservation management require a background in plant, animal, soil, and ecological sciences and a knowledge of the people who use rangelands. Excellent plant identification skills are required. A strong, demonstrated ability in oral and written communications is necessary, because range specialists must deal with a variety of people, from citizens' action groups to politicians, and from national forest visitors to the scientific community.

Education Required

Rangeland management specialists must have a four-year degree in rangeland management or ecology, or in a related field that includes at least 42 semester hours in a combination of plant, animal, and soil sciences and natural resource management.

Rangeland management: At least 18 semester hours of coursework in rangeland management, including classes in such areas as basic principles of rangeland management, wildland hydrology, botany, arid land ecology, synecology (interactions among species in communities), vegetation inventories and studies, and rangeland planning.

Directly related plant, animal, wildlife, and soil sciences classes: At least 15 semester hours of directly related courses in each of three areas: plant, animal, and soil sciences. Classes in herbivory and soil morphology or soil classification are acceptable.

Related resource management studies: At least 9 semester hours of coursework in related resource management subjects, including classes in such areas as wildlife management, watershed management, natural resources or forestry, agronomy, forages, agricultural economics, and outdoor recreation management. Candidates must have a combination of education and experience, with at least 42 semester hours of coursework in plant, animal, and soil sciences and in natural resource management (as shown above), plus appropriate experience or additional education.

Pay Scale

Rangeland management specialists are hired at different grade levels. College graduates usually start at the GS-5 or GS-7 level (table 5.1). They spend their first year in training and developmental positions, and then may be promoted to the GS-9 level. Others with advanced degrees and experience may be hired at higher grade levels. Visit the US Office of Personnel Management website (appendix B) to determine current salary ranges for each grade level and step within a position.

Botanist

Job Description

USFS botanists have the responsibility of caring for plants and plant communities on national forests and grasslands. The duties of botanists include working on teams with other resource specialists to conserve and manage the plant resources on these areas. Botanists evaluate the biological implications of various construction, logging, or other projects and develop conservation strategies to maintain threatened, endangered, and sensitive plant species. They manage and conserve plant biodiversity through a variety of programs, such as air quality, fuels management, lands, mineral, range, recreation, timber, and watershed. Forest Service botanists work closely with other agencies, public interest groups, and members of the community to conserve plant resources. Plant conservation ranges from controlling nonnative species and noxious weeds to protecting threatened and endangered species. Botanists also contribute their skills to various activities, including monitoring lichen to help determine air quality, identifying native plant species for watershed restoration projects, and developing nature trails.

Background Needed

Significant knowledge about and abilities with plants, plant identification, and plant communities are required. Strong, demonstrated skills in oral and written communications are essential, because botanists must deal with a variety of people, from citizens' action groups to politicians, and from national forest visitors to the scientific community.

Education Required

The USFS provides botanists with up-to-date training and experience. To be a botanist with the USFS, you must have a bachelor's degree with a major in biological science and complete at least 24 semester hours in botany. Courses can include plant anatomy or morphology, genetics, taxonomy or systematic botany, plant ecology, and mycology. In addition, botanists need a full year of graduate-level education, or superior academic achievement, or one year of experience equivalent to a GS-5 level. A GS-9 position requires two full years of higher-level education, or an MS or equivalent degree, or one year of experience equivalent to a GS-7 level. A GS-11 post requires three full years of progressively higher-level graduate education, a PhD or equivalent graduate degree, or one year of experience equivalent to a GS-9 level.

Pay Scale

Recent college graduates may be hired at the GS-5 or GS-7 level (table 5.1). They spend two years or less in training and developmental positions, and then may be noncompetitively promoted to the GS-9 level. Botanists may also be initially hired for grade-level positions by meeting the higher education and experience requirements. Promotion opportunities at the GS-11 level and above are competitive. Visit the US Office of Personnel Management website (appendix B) to determine current salary ranges for each grade level and step within a position.

Wildlife Biologist

Job Description

The main responsibility of USFS wildlife biologists is monitoring wildlife populations and managing, protecting, rehabilitating, and enhancing wildlife habitat. The duties of individual wildlife biologists are varied and can include such projects as building waterfowl nesting islands, cutting willows for moose browse, and conducting prescribed burns for deer and wild turkey habitats. Wildlife biologists provide the technical expertise to conserve the biological diversity of national forests and grasslands and work to protect and recover threatened and endangered species. They also interact with the public, including forest users who hunt, fish, bird-watch, and vacation on national forests and grasslands. Wildlife biologists often partner with state wildlife agencies, conservation organizations, and special interest groups to manage wildlife habitat. The states have authority over wildlife populations, but the USFS has authority over their habitat; thus cooperation is essential.

Background Needed

Good interpersonal skills and a working knowledge of economics and social and political trends are desirable. Excellent technical, biological, quantitative, and communication skills are a must. The USFS provides its employees with training as needed. A strong, demonstrated ability in oral and written communications is required, because wildlife biologists must deal with a variety of people, from citizens' action groups to politicians, and from national forest visitors to the scientific community.

Education Required

Wildlife biologists must have a bachelor's degree with a major in biological science or in natural resource management, with an emphasis in biology or ecology. An MS degree helps applicants be more competitive. Wildlife biologists must complete at least 9 semester hours in mammalogy, ornithology, animal ecology, wildlife management, or research courses; at least 12 semester hours in zoology subjects (general zoology, invertebrate zoology, vertebrate zoology, comparative anatomy, physiology, parasitology, ecology, cellular biology, entomology, genetics, or research in these fields); and at least 8 semester hours in botany or related plant sciences.

Pay Scale

Wildlife biologists are typically hired at the GS-5 or GS-7 level (table 5.1). They spend two years or less in training and developmental positions, and then may be noncompetitively promoted to the GS-9 level. You also can be initially hired for a higher-level position if you meet the additional requirements for education and experience. Promotions at the GS-11 level and above are competitive, and opportunities for advancement to higher grade levels are excellent. Visit the US Office of Personnel Management website (appendix B) to determine current salary ranges for each grade level and step within a position.

Wildlife Biological Science Technician
Job Description

The work of wildlife biological science technicians often involves regular and recurring moderate risks, such as working around moving parts, carts, or machines, as well as with contagious diseases or chemical irritants. For some positions, the job may require outdoor labor in extreme temperatures and being exposed to adverse weather conditions. Employees are required to use protective clothing or gear, such as hard hats, masks, gowns, earplugs, coats, boots, goggles, gloves, or shields to moderate risks, and they must follow procedures for minimizing risks.

A supervisor or other designated authority initially provides direction for objectives and deadlines. Technicians identify the work to be done to fulfill project requirements and objectives, plan and carry out the procedural and technical steps required, seek assistance as needed, independently coordinate work efforts with outside parties, and submit only completed work. Technicians also exercise initiative in developing their own solutions to common technical and procedural problems, such as changes in priorities, a need for extended field time, and minor requirements for additional equipment or personnel. Technicians must seek administrative direction from a higher authority before procedural changes can occur, however. Performance reviews assess how technicians have resolved technical and administrative problems.

Background Needed

The position requires knowledge of the technical methods, procedures, and basic principles of the biological sciences and statistics in order to assess readings and measurements taken, tests executed, observations made, work completed, and samples collected, as well as to understand and relate the significance of the results to the higher objectives connected with the activity. Additional requirements include knowing how to operate complex equipment systems, such as those with numerous components or parts that must be calibrated and synchronized to achieve desired results (e.g., those used in highly mechanized cartographic, hydrographic, or photogrammatic surveying; pressure chamber diving experiments; sophisticated laboratory experiments on fluids).

Education Required

At least two years of college are required, covering subjects such as mammalogy, ornithology, animal ecology, wildlife management, invertebrate zoology, vertebrate zoology, comparative anatomy, physiology, parasitology, ecology, cellular biology, entomology, genetics, and botany or related plant sciences. A bachelor's degree is preferred.

Pay Scale

The pay scale typically begins at the GS-2 level and rises to the GS-4 level (table 5.1). Visit the US Office of Personnel Management website (appendix B) to determine current salary ranges for each grade level and step within a position.

Law Enforcement Officer
Job Description

Law enforcement and investigations staff for the USFS are charged with protecting the public, employees, and natural resources across 78.1 million ha of National Forest System lands. Uniformed law enforcement officers ensure that federal laws and regulations governing these lands and resources are followed. They establish a regular and recurring presence on vast amounts of public lands, roads, and campgrounds, taking appropriate action when an illegal activity is discovered. Specific duties include working cooperatively with federal, state, and local law enforcement officials; conducting informational and educational programs; enforcing drug control laws; assisting state and local agencies with search and rescue missions on National Forest System lands; assisting special agents by conducting preliminary investigations; and responding to unplanned incidents, such as environmental protests, threats to employees, and large group events.

Background Needed

It is useful for law enforcement officers to have excellent outdoor skills, including in hunting, fishing, camping, hiking, orienteering, and backpacking. Also, expertise in firearms use and with all-terrain vehicles and watercraft is needed. Past experience working in remote areas is beneficial. Excellent interpersonal skills are necessary, because law enforcement officers need to interact with visitors to the national forests without being confrontational. Lastly, law enforcement officers need to have demonstrated an ability to make proper legal and ethical decisions.

Education Required

To qualify for a position at the GS-5 level, you need either one year of directly related experience equivalent to the GS-4 level or to have successfully completed a full four-year course of study in any field leading to a bachelor's degree from an accredited college or university. One year of full-time undergraduate study is defined as 30 semester hours or 45 quarter hours. As a general rule, visit the US Forest Service website (appendix B) to check up-to-date information on educational requirements. Graduate education per se will not qualify you for a level above GS-5, but such courses may be helpful if they are directly related to the work required in the position.

Pay Scale

The pay scale for law enforcement officers typically begins at the GS-4 level and rises to the GS-5 level (table 5.1). Visit the US Office of Personnel Management website (appendix B) to determine current salary ranges for each grade level and step within a position.

Public Affairs Specialist
Job Description

Public affairs specialists play a critically important role in the management of national forest lands. USFS audiences range from US senators to county supervisors to members of local service clubs. The USFS tailors its communication efforts to reach different audiences, depending on the issue. Public affairs specialists are involved in a wide variety of natural resource projects and activities. Examples include briefing members of Congress or their staff on USFS issues, planning and implementing field briefings for a wide variety of people and groups, and developing communication plans and strategies to inform internal and external audiences. In addition, public affairs specialists work on a day-to-day basis with local and national media, sometimes on highly controversial and complex resource issues. They advise agency leadership on communication strategies and techniques, write articles for newspapers, write and deliver speeches, and represent the agency at public events. Public affairs specialists often facilitate and sometimes lead public meetings, as well as write briefing papers and develop communications plans on complex issues for a wide variety of audiences.

Background Needed

Forest Service public affairs specialists and other agency personnel work with a multitude of partners—civic, advocacy, and conservation groups; city, county, state, and federal officials; print and electronic media; industry representatives; and members of the community—to develop land use plans, on-the-ground projects, and long-term resource solutions. Good interpersonal and oral and written communication skills are mandatory.

Education Required

To begin a career as a public affairs specialist with the USFS, you must have completed a four-year course of study leading to a bachelor's degree, preferably in the communications field, and have had three years of general experience, one year of which must be equivalent to a job at least at the GS-4 level. A GS-7 position requires completion of one full year of graduate-level education and one year of specialized experience equivalent to a post at least at the GS-5 level.

Pay Scale

Recent college graduates may be hired at the GS-5 or GS-7 level (table 5.1). Others with advanced degrees and experience may be hired at higher grade levels, and there is substantial opportunity for advancement. Visit the US Office of Personnel Management website (appendix B) to determine current salary ranges for each grade level and step within a position.

NATIONAL PARK SERVICE
Ben Bobowski, Jillian Gericke, Ben Anderson, and Apryle Craig

Agency: US Department of the Interior, National Park Service

The National Park Service (NPS), which was created by the Organic Act of 1916, is within the Department of the Interior. Its mission is "to conserve the scenery and the natural and historic objects and the wildlife therein and to provide for the enjoyment of the same in such manner and by such means as will leave them unimpaired for the enjoyment of future generations." The NPS has more than 400 units, which include historical parks or sites, national monuments, national parks, battlefields or military parks, preserves, recreation areas, seashores, parkways, lakeshores, and reserves. The NPS had approximately 307 million visitors in 2015, and about 400,000 volunteers assisting the various units in achieving their mission. Typical types of positions within the NPS can be seasonal, career seasonal, permanent, student, and volunteer.

Biologist, wildlife biologist, and ecologist are general titles that describe wildlife-oriented professional positions within the NPS. For those who have just finished an undergraduate degree and are interested in a wildlife career, work as a biological science technician is usually the entry point within the NPS. It is important to note, however, that there are numerous jobs within the NPS available for someone interested in resource stewardship, including inventory and monitoring data managers, integrated resources program managers, cultural resources program managers, natural resource program managers, wilderness coordinators, GIS specialists, historians, cultural anthropologists, archeologists, biologists, biological science technicians, entomologists, range technicians, foresters, forestry technicians, soil scientists, biological science student trainees, ecologists, botanists, horticulturists, rangeland management specialists, fishery biologists, wildlife biologists, historical landscape architects, historical architects, museum curators, physical scientists, hydrologists, geologists, librarians, and archivists. Employees in many of these positions begin as entry-level technicians, but some jobs are offered following a graduate degree. Breaking into the NPS can be challenging, but the job can be worth the effort.

Biological Science Technician
Job Description

Position descriptions within the NPS are either standardized and are referred to as benchmark position descriptions, or they are written specifically for a park unit or central office. When you apply for a position, you may have an opportunity to contact the hiring official and ask for a copy of the job description. This can be helpful to learn more about the job and develop your resume. Alternatively, if the position description is difficult to obtain, you can search on the Internet for the benchmark position description related to the job you are applying for or use the occupational questionnaire in the job announcement to better understand the job for which you are applying. Below is an amended excerpt from the entry-level benchmark position description for biological science technicians (GS-404-05).

The purpose of the position is to perform routine and uncomplicated biological science tasks common to natural resource management. Technician work is associated with and supportive of a biologist, and because of this, requires less than full knowledge of the field involved. If a professional biologist does not directly supervise a technician, then technical direction is obtained from a specialist in a central office or center. Specifically, biological science technicians collect, record, and conduct basic analyses of field data.

Biological science technicians assist in the preparation of reports, plans, and guidelines. Draft project reports may include literature research, descriptions of methods, the preparation of graphs and charts, and a summary of findings to aid biologists in interpreting complex data sets (e.g., summarizing observations gathered during vegetation monitoring efforts, bird studies). Technicians provide logistical support and area orientation for a contract (e.g., on occasion, those in charge of a large scientific effort may contract out portions of their study to assemble, install, and calibrate complex equipment and instrumentation). Biological science technicians may assist with the site-specific location of a study and the placement of equipment, or work with cooperating scientists to provide technical information, ensure quality control, and solve logistical or operational problems within the scope of their knowledge and authority. Technicians install, operate, calibrate, inventory, and maintain resource management equipment (e.g., tools; traps; sampling, monitoring, photographic, and laboratory equipment), and order replacement parts, new equipment, and supplies.

Biological science technicians also provide advice, assistance, and training to students, student conservation assistants, and volunteers involved in resource management projects and field research; use a variety of computer programs, such as data management and analysis programs, to compile, reduce, store, retrieve, analyze, and report data and other resource management information, including natural science research and long-term monitoring projects; and develop and provide interpretative information to other employees and visitors about the natural resource management program.

Optional specializations for biological science technician positions within the NPS may include different disciplines.

Wildlife: Positions involving research, conservation, and management of wildlife; wildlife control; or the determination, establishment, and application of the biological facts, principles, methods, techniques, and procedures necessary for the conservation, management, and control of wildlife.

Fisheries: Positions involving research, conservation, production, and management of fishes and fishery resources; or the determination, establishment, and application of biological facts, methods, techniques, and procedures necessary for research, conservation, and management of fishes and other aquatic animals (e.g., crustaceans, mollusks) where the work does not principally involve management responsibility for the operation of fish hatcheries.

Plants: Positions carrying out field studies on plants and plant ecology; inventorying and monitoring exotic (non-native) plants; leading field crews in exotic plant eradication; protecting plants by building and maintaining fences; conducting grazing monitoring studies; assisting researchers in vegetation and soils mapping; developing improved methods of controlling weeds and plant diseases; and similar work that usually deals with higher plant life.

Insects: Positions involving the classification, anatomy, physiology, habit, and life history, as well as response studies, of insects and related animal forms (e.g., spiders, mites); studies of measures used to control insects harmful to agricultural crops, animals, or humans; development and testing of more-effective insecticides, formulations, and biological controls; evaluation of plant varieties for resistance to insect attacks; development of methods for conducting insect surveys, to establish criteria for forecasting outbreaks; use of parasites, predators, and diseases to control destructive insects; and use of insects to control weeds.

Soils: Positions involving the physical, chemical, and microbiological properties of soils and the similar processes that take place in the soils; response and behavior of plants to physical, chemical, and biological soil properties, as well as climatic factors; and research and development efforts, through the study of soils and water, to improve cropping, cultural, and mechanical practices used in agricultural work.

Biochemistry: Positions involving the techniques and equipment peculiar to chemical analyses of biological substances or to the study of biochemical relationships. Such work may include chemical and DNA analyses of blood, urine, and other bodily fluids and substances; procedures for testing the chemistry of food, drugs, or other material; and methods for studying physiological relationships.

Water resources: Positions involving watershed management and subsurface water. These duties include monitoring water quality and recommending methods and programs to maintain quality, and monitoring and recommending mitigation actions for land uses.

Air quality: Positions monitoring and recording air quality data; monitoring and maintaining testing and data collection equipment; and coordinating data exchanges with the Environmental Protection Agency and the state.

Forestry: Positions involving forest rehabilitation, reforestation, propagation, and planting. These activities include conducting resource inventories; monitoring a park's hazard tree and disease programs; and monitoring and making recommendations for managing disease and insect infestations, especially by nonnative species.

GIS: Positions involving mapping, simple data analyses, and some data development using GIS and related equipment.

Backcountry and wilderness management: Positions monitoring, collecting, and organizing field data on ecological impacts that occur at backcountry campsites and other sites; and developing field reports and recommendations for management.

Firefighting: Positions involving wildland fire suppression activities. Employees performing this type of work are required to maintain their firefighter qualifications.

Background Needed

Previous experience, both in the depth and diversity of your skills, makes you competitive for NPS positions. Make sure to highlight those experiences that qualify under the optional specializations listed above.

Use all the resources available to you (job announcements, occupational questionnaires, the Internet, current agency employees) to gain information about the skills needed for the position in which you are interested. Requirements for NPS jobs can and do change, so it is important to do your homework before you apply. Also be sure to develop a solid resume (chapter 6). After applying for numerous jobs and being qualified but not selected, I [BB] went to the Human Resources Office and talked with the person who reviews applications. After a few minutes of visiting, it was clear that my resume was good but not great. I set up an appointment to learn more. The following week, I spent more than an hour learning tips on how to present myself better in a resume, and that made all the difference. In each and every job I applied for since, I have always received an interview. I learned that investing in understanding the application process, from beginning to end, allowed me to better communicate with the diversity of people who will look at my application in the future.

In addition, consider personally developing what the NPS defines as "the essential competencies for natural resources stewardship," which include seven components:

1. *Scientific knowledge*: In-depth knowledge of current ecological and scientific principles, with an ability to evaluate research results that can be used to resolve diverse and complex natural resource issues in parks.
2. *Scientific method*: Knowledge of and an ability to apply sound scientific approaches and appropriate methods to resolve natural resource management issues, including the design, conduct, and evaluation of inventory, monitoring, and research projects.
3. *NPS resource stewardship*: An ability to effectively apply laws, policies, regulations, and guidelines, in concert with scientific knowledge, to protect park resources and ecological systems.
4. *Planning and compliance*: An ability to apply scientific knowledge to identify and define natural resource issues, and to develop and evaluate alternative management strategies.
5. *Professional credibility*: Demonstrated expertise in and contributions to science and scientific endeavors, which are recognized by peers in governmental agencies and the academic community as providing a strong foundation and leadership in the natural resource work performed.
6. *Communication*: An ability to effectively communicate complex, technical, or controversial natural resource information to diverse audiences through a variety of media.
7. *Project and program management*: An ability to apply a knowledge of scientific concepts and NPS programs to plan, implement, and administer natural resource projects.

Education Required

In most circumstances, a bachelor's degree (or evidence that you are working toward this degree) in disciplines such as wildlife ecology, wildlife management, or a related field from an accredited college or university is needed for an entry-level position. As you move forward in your career, advanced degrees make you more competitive. Knowledge of the established practices, procedures, and techniques of at least one of the biological sciences is necessary. Familiarity with routine NPS natural resource management practices, methods, and procedures, and a basic understanding of related regulations, are expected. Knowledge of data-collection techniques, including the installation, operation, and maintenance of testing and sampling equipment, also is expected. An ability to follow detailed sampling and laboratory procedures, train others to accurately follow procedures, and recognize departures from established quality control and quality assurance guidelines is needed. A basic understanding of natural resource management principles and techniques to support, understand, and relate results to a broader natural resource function is expected.

Pay Scale

Entry-level biological science technicians just completing an undergraduate degree are eligible for jobs at the GS-4 or GS-5 level. Each time you enter a new grade, you typically begin at step 1 (table 5.1). It is important to note that being qualified for a job is dependent on a combination of education and experience. As you enter a workforce where many have an undergraduate education, consider the relevant and diverse job experiences you have that may help you compete more strongly for any given position.

NATURAL RESOURCES CONSERVATION SERVICE
Kent Ferguson and Steve Nelle

Agency: US Department of Agriculture, Natural Resources Conservation Service

The Natural Resources Conservation Service (NRCS) is an agency in the US Department of Agriculture that is charged with providing technical assistance to private landowners. Its primary mission is to address soil- and water-related issues that impact private lands through a conservation planning process with groups and individuals. The NRCS has a long history of conservation planning. During the Dust Bowl of the 1930s, the Soil Erosion Service was formed to combat this issue, which was rampant in the country at that time. That agency was later transformed from the Soil Erosion Service to the Soil Conservation Service and now is known as the NRCS. The name was changed to give the agency an identity that better describes its mission.

Wildlife management assistance has been a growing function within the agency for several decades. Nationally, there are about 300 technical staff positions that are directly dedicated to wildlife or biology. This does not include state-level biologists or upper-level biologists at the regional or national level. The staffing of biology positions varies greatly from state to state. Some states employ only one biologist, while others employ at least 20. It should be noted that NRCS biologists do not usually work directly with wildlife. Their emphasis is on habitats and habitat management, typically in association with farming, ranching, timber production, or other types of agriculture. The NRCS would be an ideal job opportunity for students who would like to work directly with wildlife habitat. Students who enjoy interacting with people and have a strong interest in wildlife habitat and agriculture are good candidates for NRCS employment.

Wildlife-related positions within the NRCS include rangeland and wildlife management specialists (series 0401) and wildlife biologists (series 0486). Other positions also may provide a significant wildlife management emphasis, depending on their location. Positions where wildlife may be an important secondary task include soil conservationists, wetland specialists, and foresters.

Rangeland and Wildlife Management Specialist
Job Description

A job as a rangeland and wildlife management specialist is an entry-level position, with an emphasis on basic training in range, wildlife, soil, and water conservation activities. Specialists also assist in the planning of conservation practices on farms and ranches. The position typically requires about 70% of an employee's time in the field, and the remainder working in the office. Specialists must be able to make accurate plant identifications and evaluations; determine plant community characteristics; and understand basic soil/plant relationships, responses to grazing management, and productivity on grazing lands. In addition, they must develop resource inventories, including the identification of ecological sites, ecological site conditions and trends, forage production, wildlife habitat evaluations, watershed qualities, and other related resources, as a step in developing conservation plans. Rangeland and wildlife management specialists also assist landowners in the development of practical grazing management plans, including planned grazing systems, stocking rates, forage utilization, fence construction, water development, range seeding, brush management, and other supplemental conservation practices

that may be needed for grazing land treatments. The evaluation of management practices and follow-up treatment alternatives are also an important part of this process. Advising ranchers and farmers on the application and maintenance of conservation practices is an integral aspect of this job. Very common duties are assisting in decisions on species of grasses to plant and brush species to control, and on pond construction. Employees in this position also receive field experience in all phases of the soil and water conservation program, as well as in the relationship of range and wildlife conservation to all land uses, with an emphasis on an interdisciplinary approach to resource use and management. This position involves the administration and technical aspects of the implementation of the Farm Bill program. Rangeland and wildlife management specialists must support civil rights activities, including affirmative action and program delivery, so that equality is demonstrated to all employees, applicants, individuals assisted, and the general public. Duties within the scope of this position involve the operation of a motor vehicle on public and private roads, so a valid state driver's license and a US Department of Agriculture identification card are required. The NRCS provides the identification card as a security measure. Specialists are required to spend many hours in the field, on various types of terrain and topography, with other individuals. Many hours are also spent in the office, preparing reports and documents for the planning process. Hours of operation are generally 0800–1700, Monday through Friday. There are several schedules of work duties that are available to employees in this position, but work schedules are governed by the local field office supervisor and that office's workload.

Background Needed

This position requires a diverse background. Rangeland and wildlife management specialists must be able to work with individuals and groups on a one-to-one basis. An ability to communicate effectively with a client is very important in completing the conservation planning process and implementing it. A farm and ranch background is highly desirable, but not required. Specialists need to have an ability and the desire to help individuals learn resource management, that is, the implementation and maintenance process that is required to plan and apply conservation practices. This experience can be acquired through on-the-job-training and exposure to the planning process. Rangeland and wildlife management specialists need to be very competent in computer skills and in written and oral communications. They must be able to perform physical activities in the field as well as work in an office environment.

Education Required

This job requires a bachelor's degree in range management or a related discipline, which includes 42 semester hours in a combination of plant, animal, and soil sciences and natural resource management. At least 12 semester hours must be in upper-division classes (junior or senior level). Of these 42 hours, at least 18 semester hours must be completed in range management, including courses in such areas as basic principles of range management, range plants, range ecology, range inventories and studies, range improvement, and ranch or rangeland planning. In addition, at least 15 semester hours of directly related classes in plant, animal, and soil sciences must be completed, including at least one course in each of these three areas. Classes in plant taxonomy, plant physiology, plant ecology, animal nutrition, livestock production, and soil morphology or soil classification are acceptable. Lastly, at least 9 semester hours of coursework are required in related resource management subjects, including classes in wildlife management, watershed management, natural resource or agricultural economics, forestry, agronomy, forages, and outdoor recreation management. The above list contains the minimum educational requirements for a rangeland and wildlife management specialist position. A deficiency in required coursework in any category can prohibit a person from being employed.

Pay Scale

Salaries are generally at the GS-5/7 pay rate (table 5.1). New employees with higher graduating GPAs generally are hired at the GS-7 rate. Promotions and pay increases normally are on a yearly basis if job performance is satisfactory, and if Congress does not freeze wages. The human resources staff reviews health, life, and long-term insurance and retirement plans during an employee's first week on duty.

Wildlife Biologist
Job Description

The actual day-to-day job description for NRCS biologists varies from position to position, but the following information is applicable to many positions. Most biologists serve on a staff with other natural resource specialists. This technical staff is usually headquartered at an area office that serves anywhere from 10 to 50 NRCS field offices. The field office is usually aligned with county boundaries and works through the local Soil and Water Conservation District. The territory of most biologists is large, and travel consumes a lot of time. Duties and daily routines are varied, based on regional wildlife issues and current Farm Bill programs.

Generally, the duties of NRCS wildlife biologists can be broadly categorized as having to (1) train NRCS employees in biology and wildlife management, (2) prepare technical materials to assist field office staff and landowners, (3) direct wildlife management assistance to landowners, (4) stay abreast of the latest research and integrate this into the NRCS's work, and (5) understand the Farm Bill and how it affects landowners and NRCS programs.

Training NRCS employees includes formal classroom sessions, organized field trips, and informal on-the-job learning. Much of this takes place on a one-to-one basis while performing other related duties. Biologists need to have strong communication skills and be able to develop and deliver effective training to others. Biologists must have excellent writing skills, be able to understand complex ecological relationships, and be able to translate them clearly and concisely in writing to landowners and other NRCS employees who may have limited ecological knowledge. Popular articles, technical notes, symposium proceedings, practice standards, trip reports, and field trial results are some of the types of writing expected of NRCS biologists.

Direct assistance to landowners is one of the more rewarding aspects of an NRCS biologist position. It is also a job that requires a high level of competence, confidence, and an ability to communicate well. Helping landowners integrate wildlife management into their overall land management objectives is challenging and gratifying to the biologist and beneficial to the landowner. Most of the landowners the NRCS works with are not interested primarily in wildlife. Wildlife is often a secondary goal. Successful biologists find ways for landowners to incorporate good habitat management into their agricultural operations. For example, biologists might suggest herbicides that are effective for the target species yet have reduced side effects on nontarget forbs and shrubs. Another example is to encourage landowners to improve habitat by fencing off ponds and riparian areas, which allows managed seasonal grazing and restricts the access cattle have to water to optimized lanes.

NRCS biologists must stay abreast of the latest science and trends in wildlife management, natural resources, and agriculture by reading scientific journals, research reports, newsletters, and popular articles and by attending relevant workshops, conferences, and seminars. NRCS biologists are also expected to cultivate and maintain good working relationships with other wildlife and agricultural professionals in the region. This may include the state wildlife agency, state conservation commission, US Fish and Wildlife Service, US Forest Service, Bureau of Land Management, US Army Corps of Engineers, Agricultural Extension Service, Farm Service Agency, Agricultural Research Service, and colleges and universities.

NRCS biologists should also cultivate good relations with many diverse private conservation and wildlife organizations, such as the National Wild Turkey Federation, Environmental Defense Fund, Audubon Society, Nature Conservancy, Quality Deer Management Institute, Ducks Unlimited, and many others. Biologists are expected to work with many different local, regional, and state organizations that are involved in wildlife conservation on private lands.

Much of the work of NRCS biologists relates to the current Farm Bill, which is passed by Congress every five or six years. NRCS funding is tied to the Farm Bill, and agency priorities are set by Farm Bill provisions. Fortunately, in recent years the Farm Bill has included increased funding and an emphasis on wildlife conservation. A good understanding of the conservation provisions of the Farm Bill is essential for this position.

The work of NRCS biologists lies at the intersection of biology, agriculture, and human nature. Simply having strong biological skills is not enough to be effective in this position. You must also understand local agriculture and be able to work effectively with many different kinds of people. One of the challenges of this position is working with individuals who disagree with you and may be antagonistic, offensive, and critical. Being able to stay calm, professional, and patient are important traits.

The percentage of time spent in the field and in the office varies from job to job. Most students who pursue a career in wildlife management prefer to spend time outdoors and often dislike office work. Within the NRCS, however, it is essential that employees keep up with required paperwork and office work, including planning, scheduling, written and electronic correspondence, progress reports, and staff conferences. The amount of time spent on administrative and bureaucratic tasks within the NRCS should not be underestimated. Few people enjoy this aspect of the job, and it consumes more time than you might anticipate. For a typical five-day workweek, the time when you are actually in the field generally ranges from one to two days per week. The remainder is spent in the office, traveling, in meetings, or other indoor functions that may or may not relate directly to wildlife management. An ongoing challenge for NRCS biologists is to be able to work effectively within a bureaucracy without becoming a bureaucrat.

Background Needed

In most cases, biologists in the NRCS are not hired directly out of college. Rather, NRCS employment usually begins as a GS-7 soil conservationist or rangeland management specialist. After several years of experience in one of these positions at the GS-7 and GS-9 levels, and after gaining a working knowledge of how the NRCS operates, individuals may seek biologist positions when vacancies occur. Most of these start at the GS-11 level. NRCS employees commonly work 5–10 years in other positions before becoming biologists. NRCS biologists are seldom hired from outside the agency.

Other types of experience will help you be more qualified and competitive for the relatively few biologist positions in the NRCS. Knowledge of and experience in range management, forestry, agronomy, soils, wetlands, water quality, fisheries, and agriculture are noticed by selecting officials. Active membership and involvement in professional societies, such as the Society for Range Management, The Wildlife Society, or others, is often viewed favorably.

NRCS biology jobs require a strong ability to work with a variety of individuals. People skills are just as important as technical biological skills. NRCS biologists are often in a position of needing to gently persuade, encourage, and motivate landowners to adopt conservation practices and techniques that are new to them. Landowners usually respond well to the show-and-tell method of teaching. For example, walking around with them and pointing out what quail nest cover, loafing cover, or brooding cover looks like is much more effective than just telling them what to look for or handing them printed material. Biologists also must be able to work effec-

tively with NRCS field office staff members who may have little knowledge about or interest in wildlife management. This is one reason why prior experience at a local field office is an important prerequisite for biology positions.

Education Required

For a wildlife biologist position, a degree in biological sciences or the equivalent is required. This includes a minimum of 30 semester hours in the following categories: (1) at least 9 semester hours in such wildlife subjects as mammalogy, ornithology, animal ecology, wildlife management, or research courses in the field of wildlife biology; (2) at least 12 semester hours in zoology, in such subject areas as general zoology, invertebrate zoology, vertebrate zoology, comparative anatomy, physiology, genetics, ecology, cellular biology, parasitology, entomology, or research courses in such subjects, and (3) at least 9 semester hours in botany or related plant sciences. For a more general biologist position, a degree in the biological sciences, agriculture, natural resource management, chemistry, or a related discipline is all that is needed, without the specific coursework requirements listed above.

Because NRCS biologists usually start out as soil conservationists or rangeland management specialists, it is important to have taken the right courses to qualify for these starting positions. Classes in agronomy, soils, plant taxonomy, animal science, animal nutrition, range management, range plants, range ecology, watershed management, wetland science, and forestry help meet the educational requirements for these entry-level positions. Students should work with their academic advisors before their junior year of college to ensure that the proper coursework is scheduled.

Within the NRCS, postgraduate degrees normally do not confer a material advantage in employment. Experience gained on the job is usually considered of greater value than additional formal education.

Pay Scale

The NRCS follows the federal pay scale. New employees fresh out of college normally begin at the GS-5/7 level (table 5.1). With a year or two of experience and good job performance, employees are usually promoted to the GS-9 level. These are generally nonbiology jobs at the field office, where important experience is acquired. After several years at the GS-9 level, employees have usually gained enough experience to apply for biologist positions, which often start at the GS-11 level.

WILDLIFE SERVICES AND THE NATIONAL WILDLIFE RESEARCH CENTER
Kurt VerCauteren and David L. Bergman

Agency: US Department of Agriculture, Wildlife Services

Wildlife Services, a federal program within the US Department of Agriculture, is tasked with helping to resolve human/

wildlife conflicts and reduce wildlife damage, such as predator issues with livestock, wildlife damage to crops, bird strikes at airports, diseases (e.g., chronic wasting disease, bovine tuberculosis, avian influenza), as well as to stop invasive species, such as feral hogs and brown tree snakes, from increasing their distributional range. Wildlife Services collaborates with numerous partners, including state and federal agencies, land grant universities, and others, to carry out this work and maintain healthy, stable wildlife populations in harmony with human populations. Employees focus primarily on human/wildlife conflicts involving terrestrial vertebrates, including birds, mammals, amphibians, and reptiles. Wildlife Services employs more than 1,500 wildlife experts and support staff working in all 50 states and several US territories. The National Wildlife Research Center, the research arm of Wildlife Services, is headquartered in Fort Collins, Colorado, with field stations located in Florida, Hawaii, Mississippi, North Dakota, Ohio, and Oregon. National Wildlife Research Center employees develop and improve tools and various methods—such as repellents, vaccines, contraceptives, toxicants, scare devices, traps, barriers, and surveillance and monitoring methods—for use in wildlife damage management and disease management. The benefits of working in Wildlife Services include frequently interacting with wildlife, playing a vital role in wildlife work that is meaningful to society, having access to cutting-edge technology, and working at a job that mixes active outdoor fieldwork with office responsibilities. Involvement at the National Wildlife Research Center leads to participating in the latest developments within the fields of wildlife damage management and disease management. Common positions within Wildlife Services include biological science technicians, wildlife biologists, and biologists. The same positions (with differing responsibilities) exist at the National Wildlife Research Center, which also employs PhD-level research wildlife biologists and research biologists to lead investigatory studies.

Biological Science Technician
Job Description

Biological science technicians (typically called technicians) in Wildlife Services assist in on-the-ground management activities, such as trapping and removing animals that are causing damage, assessing bird-strike risks at airports, and collecting biological samples to be tested for disease. Technicians working for the National Wildlife Research Center assist in planning laboratory and field studies to assess the impact of various wildlife species on agriculture, property, human and livestock health and safety, and natural resources. Technicians also help evaluate the efficacy and safety of various methods to reduce such impacts. They contribute to planning and conducting a wide range of laboratory and field studies (e.g., telemetry studies, behavior and diet analyses, pesticide efficacy and risk assessments, genetic analyses). Technicians adhere to the Animal Welfare Act and the National Environmental Policy Act; establish a good rapport and close cooperation with fed-

eral, state, and private land management agencies in carrying out various duties; and adapt and use recording forms and devices to gather test data, tabulate data, and prepare basic summaries of their activities and accomplishments. The data collected are used in analyses and written reports and manuscripts.

A typical technician with Wildlife Services is always part of a team, but the job also requires an individual to be very independent, comfortable, and competent when working in the field, often in remote locations and at odd hours. These positions often entail variety, such as assessing wildlife hazards at airports, trapping coyotes on sheep range, collecting avian influenza samples from waterfowl and passerines, and deploying toxicants for starlings at dairies, all in the same pay period. National Wildlife Research Center technicians play several roles related to conducting and reporting research findings. For example, they may be in the field for two weeks in Texas or Michigan, using remote cameras and telemetry to collect data on animal movements, and spend the next two weeks in the office, pouring through images and entering the data to be analyzed into spreadsheets.

Background Needed

Previous experience in wildlife damage management, including humanely capturing and handling birds, mammals, and reptiles using mist nets, cannon nets, leg-hold traps, live traps, and other capture devices is beneficial. Previous experience in initiating and conducting laboratory and field studies with wildlife, including the use of telemetry, GPS, and GIS; handling wildlife; and acquiring the skills associated with bleeding, immunization, chemical immobilization, euthanasia, necropsy, and feeding and caring for captive wildlife is helpful in obtaining a position as a technician. Demonstrated abilities to communicate effectively with a wide range of personnel, including peers, administrative support staff, project leaders, administrators, and stakeholders, and to gain cooperation from other agencies, landowners, and the general public are essential. Involvement in research as an undergraduate is useful, because previous experience in preparing study protocols, data summaries, draft reports, and manuscripts is expected of a technician. In addition, knowledge of computer programs, such as R, SAS, Word, and Excel, and a working knowledge of good laboratory practices, the Animal Welfare Act, quality assurance, and the Environmental Protection Agency's test guidelines all are important in working for the National Wildlife Research Center. Demonstrated abilities to work independently, with minimal supervision, as well as to work on a team are always positive accomplishments.

Education Required

Biological science technicians must successfully complete a full four-year course of study leading to a bachelor's degree from an accredited college or university, with major studies, or at least 24 semester hours, in any combination of scientific or technical classes, such as biology, chemistry, statistics, entomology, animal husbandry, botany, physics, agriculture, and mathematics. At least 6 semester hours of wildlife courses, or equivalent combinations of successfully completed education and experience, may be used to meet the total qualification requirements. Classes in wildlife damage management, human conflict, wildlife techniques, and wildlife disease, as well as courses where hands-on experience is emphasized, are particularly useful. Graduate courses or the successful completion of an MS or equivalent graduate degree in wildlife can result in being hired at a higher GS level (GS-7 or GS-8).

Pay Scale

Positions typically range from GS-5 through GS-8 pay grades (table 5.1). Further information about salaries and benefits can be found in the "Federal Pay Grade System" section of this chapter. Often, fewer benefits are provided for temporary or term (lasting for a single fixed period of time) positions.

Wildlife Biologist / Biologist
Job Description

Wildlife biologists, also referred to simply as biologists, perform a variety of technical procedures to capture, restrain, mark, and transport wildlife; collect biological samples from various species of birds and mammals; perform euthanasia and necropsy procedures; and much more. Within Wildlife Services, the positions can be either variable or specific and structured, often depending on where the job is located. In general, biologists work independently and are responsible for several ongoing wildlife management programs. They also may supervise technicians or other biologists. Besides performing duties in the field similar to technicians, they have more responsibility and often are the point of contact with cooperators, stakeholders, and state agency personnel. They also have administrative duties associated with the programs they manage. A typical month for a biologist might entail a week of travel to conduct wildlife hazard assessments at airports; another week of travel to assist state agency personnel in collecting data associated with a disease outbreak in a wildlife species; and two weeks in the home office and lab completing administrative and reporting tasks, cleaning up equipment from the most recent fieldwork, preparing equipment and ordering supplies for the next trip, and processing disease samples. In the National Wildlife Research Center, biologists assist in the design, planning, and conduct of field, pen, and laboratory studies to develop and evaluate tools and techniques for reducing damage by birds and other wildlife to agricultural crops, resolving nuisance problems, reducing wildlife / aircraft collisions, and reducing threats to human health and safety. They help with the planning, design, and methodology for experiments, as well as with data collection and analyses, and written and oral presentations of the findings. A typical month for biologists with the National Wildlife Research Center may involve a week or more of travel to a study location, working independently or with colleagues to collect data and ensure the success of a study that, for example, is evaluating a novel fencing strategy to keep deer from damaging crops or assessing the uptake of novel rabies vaccine baits by raccoons. For the rest of the month, they might be at the home office, conducting studies to, for example, determine the efficacy of a new rabies vaccine in skunks, which would involve working with the animals in pens and processing blood samples in the lab. During this period, they also would be analyzing data, drafting manuscripts, aiding in the design of new studies, and preparing equipment and supplies for upcoming fieldwork.

Background Needed

Beyond the requirements needed to be a technician, if you wish to be hired as a biologist, you need a minimum of one year's experience with statistical software, GIS, and GPS, as well as laboratory and field experience in wildlife research (e.g., an independent graduate research project). You should have experience capturing and handling wildlife and dealing with wildlife damage management, as well as have participated in wildlife damage, research, and management workshops. Good written and oral communication skills are required.

Education Required

Basic educational requirements include the successful completion of a four-year degree from an accredited college or university in wildlife or biological science that includes at least 9 semester hours in such wildlife subjects as mammalogy, ornithology, animal ecology, and wildlife management, or research courses in the field of wildlife biology; and at least 12 semester hours of zoology classes, in such subjects as general zoology, invertebrate zoology, vertebrate zoology, comparative anatomy, physiology, genetics, ecology, cellular biology, parasitology, and entomology, or research courses in such subjects (excess credit hours from courses in wildlife biology may be used to meet

A wildlife services biologist prepares to fly an unmanned aerial vehicle to gather imagery to assess feral hog damage to a corn crop (photo: Kurt Vercauteren)

the zoology requirements, where appropriate); and at least 9 semester hours in botany or related plant sciences. Additional graduate classes, an MS degree, or a PhD in wildlife will qualify a person for a higher GS level (the GS-7 to GS-11 range, depending on educational level). Additionally, you should have training or coursework in designing and analyzing management programs or research studies. You also need to demonstrate proficiency in technical writing as well as an ability to prioritize tasks and independently carry out assignments. Further, knowledge of the Animal Welfare Act, the requirements of good laboratory practices, and the chemical immobilization, euthanasia, and necropsy of wildlife are beneficial.

Pay Scale

The pay grades for these positions typically range from GS-7 through GS-11 (table 5.1). Further information about salaries and benefits can be found in the "Federal Pay Grade System" section of this chapter. Often, fewer benefits are provided for temporary or term positions.

Research Wildlife Biologist / Research Biologist
Job Description

Research wildlife biologists and research biologists (typically called researchers) are positions that are available only within the National Wildlife Research Center, not Wildlife Services. These individuals conceive, develop, plan, obtain funding for, and conduct research to manage important and complex issues related to a wide variety of human/wildlife conflicts. Depending on the assignment, the studies might involve livestock predation, crop depredation, depredation to commercial aquaculture or natural fisheries, transmission of diseases to humans and livestock, damage to property and natural resources, predation on threatened or endangered species, or aviation/wildlife strike hazards. This research typically leads to an increased understanding of the biology and impact of various species that conflict with human interests, and the discovery and development of tools and methods to manage such conflicts. Researchers apply their specialized expertise (e.g., wildlife biology / ecology, physiology, quantitative statistics, epidemiology, reproductive physiology, virology) to devising and conducting research. Assignments provide opportunities for laboratory, pen, and field studies on birds, carnivores, rodents, and ungulates. Researchers develop and keep up to date on their knowledge of biology and ecology relating to their specific areas of expertise. More-senior researchers often supervise other researchers, biologists, and technicians. They also collaborate with researchers from other agencies and from universities. A typical month for researchers may entail spending time in the field with biologists and technicians to set up a study; then leaving them on-site and heading off to attend a professional conference to present research results and discuss ideas with colleagues; followed by devoting time in the office to initiate and coordinate new studies, work toward publishing completed efforts, and address a multitude of administrative and bureaucratic issues to keep the research team functioning at a high level.

Background Needed

Minimum requirements include one year's experience using computer programming and simulation modeling, GIS in relation to wildlife research, and spatial statistics; employing wildlife field techniques (including capture and immobilization, as well as noninvasive methods); designing and developing wildlife studies; writing grant proposals; publishing research results; administering budgets and managing personnel; developing strategic plans; and preparing and giving public presentations.

Education Required

Basic requirements for researchers in the National Wildlife Research Center include successful completion of a four-year degree from an accredited college or university, with a major in wildlife biology, zoology, or botany that consisted of at least 30 semester hours of classes in the biological sciences and 15 semester hours in the physical, mathematical, and earth sciences. This coursework must have included at least 9 semester hours of training applicable to wildlife biology in such subjects as mammalogy, ornithology, animal ecology, wildlife management, and population dynamics, or related studies in the field of wildlife biology; and at least 12 semester hours in zoological subjects, such as invertebrate zoology, vertebrate zoology, comparative vertebrate anatomy, embryology, animal physiology, entomology, herpetology, parasitology, and genetics; at least 9 semester hours in the field of botany and related plant sciences; and at least 15 semester hours of training in any combination of at least two other areas of study, such as chemistry, physics, mathematics, statistics, soils, and geology. In addition, the successful completion of a PhD or an equivalent doctoral degree in the fields of epidemiology, biology, ecology, veterinary sciences, environmental health sciences, human health sciences, wildlife management, ornithology, or other comparable disciplines is required. Advanced coursework in population and community ecology, statistical methods (e.g., spatial, Bayesian, model development and selection), and GIS improve your marketability for these positions.

Pay Scale

Pay grades typically range from GS-11 through GS-14 (table 5.1). Further information about salaries and benefits can be found in the "Federal Pay Grade System" section of this chapter. Often, fewer benefits are provided for temporary or term positions.

FISH AND WILDLIFE SERVICE
Anonymous (adapted by Scott E. Henke from the USFWS website, www.fws.gov)

Agency: US Department of the Interior, US Fish and Wildlife Service

A 1940 reorganization plan for the Department of the Interior consolidated the Bureau of Fisheries and the Bureau of Biological Survey into one agency, known as the Fish and Wildlife Service. The mission of the US Fish and Wildlife Service (USFWS) is to work with others to conserve, protect, and enhance fish, wildlife, and plant populations and their habitats for the continuing benefit of the American people. The objectives of the USFWS are to (1) assist in the development and application of an environmental stewardship ethic for society, based on ecological principles, scientific knowledge of fish and wildlife, and a sense of moral responsibility; (2) guide the conservation, development, and management of the nation's fish and wildlife resources; and (3) administer a national program to provide the public with opportunities to understand, appreciate, and wisely use fish and wildlife resources. There are two bureaus within the USFWS: the Bureau of Commercial Fisheries, and the Bureau of Sport Fisheries and Wildlife, both headquartered in Washington, DC. The latter is divided into five regional offices, based in Portland, Oregon (Pacific region); Albuquerque, New Mexico (southwest region); Minneapolis, Minnesota (northcentral region); Atlanta, Georgia (southeast region); and Boston, Massachusetts (northeast region). The functions of the USFWS are to (1) enforce federal wildlife laws, (2) protect endangered species, (3) manage migratory birds, (4) restore nationally significant fisheries, (5) conserve and restore wildlife habitats (e.g., wetlands), (6) assist foreign governments with their international conservation efforts, and (7) through the Wildlife Sport Fish and Restoration Program, distribute the millions of dollars that are collected in excise taxes on fishing and hunting equipment to state fish and wildlife agencies. The USFWS manages the 60.7 million ha National Wildlife Refuge System, consisting of more than 551 national wildlife refuges and thousands of small wetlands and other special management areas. Under the Fisheries and Aquatic Conservation Program, the USFWS also operates 70 national fish hatcheries, 65 fishery resource offices, and 86 ecological services field stations. The vast majority of fish and wildlife habitats are on nonfederal lands. Programs such as Partners for Fish and Wildlife, Partners in Flight, and the Sport Fishing and Boating Partnership Council, as well as other partnership activities, are the main ways the USFWS fosters aquatic conservation and assists in voluntary habitat conservation and restoration.

The USFWS has approximately 9,000 employees at facilities across the United States. Many of the positions with the USFWS are open only to current USFWS employees serving under career or career-conditional appointments (the latter is for a three-year period measuring an employee's interest in and the government's ability to provide a federal career). Most current employees with the USFWS entered through one of the two Pathways Programs: the Internship Program and the Recent Graduates Program. The Internship Program is for current students. This replaces the Student Career Experience Program (SCEP) and Student Temporary Employment Program (STEP). The new Internship Program allows students in high schools, colleges, trade schools, and other qualifying educational institutions to have paid opportunities to work in agencies and explore federal careers while completing their education. The Recent Graduates Program provides developmental experiences in the federal government. It is intended to promote possible careers in the civil service to individuals who, within the previous two years, graduated from qualifying educational institutions with an associate's degree, bachelor's degree, MS, professional degree, or PhD, or received a vocational or technical degree or certificate from a qualifying educational institution. To be eligible, applicants must apply within two years of completing a degree or certificate. Veterans precluded from doing so, due to their military service obligation, can apply within six years after earning a degree or certificate. Once you successfully finish a stint in one of the Pathways Programs, you may be eligible to have that experience converted to a permanent position with the USFWS.

Biologist
Job Description

Biologists with the USFWS plan and conduct investigations on the impacts of various land and water development projects (primarily transportation projects) on the fish and wildlife resources of an area. Biologists review and report on permit applications under the Corps of Engineers, Environmental Protection Agency, or state permit programs. In addition, they plan and conduct inspections of existing projects to ensure that legal standards and other requirements are met and implemented in a manner most beneficial to fish and wildlife resources. Biologists provide input on land habitat restorations, including landowner contacts, mapping, surveying, staking, construction monitoring, and seeding, as well as prepare management plans for restored areas. They assist in the consultation process, pursuant to Section 7 (interagency cooperation) of the Endangered Species Act, and help in preparing and reviewing biological implications of environmental assessment and impact statements or comprehensive resource planning reports in order to evaluate environmental consequences of proposed federal actions.

Biologists assist in developing comprehensive fish and wildlife management plans to ensure the conservation, protection, and enhancement of fish and wildlife and their habitats over a geographic area having a variety of habitat conditions. This requires an ability to develop, coordinate, or review plans that may encompass any and all programs that affect fish and wildlife and their habitat conditions, including, but not limited to, fire management, moist soil management, cooperative farming, wetland management, water quantity and quality, timber management, forestry management, and grassland management and restoration. Biologists gather, organize, and interpret biological, ecological, pathological, public use, or other pertinent information to assist with the implementation of management plans, studies, and investigations that are required for species propagation, production, resource protection, ecological factors, public information, and other aspects of natural resource management; and draft and prepare reports with recommendation for changes in, the elimination of, or improvements to operations and program plans.

When in the office, biologists respond to written, telephone, and in-person inquiries from the public and the news media, obtaining and providing requested information within the established guidelines of the office and in conformity with agency and departmental policy. They also perform data analyses and write reports to evaluate their findings and make broad recommendations. In addition, biologists attend public meetings and hearings to discuss proposed or existing development projects.

Background Needed

USFWS biologists must have demonstrated their knowledge of ecology, fish and wildlife biology, and resource management; their grasp of federal and state environmental laws, regulations, and regulatory processes; and an ability to effectively communicate orally and in writing with fellow employees, other agency representatives, the scientific community, and the public. Biologists must either have or have the ability to obtain a valid driver's license.

Education Required

A biologist position typically requires the successful completion of a four-year course of study from an accredited college or university, leading to a bachelor's or advanced degree that included a major field of study in the biological sciences, agriculture, natural resource management, chemistry, or related disciplines appropriate to the position. Copies of college transcripts must be submitted with each application, to document your educational accomplishments.

Pay Scale

The starting salary for biologist positions typically is at the GS5/7/9 level (table 5.1). Information on salaries and benefits can be found in the "Federal Pay Grade System" section of this chapter. Often, fewer benefits are provided for temporary or term positions.

Wildlife Refuge Manager
Job Description

The duties of wildlife refuge managers include, but are not limited, to (1) planning, directing, and administering National Wildlife Refuge System lands in accordance with established laws, authorities, legislative mandates, and USFWS policies; (2) supervising professional, technical, administrative, maintenance, and clerical staff for the refuge; and (3) conducting administrative operations that involve short-, medium-, and long-range planning associated with all refuge operations, fiscal requirements, administration, personnel, property management, records management, historical document and museum item protection, public relations, and technical assistance to local, state, and other federal agencies. In addition, refuge managers formulate comprehensive management plans, determine missions, resolve multifaceted management problems and landowner disputes, and set objectives for the refuges within the complex, as well as develop refuge policy and coordinate the technical activities of the refuge programs

with federal, state, and local governmental officials, members of Congress, national and international resource managers and specialists, and representatives of special interest groups.

Background Needed

It is critical that wildlife refuge managers have experience with the requirements of the National Wildlife Refuge Improvement Act, the Endangered Species Act, the National Environmental Policy Act, the Wilderness Act, and the Refuge Administration Act. Past habitat management experience, including grazing, fire, farming, timber harvest, and water management, is needed. Also, a demonstrated ability to administer a fire management program and knowledge of fire suppression are required. A strong ability to communicate well both orally and in writing is essential, because refuge managers often must work cooperatively with the refuge's visitor services staff and the general public.

Education Required

A wildlife refuge manager position typically requires the successful completion of a four-year course of study from an accredited college or university, leading to a bachelor's or advanced degree that includes a major field of study in the biological sciences, agriculture, natural resource management, chemistry, or related disciplines appropriate to the position. Copies of college transcripts must be submitted with each application, to document your educational accomplishments. Refuge managers must be knowledgeable about a wide range of theories, principles, practices, and techniques of population and wildlife biology; species lifecycle requirements; habitat management and manipulation techniques; and scientific methods. They must also be able to conduct scientific investigations. Coursework in conservation biology, landscape ecology, plant ecology, soil science, hydrology, plant and wildlife diseases, zoology, genetics, statistics, environmental law, real estate policy and regulations, restoration ecology, forestry, and fire ecology and management are highly beneficial.

Pay Scale

Entry- or trainee-level refuge managers typically are at the GS-5/7/9 level (table 5.1). At the GS-9 level, managers typically supervise a limited staff. Midlevel refuge managers are normally at the GS-11/12 pay scale and usually are employed at stand-alone refuges or at a refuge complex. Positions for senior-level refuge managers (GS-13/14) generally are at a large stand-alone refuge or at a complex of at least two refuges. Information on salaries and benefits are in the "Federal Pay Grade System" section of this chapter.

Visitor Services Staff
Job Description

Providing visitor services on units of the National Wildlife Refuge System is a diverse, challenging, yet rewarding job. It entails varied responsibilities, including providing opportunities for a broad spectrum of the public to participate in primarily wildlife-dependent but also other recreational activities.

This includes designing the programs, helping to determine what facilities are needed to support them and what information needs to be provided about them, and monitoring how people participate in them. Visitor services' staff members often are responsible for all phases of facility development, including design, funding, contracting, construction, and maintenance. They also administer a diverse array of special uses on national wildlife refuges, from commercial film production to youth fishing derbies. It is usually the visitor services' staff members who manage volunteer programs and serve as the liaison to refuge support groups, cooperating associations, and other community and outside partners. They coordinate with local school districts in the development of meaningful environmental education programs to train tomorrow's natural resource stewards. Visitor services' staff work with concessionaires, as well as with outfitters and guides who help visitors experience refuges. Staff members manage fee programs, write press releases, analyze audiences, work with the media, and prepare information about the refuge for a variety of needs, including for the Internet and members of Congress. They communicate refuge messages within the organization. They help monitor the impact of visitors on the resource and establish carrying capacities when needed, so not only is wildlife protected, but conditions are also maintained to provide a quality experience. They deliver a high level of customer service and ensure that visits are safe. Visitor services' staff members are involved in comprehensive conservation and program planning to ensure that opportunities exist to instill a sense of wonder in refuge visitors, while the needs of wildlife remain first and foremost. People in local communities come in direct contact with the refuge system through programs managed by the visitor services' staff. Whether it's a waterfowl hunter in a duck blind at sunrise or a third grader on a field trip, staff members must be able to understand and meet the needs of both individuals. They are a vital link in the USFWS's community relations efforts. Most visitor services' employees are located at field stations. Support and leadership from USFWS's regional offices and its national headquarters in Washington, DC, however, are vital to bring consistency and quality to programs across the National Wildlife Refuge System. Visitor services' staff members play a key role in communicating important refuge system messages, creating opportunities for the public to experience wildlife, and ensuring that the USFWS mission, "for the continuing benefit of the American people," is met.

Background Needed

An ability to communicate effectively with the public and strong interpersonal skills are paramount to being successful as a visitor services' staff member. Past experience in a customer service–oriented business is useful, as is expertise developing and delivering educational presentations.

Education Required

A visitors' services staff position typically requires the successful completion of a four-year course of study from an accred-

ited college or university, leading to a bachelor's or advanced degree that includes a major field of study in the biological sciences, agriculture, natural resource management, outdoor recreation and education, or related disciplines appropriate to the position. Copies of college transcripts must be submitted with each application, to document your educational accomplishments. Knowledge of conservation, natural resource management, outdoor recreation principles, ecology, species lifecycle requirements, and habitat management and manipulation techniques is beneficial.

Pay Scale

The pay scale for visitor services' specialists ranges from GS-5 to GS-11 (table 5.1). Information on salaries and benefits can be found in the "Federal Pay Grade System" section of this chapter. Often, fewer benefits are provided for temporary or term positions.

Federal Law Enforcement Special Agent
Job Description

The Office of Law Enforcement within the USFWS is responsible for ensuring that federal conservation laws are followed, managing wildlife populations, and responding to the effects of climate changes. Obligations of the Office of Law Enforcement include: (1) combatting invasive species, (2) conserving migratory bird species, (3) preserving wildlife habitat, (4) promoting international wildlife conservation, (5) helping endangered species recover, and (6) safeguarding fisheries. The USFWS's Office of Law Enforcement focuses on breaking up international and domestic wildlife-trafficking rings that target protected animals and plants; enforcing federal migratory game bird hunting regulations and working with states to protect other game species and preserve legitimate hunting opportunities; engaging Americans as conservation partners; inspecting wildlife shipments to ensure compliance with laws and treaties and detect illegal trade; preventing potentially devastating threats to wildlife and plant resources, including illegal trade, unlawful commercial exploitation, habitat destruction, environmental contaminants, and industrial hazards; preventing the introduction and interstate spread of harmful wildlife; protecting wildlife from environmental contaminants and industrial hazards; safeguarding habitats for threatened and endangered species; and working with and training tribal, state, and federal law enforcement officers.

Federal wildlife law enforcement officers are referred to as special agents, although they are informally often referred to as fish and game wardens. Within the USFWS Office of Law Enforcement, fish and game wardens are criminal investigators who are called on to enforce federal wildlife laws within the United States. These federal law enforcement professionals are responsible for targeting crimes that undermine the nation's efforts to conserve wildlife resources.

Fish and game warden positions may involve investigating crimes that range from international wildlife smuggling to illegal game bird hunting. The job duties include collecting evidence, interviewing witnesses, conducting surveillance, organizing raids, making arrests, and preparing evidence for court. These individuals also often work undercover to infiltrate illegal wildlife-trafficking rings.

Fish and game wardens may work anywhere in the United States, from small, rural duty stations to large, multiagency offices in major cities, such as Los Angeles and New York City. It is common for fish and game wardens to work with state, tribal, or foreign law enforcement agencies, and routine federal partnerships include work with the US Customs and Border Protection, Environmental Protection Agency, and Homeland Security investigations.

Just 250 special agents work for the USFWS, making jobs in this field very competitive. Individuals who want to learn how to become a fish and game warden must first ensure they meet the minimum requirements for employment. Specifically, candidates must (1) be citizens of the United States, (2) be at least 21 years old but less than 37 (at the time of appointment), (3) possess a valid driver's license, and (4) be willing to sign a mobility agreement, stating that they are willing to accept reassignments to any location throughout their careers.

The employment process for a job as a fish and game warden also includes meeting the USFWS medical, physical, and psychological requirements, such as passing a medical examination and a number of physical fitness tests. Candidates also must participate in mandatory drug testing and psychological assessments. Qualified candidates complete an interview process. Chosen candidates for fish and game warden jobs must then undergo an extensive background investigation.

All new fish and game wardens must complete 20 weeks of formal training in criminal investigations and wildlife law enforcement at the Federal Law Enforcement Agency in Glynco, Georgia. The training program for new employees includes classes in (1) the rules of evidence, (2) electronic surveillance, (3) the use of firearms, (4) waterfowl identification, (5) crime scene identification, and (6) case report writing.

After successfully completing this training, new fish and game wardens report to their first duty stations, where they are expected to complete a 44-week Field Training and Evaluation Program, where they work closely with experienced officers who provide guidance on investigative skills and wildlife laws.

Background Needed

Previous experience in hunting, trapping, and fishing can be beneficial. Useful outdoor skills include camping, hiking, orienteering, and using all-terrain vehicles, boats, and other outdoor recreational equipment. Good interpersonal and verbal and written communication skills are very important. Previous law enforcement experience is an added advantage for the position.

Education Required

Although not mandatory, a four-year degree in criminal justice, wildlife management, or a related field is preferred. Given that only the most highly qualified applicants are chosen to complete the employment process, the majority of successful individuals seeking game warden careers choose to complete a bachelor's degree from an accredited college or university. A combination of two disciplines can be beneficial, that is, a bachelor's degree with a major emphasis in wildlife management or criminal justice and a minor in another field.

Pay Scale

New fish and game wardens begin their careers at a GS-7, GS-9, or GS-11 level, depending on their education and experience (table 5.1). The top performance level for this career path is GS-12, with exceptional individuals achieving a senior special agent position, which is at the GS-13 level. In addition to their base salaries, fish and game wardens also can expect to earn law enforcement availability pay, which is 25% above their base pay.

GEOLOGICAL SURVEY
Craig Allen and Joseph Fontaine

Agency: US Department of the Interior, US Geological Survey

The US Geological Survey (USGS) is a scientific organization that uses core systems in this field to provide impartial information on the health of ecosystems and the environment, natural hazards, natural resources, and the impacts of climate and land-use changes. The USGS serves the United States by providing timely, relevant, and reliable information to describe and understand the Earth; minimize the loss of life and property from natural disasters; manage water, biological, energy, and mineral resources; and enhance and protect our quality of life. The USGS employs the best and brightest experts, who bring a range of earth and life science disciplines to bear in solving problems. By integrating these diverse areas of expertise, the USGS is able to understand complex natural science phenomena and provide cogent and useable products that lead to solutions. The USGS has more than 10,000 employees, and there are many job options and position descriptions within the agency. The scientists, technicians, and support staff in the USGS work in more than 400 locations throughout the United States.

USGS employees include biological science technicians, research wildlife biologists, wildlife biologists, and supervisory wildlife biologists. Each of these jobs have different, but not necessary exclusive, requirements, which are specified for any open post. Research and supervisory positions, which generally require advanced degrees, are not covered here. Technician and biologist jobs frequently have overlapping requirements, with biologists often needing to have some specialized training and experience, which may be gained by first working as a technician. In addition to wildlife biologists, the USGS also hires ecologists, fisheries biologists, GIS analysts, and a wide range of similar positions in the climate, earth, and water sciences.

Biological Technician
Job Description

In the biological sciences, positions range from entry-level technicians on temporary (term) contracts to senior executives determining the agency's focus and its policies. The USGS is a large and complex organization, and its structure has changed over time. Current mission areas include climate and land-use changes, energy and minerals, environmental health, ecosystems, water, and natural hazards. Biologists may find employment in any of these specific areas, as well as jobs where there is considerable overlap among them. Given the diversity of the USGS's mission and the range of positions available, there is great variation among job descriptions, expectations, and duties. The USGS also provides significant opportunities at a wide range of skill and pay levels. Generally, positions are associated with USGS science centers (e.g., Patuxent or Northern Prairie Wildlife Research Centers), which are dispersed throughout the United States. Each center incorporates a comprehensive group of scientists, technicians, and support staff focused on conducting integrated research to fulfill the Department of the Interior's responsibilities to the nation's natural resources. In addition to science centers, the USGS also includes cooperative research units, focusing on fish and wildlife, which are currently found in 38 states and are associated with land grant universities. Employees in these units conduct natural resource research, in conjunction with state and federal partners, and train graduate students in wildlife, fisheries, and allied disciplines (human dimensions, global changes, landscape ecology).

A multitude of entry-level positions at the USGS are available to recent graduates with bachelor's degrees, but they are often only for limited terms. These employees provide technical help for long-term projects by collecting data in support of USGS mission areas. Entry-level positions may be office or field oriented, but they frequently include both types of work. Biological technicians in primarily field-based positions can expect to help USGS scientists in collecting basic field data, which could range from behavioral observations of animals, to estimating animal population densities, to maintaining automatic data-recording devices (e.g., acoustic detectors for bats, temperature recorders in bird nests) in remote locations. Fieldwork is frequently complemented by office tasks, such as compiling or managing collected data and writing reports that summarize results. All federal job applications must be submitted on the USA Jobs website (appendix B), the online repository where all such posts are listed. An example of a typical temporary position for a biological technician with the USGS would include duties such as collecting basic data for field investigations; making observations in which biological conditions are readily identifiable; reporting observations in field notes for use by higher-grade-level employees; searching technical sources for information and publications on designated topics and preparing summaries for reference use; assisting biologists at higher grade levels by studying and preparing

reports on wildlife biology investigations; preparing samples and performing data analyses in an office or laboratory; drafting assigned portions of reports; preparing graphs and charts; preparing correspondence pertaining to technical aspects of the work; assisting in planning, organizing, and implementing wildlife biology investigations that affect habitat conditions; and preparing reports on progress and the results of studies. The percentage of time spent on any particular task varies widely, but this generally is indicated in the description of a specific open position. Employees often receive training in gathering, organizing, and interpreting biological, ecological, pathological, and public use data, as well as other information pertinent to ongoing research studies and scientific investigations.

In an entry-level field position like the one in the example, duties routinely include days in the field (or occasionally longer time periods when field locations are very remote), often in difficult-to-access locations. Field sites may be distant and isolated, and it may take considerable time to reach them. Remote field camps are sometimes established when daily trips from the office are impractical. Field technicians have supporting roles, assisting scientists or graduate students in the collection of research data. There is enormous variation in what this means on a day-to-day basis. One example is a field research project seeking to understand the factors limiting population growth in an isolated, declining population of bighorn sheep. A technician's job might be to locate the animals by radio-telemetry, and then observe their behavior and collect data on their diet by conducting vegetation analyses. Such work can be physically demanding, since bighorn sheep are found in steep terrain, far from any development or infrastructure. Telemetry works best with few physical obstructions between the receiver and the animals wearing transmitters. Therefore, a technician may need to climb to the highest point in the landscape to get an initial fix on the animals' location or, in some cases, even fly over the area in a specially equipped telemetry plane. Work in the field can be tedious and repetitive, and this is frequently the case with field technician jobs. Experiencing remote wilderness locations and working with wild animals, however, can be highly rewarding. For instance, our hypothetical field technician working with mountain sheep may even be able to aid in the capture, handling, and radio collaring of wild bighorn sheep.

Background Needed

For entry-level positions in the USGS, there are a number of types of training that an applicant might need to have already acquired (e.g., CPR, defensive driving), but in general, position requirements are based on a combination of education and experience. In many cases, job seekers only need to have completed a few years of post–high school education to qualify for a position, but competitive applicants often have hands-on experience. Volunteer or previous work experience conducting plant surveys, handling animals, or entering data is often beneficial. In our example of a technician working with bighorn sheep, applicants with telemetry expertise or experience recording behavioral data most likely would be preferred over someone who simply had a wildlife degree.

Education Required

The educational requirements at USGS, as with all federal positions, depend on the GS level of the position. In general, most entry-level temporary positions in wildlife biology for the USGS require a bachelor's degree.

Pay Scale

Biological technicians with a bachelor's degree typically are hired at the GS-5 level (table 5.1). Often, fewer benefits are provided for temporary or term positions.

TRIBAL GOVERNMENTS
Jonathan H. Gilbert

Agency: US Department of the Interior, US Bureau of Indian Affairs, Tribal Governments

Often when people hear the job title "tribal wildlife biologist," they think the employee is a member of a Native American tribe who happens to work as a wildlife biologist. Anyone who works for a Native American tribe as a biologist, however, can fill that position. Currently, there are 326 American Indian reservations in the United States, each of which are managed by a Native American tribe under the US Bureau of Indian Affairs. Not all of the 567 recognized tribes have reservation land, but the collective total for these lands is approximately 22.7 million ha. Writing a short description of a typical tribal wildlife biologist position is the same as writing a short description of a typical Native American tribe—it is nearly impossible. Job descriptions for biologists working for tribal governments are as diverse as the tribes themselves. Tribal biologists work in a wide array of environments, with a diverse set of plant and animal species, in the field, in the office, independent of state counterparts, or in concert with those colleagues. My job, for example, is as a technical advisor assisting the Ojibwe tribes in the implementation of their off-reservation treaty hunting and trapping rights. Native American tribes are located throughout North America, inclusive of Alaska and Hawaii, in locales ranging from the arid regions of the Southwest to the high mountains in the Northwest, and from the lakes of the Great Lakes region to the ocean coasts. In addition, tribal biologists work in a variety of political environments. All tribal biologists are employed by tribal governments and must be adept at understanding and navigating their way through these specific sets of governmental processes. Tribal biologists may work on individual reservations for a single tribe, in which case they may have one set of roles and responsibilities. Or they may work off the reservation as part of an intertribal organization, in concert with state and federal wildlife agencies, so they wear multiple hats and have diverse responsibilities.

Wildlife Biologist
Job Description

As a result of the wide variety of environments and diversity of tribes, there are many different job duties tribal wildlife biologists can have. No day is typical. Each one provides challenges and opportunities, with an occasional dose of surprises. Some tribes manage hunting seasons for either tribal members or nonmembers. Tribal biologists' duties cover much of the work in setting harvest limits, administering the hunts, and monitoring and evaluating the harvest. Other tribes have significant endangered species projects. Still others seek to integrate their forestry and wildlife programs and manage them within an ecosystem perspective. Although I've emphasized the diversity inherent in this position, there are at least three common elements: the advancement of tribal sovereignty, culturally appropriate plans, and effective communication skills.

My job, for example, is to assist the Ojibwe tribes. I work with them to implement their treaty-reserved hunting, trapping, and gathering rights. The advancement of tribal sovereignty and tribal jurisdiction is a key element in this job. Tribes wish to function as governments. Therefore, much of my job is geared toward securing the tribe's ability to enforce hunting, trapping, and gathering regulations for members who are exercising these rights on public lands off their reservations. The rules tribal members must follow are different from the rules non-Indians must follow. This is sometimes a controversial part of my job. For example, tribal members in Wisconsin may hunt white-tailed deer with rifles from early in September through the end of December, a much longer period than the nine-day hunting season non-Indians in that state enjoy. This extended season caused much controversy initially, but, through the accumulation of harvest data, the tribes were able to demonstrate that the number of deer they took, even with more than 110 days of hunting, was far lower than the nine-day harvest result during the gun season by state-licensed deer hunters. Other positions in tribal natural resource agencies may not have as explicit a role for tribal sovereignty, but be assured, sovereignty is an underlying aspect of every tribal wildlife job.

My primary duty is to provide biologically sound and culturally appropriate advice to the tribes. Note that two areas of expertise are required: biological training and cultural sensitivity. Any person interested in being a wildlife biologist should have the necessary training in natural resources, ecology, and biology. But a unique aspect of working for a tribe, or an intertribal organization, is to infuse wildlife biology with the cultural lessons of the tribe. This is something that is rarely learned in school. It is knowledge that only comes from working with, listening to, and respecting the people in the tribe. It takes time to develop this ability, and it only comes when there is a level of trust established between yourself and tribal members.

An important skill for tribal wildlife biologists is to establish and foster this trust within the community. One day I was on an outing with some of my Indian friends when I noticed a very evident browse line on some of the vegetation. Wanting to impress them, I started to tell them about these lines, which indicate heavy browsing pressure from white-tailed deer. They asked me if I was sure there were lots of deer in the area. Wanting to continue to impress them, I said, "Yes, just look at the deer pellets all around," and pointed to what I thought was a group of them. As we inspected this more closely, we saw that it was a pile of hemlock cones, looking just like a clump of deer droppings. Well, that got a laugh, and the joke spread around the community. More than 25 years later, I still hear about it. How could such a silly mistake earn their trust? Well, the Ojibwe love humor, but more importantly, it showed that even though I was young and naïve, I was willing to own up to my mistake (chapter 3). I could accept being the brunt of a joke and laugh along with them. This particular story has earned me plenty of good will in my job.

Excellent communication skills are critically important for this position. It is essential that tribal biologists be able to communicate effectively within and among various groups from both a biological perspective and a cultural one. For example, if I were to meet with tribal deer hunters about a deer management program, I would need to be able to listen to and understand how their culture views deer. I would then need to be able to put this into an ecological setting, using my best skills, explaining the cultural perspective to the scientists and the science to the deer hunters.

A large part of my time is spent attending meetings. I interact frequently with my state and federal counterparts. Much of what we discuss involves the management of harvested species. We evaluate the population status, examine the habitat, set harvest limits, and report harvest results. Other meetings involve planning efforts, whether this is writing a management plan for individual species or forestry plans for national forests. I frequently get together with tribal leaders to explain natural resource issues or controversies. I monitor the harvests of white-tailed deer, American black bears, fishers, river otters, and bobcats, and I write reports documenting these harvests.

Research is another part of a tribal biologist's job. For example, I have a long-term research program examining the forest carnivore community (bobcats, fishers, martens). This keeps me mentally sharp and is one of my keen interests, so I'm highly motivated. I write and publish research articles and attend professional conferences, in order to disseminate my research results and support my professional society.

Background Needed

Individuals wishing to work as tribal biologists should seek cultural awareness training as well as traditional ecological knowledge. Locate university courses that offer cross-cultural training. Practical experience can be obtained by joining the Peace Corps. I spent five years working with people from a different culture: two years teaching biology in Fiji, in the South Pacific, and three years working in national parks in Ivory Coast, in West Africa. There, I learned to speak several languages, as well as how to interact with people who know nothing about being an American. This cross-cultural training qualified me for my current job and put me ahead of my competitors when I applied for the position.

Education Required

The educational requirements for an entry-level position working for tribal governments vary. My employer requires either a bachelor's degree with five years of relevant work experience or an MS degree in natural resources or related fields. It is rare to find an applicant with the former set of qualifications, so most of the biologists we hire have earned an MS degree.

For tribal wildlife biologist positions, I would refer to the classes suggested by The Wildlife Society's certification requirements. This is a broad list of courses that are designed to provide sufficient training for entry-level biologists. Tribal biologist jobs require diverse skills, and a well-rounded education in the wildlife sciences will serve you best.

Pay Scale

There is no set pay scale for tribal biologists. The salaries differ widely, depending on the tribe and the job responsibilities. An entry-level position with the above qualifications often pays between $35,000 and $45,000 per year. Most tribal jobs come with health insurance and access to retirement plans, such as a 401(k) plan.

Tribal biologists participate in many different types of projects, including capturing and radio collaring animals. Here, biologist Jonathan Gilbert examines a radio-collared American marten as part of a study on their home range and habitat use. Martens (*waabezheshi* in Ojibwe) are considered to be a clan animal and thus are important in the Ojibwe culture (photo: Jonathan Gilbert).

GOVERNMENT AGENCIES: STATE ORGANIZATIONS

DEPARTMENT OF PUBLIC HEALTH
Emily W. Lankau

Agency: State Department of Public Health

One function of a state department of public health is to prevent and control zoonotic diseases (those transmitted from animals to humans) in human populations by analyzing disease-surveillance data, performing outbreak investigations, partnering with other state agencies and health care professionals on disease control programs, and educating the public on how to prevent zoonotic diseases. Official job titles for this type of work may vary among states and could include zoonotic disease specialists, zoonotic epidemiologists, environmental epidemiologists, public health program specialists, or public health scientists.

Zoonotic Disease Specialist
Job Description

State zoonotic disease specialists work for a state's department of public health. While the exact names of these departments vary considerably (e.g., Arizona Department of Health Services, Georgia Division of Public Health, Maryland Department of Health and Mental Hygiene, Michigan Department of Community Health, Texas Department of State Health Services), all states have a governmental unit whose mission is to protect the health of the human population in that state by monitoring and responding to disease events and educating people on how to prevent illnesses. Unlike a clinic or hospital, where the focus is on providing individualized medical care to people who are sick, public health departments apply data on disease occurrence patterns to programs and activities that will reduce the number of their state's residents who get sick. State public health departments employ people with many different skill sets, from microbiologists and data analysts to policy and communication experts, because monitoring and preventing diseases at the population level are complex tasks that require input from many disciplines. State zoonotic disease specialists are members of such multidisciplinary public health teams. They provide specialized skills and training in epidemiology and in how infectious diseases can be transmitted.

This is a job that attracts generalists who enjoy working on a diversity of health concerns, including rabies; vector-borne diseases (those transmitted by an intermediate species that is not affected by the disease), such as West Nile virus; and enteric infections (those affecting the intestines) that are associated with exposure of affected food, water, and animals, such as salmonellosis and cryptosporidiosis. Zoonotic diseases that may be encountered include a number of wildlife-linked diseases, particularly rabies, but also those tied to companion animals (e.g., dogs, cats) and with livestock and poultry. The breadth of zoonotic diseases covered in a particular health department position may depend on the size of the state's human population, caseloads related to different zoonotic disease concerns within that state, and available funding for personnel. Some public health departments use the job title "zoonotic disease specialist" for employees who handle all zoonotic diseases, while others hire specialists in particular areas (e.g., vector-borne diseases, food-borne diseases).

Zoonotic disease specialists are scientists, and a major part of their job is to provide guidance and factual information on zoonotic diseases to other state agencies, local public health departments, and health care providers. They are also responsible for the collection and analysis of surveillance data, in order to document patterns of disease occurrence and assist public health partners in understanding the risk factors that might result in an illness. For example, zoonotic disease specialists might monitor bird and mosquito populations near urban centers for West Nile virus, to help the public health department target mosquito control efforts more efficiently. These specialists also provide leadership and technical support for outbreak investigations and emergency responses, and they may serve as the public health department's liaison with federal partners at the Centers for Disease Control and Prevention (CDC) during an outbreak that affects multiple states at the same time. In addition, zoonotic disease specialists are responsible for the dissemination of information to the broader professional community and to the public at large by writing reports and scientific publications and by making public presentations.

Zoonotic disease specialists must be able to flexibly transition among a variety of roles within a public health team, serving as scientists or data analysts on one project and youth educators or meeting facilitators on the next. Therefore, strong interpersonal and communication skills are vital. For instance, zoonotic disease specialists often serve as liaisons or facilitators for partnerships among a state's public health department and its other agencies, including the departments of agriculture, natural resources, and wildlife, as well as local health departments, doctors, and nurses. They must be comfortable working across different scientific cultures to bring people together for a common goal—disease prevention.

No day is typical. Daily activities depend on current disease concerns within the state, and priorities may shift rapidly when an emergency situation arises. The majority of the work is done in the office. Some days may focus on routine activities, such as managing incoming surveillance data, communicating with partners by email and phone, and writing or editing papers or reports. Zoonotic disease specialists also know that a routine day could be disrupted at any moment by an emerging disease issue in their state or in a neighboring state. For this reason, people who work in these positions tend to be very comfortable with uncertainty and are able to refocus their energies quickly as new situations arise. While the job is primarily office based, some fieldwork may be required during disease events, in order to visit with local public health partners and support the coordination of a response to an out-

Zoonotic disease specialists monitor numerous wildlife diseases, such as West Nile virus, by obtaining blood samples and checking disease titers (measurements of antibody levels) (photo: Christine Hoskinson)

break. Typically, local public health departments handle the majority of field-based work with the general population, although the relationship between state and local public health departments does vary among states. Local public health departments at the district, county, or city level may have their own zoonotic disease specialists, who support public health activities with job descriptions similar to the state-level positions, but often with a more active field component to their work requirements.

Background Needed

Zoonotic disease specialists should have a strong background in general biology and infectious diseases, an interest in human health care, good people skills, and strong verbal and written communication abilities. An important preparatory step for this position is experience in teaching and public speaking or in facilitating meetings. Students interested in this type of work should consider volunteering with scientific or environmental outreach groups, in order to develop skills in explaining complex topics to people of different ages and with varying levels of scientific understanding. Advanced computer skills, including data management in database programs (e.g., Excel, Access) and data analysis (e.g., using EpiInfo, SAS, Stata, JMP, R), are especially important. Familiarity with GIS technology and an ability to make maps are useful attributes, although many departments have technical experts in informatics and GIS mapping to provide support in this area.

Education Required

Entry-level zoonotic disease specialist positions in state public health departments typically require a master of public health (MPH) degree or an epidemiology or infectious disease–related MS degree. Supervisory positions or those requiring advanced expertise in epidemiology may be filled through the promotion of experienced MPH-level scientists, public health

nurses (BSN or MSN), or doctoral-level scientists holding a medical degree (MD or DO), veterinary degree (DVM), or PhD.

MPH programs are primarily course based, consisting of approximately 30–40 semester hours of classes and a small capstone project. Required courses for most of these programs include epidemiology, public health, infectious diseases, biostatistics, and public health policy. MPH programs vary in their level of specialization. Some are very broadly based and focus on general public health and epidemiology skills, while others have their students select an area of concentration, such as biostatistics or environmental health, for more-specialized training. Many programs encourage students to complete at least one health-related internship with a federal, state, or local public health agency, in order to gain practical experience and begin building a career network. Networking is critical for successfully finding a job in the public health field, and student internships often form the beginning of long-term relationships between students and mentors that can lead to job offers.

A capstone project for an MPH degree is a small project that demonstrates the student's skills and understanding of public health practice. These are often completed during at least one internship with a public health agency, but they also may be original research projects facilitated by an academic advisor at a university. Capstone projects usually reflect a particular student's interests and the skill set that person would like to develop during his or her training. For example, an MPH student interested in rabies could complete the capstone requirement by reviewing and synthesizing the information from literature on the epidemiology and control of rabies in skunks; analyzing existing rabies-surveillance data in skunks while serving as an intern with the state public health department; or working with a university-affiliated field biologist to study wild skunk behavior and habitat use in urban settings, to better target rabies control efforts.

After completing an MPH or other graduate or professional degree, many people then pursue a fellowship in public health as an entry-level pathway to a government-service career. For example, the Council of State and Territorial Epidemiologists has a two-year applied epidemiology fellowship program for MS or PhD graduates in epidemiology or a related field. This program places recent graduates in state, territorial, or local health departments, so they can learn applied epidemiology skills on the job. The intent is to help fellows obtain permanent positions in the public health field after completing the program. There are also a number of other public health fellowship programs through the CDC and partner organizations that can facilitate entry into the public health workforce by providing on-the-job training in areas such as general public health, informatics, health policy, and global public health. Learn more about these opportunities at the CDC's public health fellowships website (appendix B). All of these programs, which typically accept applications annually, are highly competitive.

Pay Scale

Entry-level salaries for this position can vary considerably across the United States, depending on specific job responsibilities and the cost of living at a particular location. Full-time annual incomes can be between $35,000 and $60,000, with a potential for promotion beyond the upper end of this range after you gain sufficient work experience or acquire additional education beyond the MS level. Many states have salary-graded epidemiologist positions (e.g., epidemiologist I, epidemiologist II, epidemiologist III), where the hiring level and compensation are based on the amount of experience a person has beyond the minimum educational requirements. Entry-level zoonotic disease specialists typically begin at the lower end of a salary range and can be promoted to higher pay grades with three to five additional years of experience. State agency employment also includes benefits, such as health and life insurance and a retirement plan.

Zoonotic disease specialist positions are usually for full-time work (40 hours per week). Specialists may also have additional on-call duties when they are responding to health emergencies occurring outside of normal business hours, or have extra paid overtime duties when dealing with health emergencies. Compensation may be described in position announcements as an hourly rate, a monthly rate, or a total annual salary.

State-level zoonotic disease specialist positions in public health can be supported by federal funding programs (e.g., Epidemiology and Laboratory Capacity Cooperative Agreements with the CDC). These positions also may be filled by federal employees who are appointed to jobs at the state level. Employees in such posts support state and local public health activities and act as liaisons with the federal public health system. Federal or federally funded epidemiologist positions may have slightly higher salaries and different benefits from the state-funded jobs. These federally funded positions may be temporary or permanent.

DEPARTMENT OF TRANSPORTATION
John H. Young Jr.

Agency: State Department of Transportation

The Department of Transportation (DOT) was established by an act of Congress in 1966. The mission of the DOT is to serve the United States by ensuring a fast, safe, efficient, accessible, and convenient transportation system that meets national interests and enhances the quality of life for the American people, today and into the future. While we tend to view the DOT as dealing with highways, it is also responsible for railways and airports. The Federal Railroad Administration focuses on assistance programs to expand the railway system in the United States, as well as on promoting rail safety and certifying that rail projects are in compliance with environmental regulations, such as the National Environmental

Policy Act. The Federal Aviation Administration works to ensure airline safety and reviews airport development projects to secure compliance with the National Environmental Policy Act, noise restrictions, and other state and federal regulations. The Federal Highway Administration is responsible for establishing rules for building safe roads, but it is the responsibility of state and local government to build and maintain the nation's highways, with funding from the Federal Highway Administration.

Environmental Specialist
Job Description

Environmental specialists are concerned with developing, analyzing, evaluating, and modifying environmental programs, policies, and procedures. They use their knowledge of the natural sciences to protect the environment and human health by assessing the risks that new construction projects pose to the environment and making recommendations on how to minimize the environmental impact of projects.

Environmental specialists assist in preparing necessary documents under the National Environmental Policy Act to demonstrate compliance with federal agencies' regulations and procedures and related federal, state, and local environmental laws and ordinances. Natural resource statutes that apply include but are not limited to Section 7 (interagency cooperation) of the Endangered Species Act; Migratory Bird Treaty Act; Bald and Golden Eagle Protection Act; Clean Water Act, Sections 303(d) (impaired water resources), 401 (wetlands), 402 (storm water runoff), and 404 (fill material in water); Fish and Wildlife Coordination Act; Farmland Protection Policy Act; and Coastal Barrier Resources Act. In addition to national regulations, state and regional ones also apply. For example, in Texas, other codes that may apply to a project, depending on where the project is located within the state, could include the Texas Council on Environmental Quality's Construction General Permit, the Recharge / Contributing Zone rules of the Edwards Aquifer Authority, and any city ordinances (e.g., Trinity River Corridor Development Permit, required by the City of Irving). Each project that is reviewed by an environmental specialist is evaluated for compliance with these and other rules. An in-depth understanding of state and national regulations, biology, and the ecology of species is essential for environmental specialists, in order to determine when and if a regulation applies and whether mitigation/conservation actions meet the various requirements. Once it is determined that an act or a code applies to a project, specialists then evaluate whether the necessary mitigation or environmental protection actions meet the regulatory mandates.

For example, replacing a bridge over a stream may or may not involve coordination with the US Fish and Wildlife Service and a state natural resource agency (e.g., Texas Parks and Wildlife) under the Fish and Wildlife Coordination Act. If the bridge design spans a body of water and does not involve impounding or diverting the flow, deepening the channel, or modifying the bank or bed, then coordination is not required. If it does involve one or more of the above, however, then coordination with these agencies is required. After determining if any statutes or ordinances apply, environmental specialists evaluate the proposed work and coordinate with natural resource agencies to determine a course of action that reduces, minimizes, or avoids impacts of the bridge replacement project to the water body in question. The review of environmental documents for compliance with state and federal regulations is a major part of the work for this position, as 50% of environmental specialists' time is spent in evaluating projects and staying current with changes in regulations.

Project scoping accounts for another 10%–20% of their time. Scoping involves assessing natural resources and their status (presence/absence, increasing/decreasing populations) within the project limits. Environmental specialists must identify and foresee potential issues for species and the environment. Environmental specialists may perform or oversee the fieldwork required to determine species presence/absence, assess possible impacts to species or habitats, and make recommendations to minimize, mitigate, or eliminate environmental impacts. Thus an extensive background in habitat restoration is desirable. When impacts to a special or protected habitat (e.g., wetlands, longleaf pine forests) are unavoidable, knowing what recommendations to make to restore or replace the function of those habitats is critical.

Environmental specialists develop and administer contracts for consultants and contractors, as well as budgets, in areas related to their field of expertise. Contract administration requires coordinating and overseeing the work of consultants, developing work authorizations and fee schedules, negotiating the scope of a contract, monitoring performance, reviewing invoices to identify discrepancies, and ensuring that the final reports for projects meet the necessary standards. Contract administration is a vital part of this position, accounting for about 10%–20% of the specialists' time.

Many of the regulations that must be applied to projects require environmental specialists to coordinate with state (e.g., Texas Parks and Wildlife Department) and federal agencies (e.g., US Fish and Wildlife Service, National Marine Fisheries Service, Natural Resources Conservation Service). Coordination includes evaluating projects; informing agencies of any potential project impacts; detailing how those impacts may be mitigated, reduced, or eliminated; and then discussing and working out required actions to address the impacts. Negotiation and communication skills are essential. Poor efforts in this area by environmental specialists can lead to excessive costs, resulting in the cancellation or postponement of a project, or can cause harm to an endangered species if mitigation measures do not offset project impacts.

Environmental specialists may become involved in conducting or supervising research projects. They must be knowledgeable about techniques for data collection, compilation, and analysis; capable of summarizing and communicating findings; and able to recommend a course of action. De-pending on the amount and variety of species expertise and knowledge the specialists have, managing research projects may account for 10% or less of their time. As an example, the Texas DOT recently initiated a multiyear project on bat diversity, abundance, and occupancy at a series of bridges slated for replacement. The study objectives were to characterize pre- and post-construction bat use at existing and planned bridge locations, develop design specifications for artificial roost structures, and measure and compare the available surrounding habitat with the habitat provided by artificial roost structures. Contract biologists used microclimate measurements from bat-occupied gaps on the existing bridges to determine the placement and size of artificial roost structures, in order to ensure that a comparable amount of roosting area was present and to help decide on where the structures are put on the new bridges. Environmental specialists used this information to work with design engineers to develop artificial roost structures that supplied a comparable amount of habitat to what had existed under the old bridges. Communicating and sharing the results within Texas DOT resulted in replacement projects on other bridges also providing artificial roost structures for bats.

In addition to these four main duties and responsibilities, environmental specialists may also be involved with preparing regulations, rules, policies, processes, and procedures related to their areas of responsibility. They may respond to requests for information regarding specific environmental activities and represent the department in meetings with internal and external agencies regarding program issues.

Background Needed

Experience in conducting biological surveys for plants and animals is a necessity, as is an ability to work outdoors under a variety of rigorous conditions. Skill in identifying the tracks of various species is not essential but is desirable. Applicants should have demonstrated good written and oral communication skills; published papers in scientific journals and articles in popular magazines and similar outlets; and given presentations at scientific meetings and to other organizations (e.g., local Lions Club, Rotary Club, school groups). Environmental specialist positions require an ability to work and cooperate with others and build effective working relationships with internal and external collaborators. Therefore, well-developed people skills are very important.

Education Required

A bachelor's degree in the natural, physical, or environmental sciences, natural resource management, conservation biology, or a biology-related field is mandatory. Core courses should include physics, hydrology, watershed science, sampling and analysis, statistics, environmental biology, chemistry, elements of atmospheric pollution, environmental resources management, and environmental policy. An MS degree is recommended but not required for an entry-level position.

Environmental specialist positions require particular types

Department of Transportation biologists monitor bat populations living under an overpass bridge (photo: John Young)

of knowledge, including (1) principles and methods of administering environmental protection programs; (2) environmental laws and regulations related to environmental protection; (3) ways to plan, fund, organize, administer, and evaluate environmental programs; and (4) practical information about natural resources, habitat restoration and management, and endangered species. A working knowledge of mitigation requirements pertaining to natural resources and hazardous materials management also is needed, as well as some familiarity with contract development, management, and oversight.

Pay Scale

Environmental specialist positions typically are divided into five levels. The salary for an Environmental Specialist I is $35,000–$54,000, while salaries for an Environmental Specialist V range from $59,000 to $94,500. Nationally, the 2015 salaries for environmental specialists were between $45,144 and $95,611, with a median income of $67,821. Benefits are consistent with those in other state and federal agencies.

GENERAL LAND OFFICE
Stephen M. Williams

Agency: State Land Department

The main function of a state land department is to administer state trust land. This acreage was removed from the federal domain and granted to individual states by the US Congress, for the benefit of several public institutions, the largest of which (based on the number of hectares in the grant) is the common schools, otherwise known as kindergarten through twelfth-grade education. The management and disposition of state trust land is regulated by the laws, statutes, and consti-

tution of a state. For example, in Arizona, the management and disposition of state trust land is regulated by the Arizona and New Mexico Enabling Act, the Arizona State Constitution, Arizona Revised Statutes (Title 37, State Land Code), and more than 100 years of case law from judicial proceedings. State trust land is managed to generate the highest return for its beneficiaries through the sale and leasing of this land, and by the sale of its natural products. Although individual states differ in what kinds of products are generated, the concept of managing them for state beneficiaries is a constant. The Arizona State Land Department (ASLD) is used as an example to highlight the job duties and requirements for positions in state land departments. Individual states may prioritize these duties differently and may manage other types of resources (e.g., timber). The position description, requirements, and pay scale, however, are fairly constant from state to state.

The ASLD and the system by which its trust lands are to be managed was established in 1915 by the Arizona State Land Code. In compliance with the Enabling Act and the state constitution, the code gave the ASLD authority over all trust lands and the natural products from these lands. Since its inception, the mission of the ASLD has been to manage state trust lands and resources to enhance their value and optimize the economic return for the beneficiaries, consistent with sound stewardship, conservation, and business management principles supporting socioeconomic goals for the state's citizens, today and for generations to come. The ASLD's mandate is to manage and provide support for resource conservation programs for the well-being of the public and the state's natural environment.

Natural Resource Manager
Job Description

A natural resource manager works within the Range, Agriculture, and Conservation section of the Natural Resources Division. This section administers 1,240 grazing leases on 3.4 million ha for livestock grazing, under its range function; 354 agricultural leases on 63,535 ha for raising crops, fruits, grains, and similar farm products, under its agricultural function; and 32 natural resource conservation districts across the state, which are recognized as quasi-political subdivisions of state government that develop comprehensive plans for the conservation of soil and water resources within the district.

This description focuses on the rangeland management and grazing lease administration duties of natural resource managers. The working title for this job is range resource area manager (RRAM), since each manager is responsible for a specific geographic area within the state, known as a range resource area.

RRAMs are responsible for an area encompassing approximately 690,000 ha of state trust land, with 242 grazing leases and permits. The busiest seasons historically were autumn through spring, dealing with rangeland monitoring and authorization for the number of animals that would be allowed to graze under the terms of a permit. This old paradigm has changed with the advent of various funding sources and programs that are available to assist grazing lessees with rangeland improvements. Fieldwork is yearlong now, with reviews of rangeland improvement projects taking up the formerly slack months. RRAMs have fieldwork days in every week of the year. A general rule is that every venture into the field also requires office time to document the trip and complete the action item that prompted it. In the office, RRAMs answer phone calls and emails and deal with walk-in customers (lessees or members of the public who arrive at the office without an appointment). Lessees typically need assistance completing applications and forms, while a member of the public may want to know if there is any unleased land available for grazing. The time involved in this customer contact varies, depending on the amount of background and knowledge customers have that is related to their questions, but 30 to 60 minutes per contact is typical. Some overnight travel is required, due to the size of the range resource areas and the remote locations of ranchers who hold grazing leases.

Arizona is a mixture of various types of landowners. Federal land encompasses 70% of the state's acreage and is administered by the US Forest Service, Bureau of Land Management, Department of Defense, US Fish and Wildlife Service, and 18 Indian Tribal Councils, in cooperation with the Bureau of Indian Affairs. Private and state trust land makes up 17% and 13%, respectively.

Some of the primary duties for RRAMs are developing management plans, reviewing rangeland improvements, monitoring the land, overseeing grazing authorizations and compliance, reviewing applications, and educating the public by providing information about the state land department.

Management plans: RRAMs work with grazing lessees and other federal agency partners (Bureau of Land Management, US Forest Service, and Natural Resources Conservation Service), state agency partners (Cooperative Extension Service, Arizona Game and Fish Department, Arizona Department of Environmental Quality), and others (natural resource conservation districts, nongovernmental organizations, consultants) to develop coordinated resource management plans for a ranch unit. The primary components of these plans are (1) goals and objectives; (2) management actions to be implemented, to help achieve the goals and objectives; (3) a monitoring protocol, to determine if the stated goals and objectives are being met; (4) a list of proposed improvements and a base map with all the necessary layers for land status; (5) existing improvements; (6) ecological sites; (7) habitat areas; and (8) proposed improvements.

Land improvements: By law, grazing lessees must request prior approval for the placement of rangeland improvements (e.g., water catchment devices, fences) on state trust land. Such improvements assist with proper grazing distribution, livestock control, and rangeland restoration. RRAMs help grazing lessees complete their applications and see that all of the questions are answered correctly. They conduct a field review to verify that a proposed project is on state trust land;

a cultural resource clearance, to ensure that the project does not adversely affect any cultural, historical, or paleontological sites; and a native plant survey, to verify that plants listed under the Arizona native plant law will not be adversely impacted. If cultural sites or protected native plants are found in the area covered by the proposed range improvement, RRAMs prepare conditions, placed in the project authorization, that will avoid or mitigate the impact of the improvement on these sites and plants.

At the conclusion of this fieldwork, RRAMs prepare a memo, summarizing the comments received from the specialists who were asked for their input, as well as the results of the field inspection, and making a recommendation that would include any applicable supplemental conditions. These conditions, which consider wildlife values, could consist of placing escape ramps in water troughs or constructing fences according to wildlife specifications. The memo is then routed to the section manager for review, the division director for approval and a signature, and the title and contracts section to update the lease file.

Monitoring: This is a means to determine if a management plan's goals and objectives are being achieved. Clearly stated goals and objectives provide the framework for deciding what, where, and how often to monitor. Attributes that have a known relationship to the management objectives (e.g., basal cover, canopy cover, density, production) are selected for measurement. The appropriate monitoring methods are determined. Transects are established, documented, and measured at a time of year that is advantageous for plant identification. Most herbaceous plant attributes are best measured when the vegetation reaches its peak of maturity, near the end of the growing season. Not only do RRAMs participate in the selection of the key area where transects are placed, but they also take part in the actual data collection and interpretation of that monitoring data.

Ephemeral grazing authorizations: In some areas of the Sonoran Desert, grazing lessees chose to run seasonal rather than yearlong livestock operations. These seasonal operations take advantage of the yearly grass and forb growth generated by autumn and winter rain. In a good wet year, 907 kg of air-dried plant material can be produced on the most productive ecological sites. The ASLD can issue special authorizations for extra animal unit months (AUMs), known as Additional AUM Grazing Permits, which allow a lessee to exceed the carrying capacity of the lease for six months or less, in order to make use of this ephemeral forage. RRAMs conduct field investigations for special authorizations to determine whether there is sufficient plant growth to sustain the number of livestock requested by the grazing lessee. At the conclusion of the grazing season, RRAMs calculate the actual use (the amount of AUMs of forage that was consumed) and prepare a billing memo for the grazing fees owed for that use.

Compliance: Because grazing leases and permits are contracts, with specific terms and conditions, RRAMs must be vigilant regarding any violations of the lease terms. Such violations result in a lease default being issued, through a commissioner's order. This order is an agency action that can result in an administrative hearing before the Office of Administrative Hearing and Appeals. RRAMs are responsible for assembling all the factual background concerning the default and providing testimony during any proceedings that arise from a lessee's appeal of the order. RRAMs then enter the ultimate disposition of the case into the lease file.

Application review: RRAMs have an opportunity to assist the administrators for rights of way, sales, and commercial leasing by offering comments and recommendations on applications for those proposed uses of state trust land. If potential adverse impacts to rangeland resources or the grazing lessee are identified, RRAMs can provide supplemental lease conditions to avoid or mitigate these impacts. For instance, if construction of an access road across state trust land was to occur, possible supplemental conditions could include power washing the construction equipment, to remove seeds from invasive plant species prior to arrival at the job sites; reseeding road shoulders with a mix of native grasses and forbs; and installing features such as out-sloping, berm removal, and water bars to improve surface drainage.

Information and education: RRAMs are expected to offer customers information and education about how the ASLD functions, describing what its powers are and what the limits of that authority are. RRAMs are often asked to provide relief for an aggrieved party, or just to supply basic information. Some common topics involve livestock eating ornamental vegetation in a private landowner's yard, all-terrain vehicles driving off of established roads, trash dumping, target shooting, hot air balloon launches and retrievals, and the collection of fuel wood and dead cactus skeletons for personal use. RRAMs need to have a broad knowledge of ASLD policies and procedures to be truly effective at providing the necessary responses.

If the old adage that variety is the spice of life appeals to you in looking for a fulfilling career in natural resource management, a RRAM position can provide just that. Although the duties described above may not occur daily, they are typical of what you should expect from this job.

Background Needed

In order to be a satisfactory employee as an RRAM, you must have basic knowledge about and be able to operate office equipment, work in the outdoors, use effective oral and written communications, drive a four-wheel-drive vehicle, and read livestock brands. Due to the variety of duties that must be performed, you need excellent organizational and time management skills. In addition, you must be able to identify plants, wildlife species, and livestock breeds, as well as determine the sex and age of livestock; plan, organize, and accomplish activities according to a specific work plan; interpret, analyze, and evaluate data; make independent decisions and judgments; establish and maintain effective relationships with coworkers and customers; work as an effective team member in collaborative efforts; and travel to and work in the field under sometimes adverse weather conditions.

When preparing a resume or filling out a job application, it is important to convey to a prospective employer what you have done to augment your classroom training. Highlight the hands-on experience you gained from activities in student chapters of professional organizations (e.g., Society for Range Management, The Wildlife Society), internships, volunteer work, and attendance at conferences and workshops pertaining to natural resources.

People skills are crucial. Those you encounter while in the performance of your duties form opinions about the agency, based on these contacts. Professional conduct in dealing with lessees, the public, and other agency personnel is crucial, both for effective job performance and maintenance of the agency's image. The average age of farmers and ranchers in the United States is 57. Young professionals need to take this age differential into account and realize how it affects the communication style needed to effectively interact with an older generation.

Education Required

A bachelor's degree in a natural resource field (e.g., range management, watershed management, soils, wildlife biology) is required. Essential courses include ecology and ecosystem functions, rangeland monitoring methods, map interpretation, GIS, GPS, remote sensing, environmental law and legal descriptions, statistics, and identification of historical and prehistoric cultural resources. Training often is available after being hired, however, to meet para-archaeology requirements (how to identify artifacts and sites; procedures for properly documenting them; local, state and federal laws protecting cultural resources; archaeological field methods), because the majority of potential employees lack this knowledge and experience.

Pay Scale

The salaries for jobs in the Arizona State Land Department are based on national averages for positions that have similar tasks and duties. A RRAM position, classified as a natural resource manager II, is at pay grade 19 and has an annual salary range from $33,435 to $56,964. An entry-level employee with no experience or advanced degrees would start at the lower end of the scale, while an employee with these qualifications could start on the upper end. The ASLD also offers various medical, dental, vision, life, and long-term disability insurance plans for its employees. Employees also have the option of participating in a 457(b) deferred compensation program, which allows them to build a retirement savings account by having automatic payroll deductions go to specific investments. Cost of living adjustments are provided only when the state legislature appropriates the money for them. Such adjustments have had an irregular history in recent years.

STATE EXTENSION
Billy Higginbotham

Agency: State Cooperative Extension System

While all universities engage in teaching and research, the nation's land grant universities have a third critical mission—extension. Extension means reaching out and addressing public needs with research-based information through informal programs, applied research, mass media, one-on-one contacts, websites, and social media outlets.

Today, the conservation and management of rural farm and ranch lands is especially critical for wildlife, since two-thirds of the United States is privately held and approximately 80% of our nation's wildlife is found on these lands. Extension wildlife specialists serve as an important link between research-generated information and landowners by providing educational offerings and expertise about wildlife management on these privately held properties.

Cooperative Extension Service programs are largely administered through county and regional extension offices, with an office in or near most of the nation's approximately 3,000 counties. The Cooperative Extension Service (often referred to just as *Extension*) offers landowners information on and support for their crops and livestock; helps them become better stewards of the land by assisting them with the management of the natural resources on it (including wildlife and wildlife habitat); assists homeowners in planning and maintaining their homes and improving family nutrition and quality of life; and, through 4-H and youth development programs, helps children learn skills to become tomorrow's leaders.

Today, Extension works in six major areas, through a partnership funded at the federal, state, and county levels: (1) 4-H youth development, (2) agriculture, (3) volunteer development and recruitment, (4) natural resource management, (5) family and consumer sciences, and (6) community and economic development.

Extension Wildlife Specialist
Job Description

Every state has a land grant university, but not every state employs an extension wildlife specialist. Extension wildlife specialists may have statewide responsibilities or, in states where several such specialists are employed, job responsibilities may be divided geographically or by expertise in a particular subject matter. Regardless, extension wildlife specialists often reach the public directly, either by one-on-one interactions or through a local contact, usually a county educator or a county extension agent. These local educators are extension-only employees that normally are assigned to a single county and assist the public with a host of issues. A typical county has from one to three county faculty members: an agricultural / natural resource agent, a family and consumer sciences agent, and a 4-H youth development agent. Extension wildlife specialists have opportunities to address youth audiences through a variety of 4-H programs, including the Wildlife Habitat Evaluation program, field camps, Shooting Sports programs, and Natural Resources School Curriculum Enrichment programs.

County-level faculty consider extension wildlife specialists as their go-to subject matter experts on wildlife and consult with them to obtain information or answer questions generated by their clientele at the local level. In addition, extension wildlife specialists help county extension faculty (often through county-based wildlife committees) plan educational events to address local needs regarding wildlife. In some cases, these educational programs are single events, while others may be conducted as a series of programs over several weeks or months. Extension wildlife specialists also are expected to provide in-service training to county faculty, so the latter can answer many of the questions they receive on wildlife issues.

Extension wildlife specialists coordinate educational activities with colleagues in their respective state's game and fish commission, nongovernmental organizations, other universities, and extension faculty in other departments (e.g., forestry, soil and crop sciences). They also often participate in grant-funded projects with researchers, chair or serve on graduate student committees, and serve as mentors for undergraduate students. Specific day-to-day activities might include answering questions from clientele through office visits, phone calls, and emails; supporting county extension agents; and developing county-level programming in natural resource management. They are also expected to create innovative extension educational materials directed at enhancing wildlife resources in their state. The focus of such programs include native game and nongame species and management strategies that support wildlife/livestock/agriculture interactions; collaborations with a network of wildlife biologists and other natural resource managers and scientists within state and federal agencies and nongovernmental conservation groups to address issues of common concern; the expansion of youth conservation education programs; the development of external funding sources to support extension and applied research activities in a faculty member's area of expertise; committee service for departmental, college, and university activities; general support to the land grant mission of their university; and service and leadership on committees within their professional societies. Extension wildlife specialists are also expected to design, find funding for, and conduct applied research projects to address wildlife management issues.

How do extension wildlife specialists proffer this research-based information to their clientele, when that customer base may be the general public or a specific group (e.g., a local 4-H Club interested in wildlife management projects)? One way is to disseminate information through workshops, field days, seminars, newsletters, news releases, magazine articles, office visits, phone calls, emails, websites, and social media. Extension wildlife specialists also may write a number of electronic/

printed publications that address common management issues (e.g., deer management, nuisance wildlife control, wildlife habitat management). In addition, they are often contacted by mass media outlets (radio, television, newspapers) for interviews as subject matter experts on all things pertaining to wildlife. Extension wildlife specialists must learn to become very comfortable in front of a camera or microphone, so developing excellent people skills can greatly enhance their credibility with the public.

Lastly, extension wildlife specialists must be able to document the positive impacts of their extension program to retain their jobs and be promoted. Much like university professors, who disseminate course evaluation forms to students at the end of the semester, extension wildlife specialists should develop evaluation instruments or surveys to measure the impacts of some of their major extension educational programs. Surveys typically include impact metrics, such as knowledge gained by clientele, management practices that could be adopted, economic impacts, and a customer satisfaction index (e.g., net promoter score). If extension wildlife specialists are also university faculty members, additional measures of success may include the number of peer-reviewed extension and scientific articles published, news releases and popular articles written, contacts made through websites and social media, and committee service in the department, college, university, and professional societies. These collective data are considered when a faculty member is evaluated annually or becomes eligible for promotion to the next academic rank.

Background Needed

Make no mistake about it—wildlife management is people management! If you have no desire to have frequent contact with the public to answer their questions or advise them on solving management issues related to wildlife resources, then a career as an extension wildlife specialist is not for you. On the other hand, assisting the public with a myriad of wildlife-related issues, from urban backyard wildscaping to addressing wildlife-caused livestock depredation, can be an extremely rewarding job. A successful extension wildlife specialist must possess and further develop the communication skills necessary to discuss technical wildlife issues with laypeople who normally do not have a natural resource management background. These clientele groups may be rural or urban, have little or no background with regard to wildlife, or do not have a deep understanding of specific wildlife issues. In addition, many of these issues may not have clear-cut recommendations to use in advising a client. In such cases, extension wildlife specialists have an opportunity to design and conduct applied research (often referred to as *result demonstrations*) to address specific problems. Examples of applied research might include investigating new trapping techniques for controlling feral hogs, evaluating the impacts of a wildlife habitat management technique (e.g., forest thinning, prescribed burning) on habitat quality, or field testing new census techniques on estimating the population of a species (e.g., camera census). A primary responsibility of extension wildlife specialists is not to make decisions for clientele, but rather to provide them with the research-based information they need to make an informed decision.

Successful extension wildlife specialists must develop the skills necessary to raise money to fund projects (primarily through gifts and grants), sizeable educational events, and research. There are many opportunities to work with other faculty members when submitting grant proposals, as many governmental funding entities now require an extension or outreach component that details how the information derived from a research project benefits specific clientele groups or the general public.

Extension wildlife biologists work closely with county extension agents and other agency biologists to demonstrate best-management practices at landowner education field days (photo: Billy Higginbotham)

Rest assured that no extension wildlife specialist came to his or her job with a complete skill set! Much of the background needed for these positions was acquired on the job and through professional development opportunities provided either internally (by a state extension service) or externally (by professional societies).

Education Required

A minimum of an MS in forestry, range, or wildlife science is required for nonfaculty extension wildlife specialists, and a PhD in forestry, range, or wildlife science for specialists who are faculty members at the level of assistant, associate, or full professor. Although it is not imperative to have various degrees within different fields of study (e.g., MS degree in range management and PhD in wildlife management, or vice versa), basic knowledge within multiple educational fields can be useful. You never know what the general public might ask, but often they desire immediate answers. Therefore, a variety of courses (e.g., fisheries, range, wildlife, habitat, GIS) can help prepare you for whatever questions might come your way. While a few states still have extension-only faculty positions, most offer joint appointments, where the extension component is complemented by a research or teaching appointment. The advantage of joint extension/research jobs is that these faculty members have an opportunity to obtain funding for and conduct applied research that has an immediate impact, and they can quickly share the results with clientele (e.g., landowners managing their properties for the benefit of wildlife). Another advantage is that many of the faculty members' extension experiences, gained while working with clientele, can enhance their knowledge base, which is then shared with students in the classroom. In a few instances, faculty members may have a three-way joint appointment involving extension, teaching, and research. Although these positions are common at some land grant universities, it may be advantageous for newly hired faculty to gain experience in no more than a two-way joint appointment and grow into a three-way post (if applicable) as they progress in their careers. Regardless, extension wildlife specialists with joint appointments are advised to work *smarter*, not *harder*, by combining their efforts to meet extension, research, and teaching responsibilities through an integrated approach.

Pay Scale

There are two categories of extension wildlife specialist positions, but both usually are appointments covering 12 months of the year. The first category is for entirely extension-funded positions and may or may not carry a faculty title of assistant, associate, or full professor, depending on the academic guidelines of that particular university. For jobs that are not considered to be faculty positions and do not require a PhD, beginning salaries range from $55,000 to $65,000 for a 12-month appointment and typically have some type of career ladder system in place for advancement and salary increases. The second type of extension wildlife specialist category requires

a PhD and is most likely to be considered a faculty position within the appropriate university department. Beginning salaries for these posts generally range from $70,000 to $80,000 for a 12-month appointment. Promotions and salary increases are similar to those in the "University Professor" section of this chapter. In many states, extension specialists in a faculty-level position may hold a joint appointment where, instead of being extension-only positions, they must also teach and conduct research. Joint appointments often offer a slightly higher starting salary, compared with an extension-only job.

Regardless of the type of position, being an extension wildlife specialist is an extremely rewarding career choice. No two days are exactly alike, and there are abundant opportunities to design a program that fits your expertise and skill set with a good balance of working behind a desk and days spent in the field.

STATE WILDLIFE DEPARTMENT
Joe Beach, Vernon C. Bleich, and Leonard L. Ordway

Agency: State Wildlife Department

State wildlife departments—sometimes called a Department of Natural Resources, Department of Fish and Game, or Department of Parks and Wildlife—all have one main purpose, which is consistent with the doctrine of public trust: they have a responsibility to conserve the wildlife and other natural resources of their state for the public's benefit. This responsibility is embodied as one of the seven pillars of the North American Model of Wildlife Conservation (Organ and Batcheller 2009), the first of which emphasizes that fish and wildlife are held in trust for the public by state (or federal) governments. Although an individual may own the land on which wildlife resides, he or she does not own that wildlife. Instead, wildlife belongs to all citizens.

Wildlife Biologist
Job Description

Wildlife biologists are responsible for management and conservation decisions that either are applicable to a particular geographic area or are part of a specific species management program at the statewide level. Wildlife biologists operate four-wheel-drive vehicles of various types over rough terrain, sometimes many kilometers from the nearest road, and, whether planned or unplanned, occasionally spend days and nights alone in remote locations. They are away from the office and, occasionally, their families for extended periods of time. Wildlife biologists should expect to hike or walk long distances under sometimes unpleasant weather, including excruciatingly hot temperatures or frigid conditions. Thus biologists need to be in excellent physical shape and must possess the survival skills and, in particular, the common sense that will allow them to survive in case of unanticipated events or emergencies.

Common sense is an especially desirable attribute. For example, during an interview, I [VCB] presented a prospective employee with the situation of a vehicle breakdown and a concomitant radio malfunction some 32 km from the closest human contact, during July, near Death Valley, California. In the scenario, that individual was to be gone for one week but carried food and water for a 10-day assignment and was instructed to not move his vehicle from the specified location that was to be his base camp, an area his supervisor was very familiar with. He also was to return on a particular date and was to notify his supervisor as soon as he returned. When I asked the prospective employee how, in that situation and with me as his supervisor, he would handle the problems of a broken vehicle and a nonfunctional radio, he replied, "I would walk out, because you are expecting me to notify you on a certain day that I have returned safely to civilization." My response was that temperatures in Death Valley during July routinely exceed 52° C, and that he could not possibly carry enough water to safely walk 32 km under those conditions. The candidate continued to argue that he was obligated to notify the supervisor of his safe return, and that alone was his rationale for attempting to hike out. I replied: "You are at far greater risk attempting to walk out than waiting at your vehicle. I'm aware of where your camp is, and I know you have resources adequate for an extended stay and have shade at (or under) your vehicle. I would come looking for you if I did not hear from you on the predetermined day. Thus, why would you attempt to walk out?" Unfortunately, the prospective employee's response was: "I know I am in good enough condition to make the hike, and you are expecting me to call and confirm that I have exited the area safely. If I did not notify you, you would be worried." Needless to say, this individual was not selected for that particular position.

Wildlife biologists conduct surveys of wildlife populations for many species, both game and nongame. They adhere to established protocols, but can also bring their supervisors novel ideas that could result in improvements to or modification of established protocols. Biologists often are asked to fly at low elevations, in either fixed-wing aircraft or helicopters, to conduct wildlife surveys. Hence a proclivity for airsickness is an undesirable trait. During a career spanning nearly 35 years, I [VCB] was amazed at the number of applicants who claimed to not be subject to aeropathy but had erred in their assessment. Prospective biologists are encouraged to be honest and forthcoming about their skills and limitations, and it would have been far better to know ahead of time what to expect from those aerial observers.

Wildlife biologists often interact with outdoor enthusiasts, some of whom are avid hunters, and others who are vehemently opposed to the harvest of wildlife. Therefore, an open attitude is a necessary component of the job. Biologists should expect and welcome drop-ins (e.g., other professionals, members of the public) at any time during normal office hours and should not be surprised to receive telephone calls from the public at home or on weekends. These events are especially common for biologists with an office and residence in a small town or a rural location. Biologists must respond to any such communications in a polite manner and, if they are not available then, need to return calls as quickly as is practical. Biologists must also be adaptable and able to address a variety of audiences. In another situation, I [VCB] was involved in a conversation with two houndsmen who were working for me on a mule deer and mountain lion investigation; a young graduate student working with us also was present. During our conversation, much of which centered around a location in which to hunt that day, the health of a favorite hound that had been injured in an earlier pursuit, and which tavern to meet in on Friday evening, I was interrupted by phone call from a very well-known population biologist concerning a question I had regarding metapopulation dynamics. While I was on the phone, the houndsmen left the office, but the graduate student remained. When I terminated the call, she commented: "My gosh, I can't believe you went from discussing Bowser's injuries and where to have a beer on Friday night with the houndsmen directly to discussing metapopulation dynamics with one of the top scientists in the world! You really have to wear many hats, don't you?"

Wildlife biologists compose activity reports, survey reports, technical reports, and articles for professional publications on a regular basis. They need to put the same amount of care and detail into preparing letters, memoranda, or reports as they would into writing a manuscript for publication. It has unfortunately been my [VCB] experience that individuals not interested in writing invest less effort into the preparation of reports than they do in most of their other job activities. Employers have no tolerance for a poorly prepared document. The best advice I was given was to write everything, from memos on up, as though you were composing them to be considered for publication in a professional journal. That counsel has served me well. Prospective wildlife biologists should have developed excellent writing skills while in school, not after they are hired.

Wildlife biologists use modern technology—including telemetry equipment, geographic information systems, modeling software, and statistical software—in their daily activities. No one is expected to be competent with all that is currently available, but knowledge about the utility of various programs and devices, and an ability to work with others having those skills, enhance your productivity and effectiveness. Nevertheless, familiarity with all of the technology in the world is no substitute for competence in assessing problems, developing effective responses, and implementing them in a meaningful way. An ability to think through issues and arrive at appropriate solutions or responses challenge wildlife biologists on a regular basis. You should consider this position as the opportunity of a lifetime to make meaningful contributions to wildlife conservation.

Background Needed

A strong background in ecology is essential, as wildlife biologists interact on a near-daily basis with members of the public, many of whom claim to have far more knowledge about the subject matter than a recent college graduate. Biologists have to maintain cool heads and not be easily frustrated, remain patient and persuasive, and have a desire to interact with members of the public. They must converse intelligently with various audiences about many species and on many topics. Some individuals can be very opinionated, so biologists need to approach these opportunities with the realization that members of the public interested enough to engage in a debate regarding wildlife issues sincerely care about wildlife species and their future. Hence good interpersonal skills are mandatory, as is an ability to listen attentively and remain unflappable under fire. On numerous occasions during public meetings on the topic of wildlife regulations, I [VCB] was challenged regarding the rationale for proposed regulatory actions. During one particularly memorable occasion, an attendee (who was presumed by others in the audience to be an expert on grazing practices and deer ecology) asked, "Well, how many litters do deer have each year?" I did not need to respond to the question. The other attendees suddenly became silent, and the meeting continued in the absence of further argumentative disruptions. All of the agency representatives could hardly keep from bursting into guffaws at the question, and the audience handled that response for us.

At a similar meeting regarding the regulatory process, one individual was particularly incensed about the decline in the quality of bucks that were available for harvest. He suggested that additional deer be translocated to the area to eliminate the inbreeding that was occurring. When I [VCB] asked why he was concerned about inbreeding, he stated, "The deer I shot this year didn't have any top teeth." Once again, local experts had played their cards to the benefit of agency personnel. The biggest challenge following that statement was not to laugh.

Wildlife biologists should interact in a professional manner with all members of the department (e.g., fisheries, enforcement, administrative staff) and acknowledge that personnel in each of these areas play an important role in meeting agency objectives. I [VCB] had always gotten along well with the law enforcement officers in my department, but the availability of new technology and the application of population modeling created some strife. One day, the captain in charge of the wildlife officers I had worked with on a daily basis stormed into my office and wanted to discuss the recommendations submitted to the agency's state office. I explained that we had arrived at the harvest recommendations following discussions with members of his staff, combined with results from some of the models that we were using at that time. The captain became very incensed and shouted, "That's the trouble with you biologists: garbage in, garbage out." I kept my composure, however, and the proposed regulatory changes were made. Hunters were happy with the new rules, and the captain eventually calmed down. It was a successful interaction with a professional colleague.

Educational Required

A minimum of a bachelor's degree from an accredited college or university in an appropriate field (e.g., wildlife management, conservation biology, biology, ecology, zoology, another closely related biological science) is required. Individuals possessing an advanced degree most likely have advantages over those without such a degree. Courses in population dynamics, natural history, physiology, nutritional ecology, and animal behavior are extremely useful. Although it would be desirable for any candidate to meet all of the requirements for certification as an Associate Wildlife Biologist by The Wildlife Society, this is not essential. For one thing, good technical skills do not necessarily compensate for a strong background in evolutionary biology or population ecology. In addition, I [VCB] believe that an MS is the most meaningful degree prospective wildlife biologists should obtain. Students receive good training in basic biology, conservation, and management in many undergraduate wildlife programs. At the MS level, students are applying what they learned as undergraduates, working independently on a project of their own design, experiencing the perils and pitfalls of many things going wrong (and some working well), and learning how to deal with those setbacks. These are real-world experiences that parallel what a state wildlife biologist does. Personally, I believe individuals with an MS degree in wildlife are ahead of those who only hold undergraduate degrees.

Pay Scale

Starting salaries for wildlife biologists working in state agencies are generally about $36,000 per year, with step increases raising the annual compensation up to around $50,000 per year. For example, in California, where the starting salary for a wildlife biologist is $37,000 per year, incumbents can remain in what is referred to as a *deep class*, advancing in the same position to eventually earn an annual salary of approximately $68,000 per year. Most states provide at least some contribution toward various benefits, such as dental care, vision care, and general health insurance, a couple of weeks off for a vacation, and a defined-benefits retirement plan, all of which contribute substantially to the net value of the position.

Program Specialist
Job Description

Program specialists work with complex directives and manage and lead the program staff. This includes areas such as data analysis and administration; work with upland and migratory game birds; oversight for wildlife management areas statewide; the conservation and management of waterfowl, big game, small game, invertebrates, and herptiles (reptiles and amphibians); wildlife diversity programs; public hunting; interactions with private landowners through the Farm Bill or other programs; urban wildlife and conservation outreach programs; the coordination of cross-agency natural resource conservation and natural resource management efforts; grant administration and writing; archeology; interagency working groups and joint ventures; direction given to citizen science efforts, and more. They also assist in the formulation of regulations and statutes for conservation law enforcement.

Program specialist positions are broken down into levels, typically ranging from program specialists I through VII. You could be working solo; as part of a cross-functional team; on a joint venture with professionals from other agencies, universities, and nongovernmental organizations; or as a supervisor for employees and contractors in at least one area of the major wildlife disciplines. You could deal with concerned citizens who may have competing or contradictory interests in wildlife management and conservation.

A program specialist I performs routine (journey-level) work, which involves helping with planning, developing, and implementing one agency program and providing consultative services and technical assistance to program staff, governmental agencies, community organizations, or the general public. Individuals in this position may train others, and they work under moderate supervision, with limited latitude to use initiative and independent judgment. At the other end of the spectrum, a program specialist VII performs highly advanced (senior-level) work, which involves planning, developing, coordinating, and implementing major agency program(s) and providing consultative services and technical assistance to program staff, governmental agencies, community organizations, or the general public. These specialists train, lead, assign, and prioritize the work of others. They are minimally supervised, with extensive latitude to use initiative and independent judgment. More specifically, program specialists at all levels are required to know the history, status, ecology, and survey methods used for the wildlife species being monitored. They coordinate stakeholder groups and interpret, analyze, and explain laws, policies, and procedures. Program specialists may be responsible for either a single species (e.g., whooping cranes) and its associated habitats or a group of species and their associated habitats (e.g., upland game birds, herptiles, invertebrates). Or they may provide the specialized data required by wildlife biologists to perform their duties (e.g., GIS, database administration). Specialization requires knowing and being able to uniformly and consistently interpret and enforce state and federal (and sometimes international) laws pertaining to the species for which your program is responsible; developing and delivering effective messages and programs to the public; working productively (and supervising and mentoring staff to do same) with private landowners and interested stakeholders; writing, reviewing, and editing technical documents; prioritizing workloads for yourself and your staff, while being involved with multiple projects simultaneously; gathering and synthesizing complex data and clearly communicating this information in practical terms to various audiences; and developing, monitoring, and controlling budgets. Program specialists at any level work under stressful conditions; hike to areas for site monitoring, mapping, and data collection; perform manual labor as required (e.g., lift and

Program specialists supervise the management of habitat, such as through a prescribed burn conducted by a fire ecologist, or the management of a group of species, such as upland game birds or reptiles, for a state wildlife agency (photos: Scott Henke)

carry supplies and materials); and work in remote locations under extreme and adverse weather conditions. Complying with Mother Nature's cycles instead of the human calendar and clock comes with the territory—holidays, birthdays, and anniversaries included.

Program specialists spend varying amounts of time outdoors. They interact with private landowners, interested stakeholders, and staff from other agencies, both in person and in meetings. They attend public hearings and meetings and may also be asked to speak at them. Writing and speaking succinctly and clearly, using diplomacy, questioning folks accurately and dispassionately, and actively listening and paraphrasing what you have heard are among the skills you need if you choose to pursue this career path. Always be friendly and tactful. Also remember that you are a public servant. Social media and cell phones are ubiquitous, so never say or write anything that does not reflect your employer's perspective. Program specialists are in a 24-hour-a-day profession. Writing reports or compiling data must be done, no matter what the hour or the day. At times, you may feel as though you are in front of a computer screen more often than when you were in college. Activities such as research and keeping current with various types of knowledge do not end once you receive your diploma.

Background Needed

The minimum qualifications range along a continuum of education, experience, and licensure. For some program specialist I and II positions, two years' experience in wildlife management or related work is needed, along with having a state driver's license or being able to obtain one within 30 days. For fire management program specialists, the requirements extend to an ability to obtain the National Wildfire Coordinating Group's fire fighter type II or crew chief certification within one year of employment. Being able to remain in your position is contingent on obtaining and maintaining the necessary certifications. For higher-level program specialists, more years of relevant experience are required, and sometimes advanced degrees in wildlife science, range and wildlife management, wildlife ecology, or a closely related field are preferred. You may be able to substitute an MS or a PhD degree in a relevant field for some of the necessary years of experience. Please note that if you see such categories as "preferred qualifications," "desired skills," or the like, this is generally the hiring manager's description of the experience, knowledge, skills, abilities, certifications, and licensures the ideal candidates would possess for that position. These qualifications also could be used as a road map for how you can strengthen your candidacy for similar positions in the future.

Field experience and evidence of responsible time management are requisite qualifications for successful program specialists. Cite your relevant experience, beginning as far back as necessary. Generally, the greater the total amount and the different types of fieldwork you have, the better. Volunteer for such opportunities now, to gain experience, and be sure to budget your time appropriately. Seek out graduate students and professors with whom you could get such experiences. Learn how to say no to distractions, so you can focus your time on gaining valuable field and lab skills.

Memberships in professional organizations, such as The Wildlife Society, Ducks Unlimited, and others, show your dedication to and passion for obtaining a career in wildlife. Indicate your leadership experience by citing any positions you have held (e.g., officer, committee chair) in professional organizations. Networking is important. Monies from external sources fund our profession, so the more resources you are familiar with and the greater the number of potentially willing partners you meet, the better. In addition, communicating well in writing and in person is paramount. Know your audience. For example, PhDs and research scientists may require a different level of writing or speaking than a high school group. When in doubt, brevity and simplicity suffice.

Remember to practice and follow the code of ethics and standards for professional conduct, as determined for our profession. The Wildlife Society lists these on their website (appendix B). Learn about and follow any additional standards and ethics prescribed by the agency, state, or other jurisdiction for which you wish to work.

Various means of demonstrating your value to an employer include obtaining professional certifications, such as those offered by The Wildlife Society and the Society for Range Management; teaching university-level courses; being an author for peer-reviewed publications, as well as editing and writing professional books; and identifying and judiciously using well-trained, motivated volunteers in citizen science efforts. Do not pass up opportunities to gain such types of experience. They are beneficial to your future.

Education Required

You may be able to begin your career as a program specialist I with a bachelor's degree in wildlife or a closely-related field from an accredited college or university, but to advance within the position, a relevant MS or PhD degree helps. Generally, but not always, program specialist jobs are not entry-level positions.

Sometimes a relevant advanced degree may substitute for one year (MS) or two years (PhD) of required professional experience as a program specialist. Each agency may differ slightly in how it interprets and classifies its staff in the program specialist series. For example, one agency may require five years at that agency as a program specialist IV before you are eligible to be promoted to a program specialist V, while another agency may require 10 years in service as a program specialist IV, with budget authority and other responsibilities, before you become eligible for promotion. Generally, a minimum of two years' relevant experience at each level, plus meeting or exceeding required additional criteria and not violating the ethics and standards of our profession, are all required for promotion.

Also take classes and get other training in public speaking and writing, because, as a program specialist, you constantly are addressing the public. Certifications, such as that for an Associate Wildlife Biologist, offered through The Wildlife Society, are important to obtain while completing your education.

Pay Scale

Annual salaries for a program specialist I range from $36,000 to $60,000, while a program specialist VII could earn between

$60,000 and $102,000. In addition to state retirement benefits and health insurance, full-time state employees earn paid annual leave and paid sick leave for each month of service, and allotments for both typically increase with the length of service. Paid holidays for state employees are determined at each session of the state legislature and include most major federal holidays. Flexible work schedules may be offered and are considered on a case-by-case basis.

Wildlife Technician
Job Description

Technicians have various duties: performing wildlife population surveys, maintaining scientific databases, collecting data concerning the management of vertebrate wildlife species, working at wildlife management areas, assisting with public hunts, informing the public about wildlife, providing conservation outreach activities to the public, and helping to develop wildlife management plans. More specifically, wildlife technicians trap and tag live game; maintain equipment, buildings, and grounds; operate and maintain motor vehicles, power boats, and specialized equipment; coordinate the collection, tracking, recording, and processing of biological data for wildlife studies; organize and conduct public activities, such as public hunts, tours, or presentations to groups; recommend specialized equipment and techniques to obtain wildlife specimens for study; observe and record ecological conditions and the effects of wildlife on the habitat; and assist in creating and implementing safety procedures and programs.

Technicians spend about half of their time outdoors, but much of this is taken up with maintaining and repairing equipment. Interacting with the general public is also included in some of that time outside of the office. The Boy Scout motto is good advice for a technician—always be prepared. Carry a sturdy backpack stocked with a flashlight and batteries, a hat, extra water, snacks, maps, matches in a waterproof bag, binoculars, sunscreen, and bug repellent. Interacting with the public—assisting with youth or guided hunts, attending public hearings, or meeting curious landowners—often occurs, so have an extra shirt and pair of pants and boots to change into, because you never know when these may be required. You do not want the odors of the field to distract from what you have to say. Most people are cooperative and helpful, but not always. Running to the store in your uniform or driving a truck emblazoned with the agency's logo can lead to general questions and conversations, which you must always handle courteously and professionally. Once people get to know you, they may come up to you even when you're not in uniform or on the job. Always be friendly and diplomatic. Remember, these folks pay part of your salary. Cell phones are ubiquitous, so never say or text anything you do not want those you respect to read or hear. Writing reports and compiling data must be done, no matter what the hour or the day. At times, you may feel as though you are in front of a computer screen more often than when you were in college. Activities such as research and keeping current with technology and new techniques do not end with receiving your diploma. Welcome to the real world!

Background Needed

The minimum qualifications range along a continuum of education, experience, and licensure. For some wildlife technician positions, two years' experience in wildlife management or other relevant work is required, along with having a state driver's license or the ability to obtain one within 30 days. Sometimes requirements include an ability to acquire a commercial driver's license within six months of being hired, or to obtain the National Wildfire Coordinating Group's fire fighter type II certification within one year of employment.

Memberships in professional organizations, such as student chapters of The Wildlife Society, Society for Conservation Biology, Future Farmers of America, 4-H, or a scout troop, as well as listing the type of hunting license you have or your completion of hunter education training, demonstrate your dedication to and passion for obtaining a career in the wildlife field. Previous experiences with livestock, bird and woody plant identification, breeding bird surveys, and endangered species fieldwork are useful.

Communicating well in writing and in person is paramount. In addition, know the mission of the agency. Excellent orienteering skills and an ability to remain calm, even when you think you're lost, are valuable skills.

Education Required

In several states, you can become a wildlife technician without a bachelor's degree; a high school diploma or a GED is all that is required. Realize, though, that most likely you are competing for the same technician's job against applicants with bachelor's degrees and several years of professional experience. Therefore, a bachelor's degree in wildlife management or a closely related field from an accredited college or university is very beneficial.

To advance up the career ladder, you must meet or exceed the education and experience requirements for technicians. You can also demonstrate your value to your employer through certifications, such as that for an Associate Wildlife Biologist (offered through The Wildlife Society) or the National Wildfire Coordinating Group's firefighter certification to conduct controlled burns; hunter education courses; knowledge of data analysis programs (e.g., R) and citizen science tools (e.g., iNaturalist); and an herbicide and pesticide applicator's license.

Pay Scale

State employees are typically paid once a month. Annual salaries for technician I, II, and III positions range, respectively, from $30,000 to $45,600; $38,400 to $50,400; and $43,200 to $57,000. A technician IV typically starts at a higher salary than a technician III, but the amount depends on education, knowledge, skills, experience, licensures, and certifications. In addition to state retirement benefits and health insurance, full-time state employees earn paid annual leave and paid sick leave for each month of service, and allotments for both typically increase with the length of service. Paid holidays for state employees are determined at each session of the state legislature and include most major federal holidays. Flexible work schedules may be offered and are considered on case-by-case basis.

Game Warden or Conservation Officer
Job Description

Game wardens, otherwise known as *conservation officers*, provide national-, regional-, and local community–level wildlife management and conservation services, expertise, coordination, and enforcement measures to promote, sustain, monitor, and protect wildlife populations on lands that are within the public trust. They are part of key government entities that, along with nongovernmental organizations, sportsmen, and industry, fulfill the tenets of the North American Model of Wildlife Conservation.

Conservation officer positions within the various jurisdictions are field-oriented. The duties of these commissioned law enforcement officers include protecting and conserving wildlife resources for the public's benefit. The individuals in this position are generally considered the local frontline, face, and community wildlife management contacts and experts for their respective agencies. As such, conservation officers are strategically spread out across their assigned jurisdictional territories, in order to provide as much official presence as possible to manage and protect the local wildlife resources. Based on the 2013 annual survey by the Law Enforcement Committee of the Western Association of Fish and Wildlife Agencies, member agencies reported average district sizes for conservation officers ranging from 1,000 km^2 to as much as 52,000 km^2 of land and water. The jurisdictional human populace per officer ranged from 2,500 to 135,000.

Assignments involve remote duty (when you are usually alone) and have a great degree of independence, requiring constant self-analysis of your effectiveness and efficiency in determining how, when, and where you work. Often there is minimal face-to-face contact with your supervisor. It is up to the individual conservation officers to seek out and stay abreast of agency information and directions, which they are responsible for implementing. Duties vary widely, depending on the jurisdiction and the district assignment. In addition to wildlife law enforcement, in many places conservation officers also have primary jurisdiction over boating safety and off-highway vehicle laws and regulations. Whether a jurisdictional assignment only involves wildlife law enforcement tasks or is more varied and includes a full array of wildlife and habitat management duties, conservation officers must have a full understanding of their management, research, and law enforcement responsibilities—under an overarching umbrella of informational and outreach efforts—along with a thorough grasp of environmental laws and regulations, to effectively work as the local point of contact and wildlife expert for the relevant jurisdiction.

Law enforcement is the primary function of conservation officers, and time spent on this aspect of wildlife management may vary from roughly 50% to 100%, depending on the ju-risdiction. The objective of conservation law enforcement is to provide a local presence that can inform and educate the public, deter and detect violations, and apprehend violators, as necessary. Conservation officers must have good people skills. They are generally well known by sportsmen within their local communities and are readily sought out for information and for help on jurisdictional wildlife management issues. Also, because of their remote and distant duty assignments, conservation officers may be the initial responders in law enforcement matters unrelated to wildlife. Networking and partnering with all law enforcement jurisdictions (city, county, state, and federal) at the local level is a must for conservation officers, in order to be effective and stay safe.

Conservation officers are generally considered to be responsible for law enforcement beyond the pavement, where backup is often nonexistent. Officer safety skills and an ability to control and de-escalate volatile situations are critical. Law enforcement activities include patrolling, investigating criminal complaints, making arrests and issuing citations, applying for and serving warrants and subpoenas (court orders), preparing case reports, testifying in court, receiving training and supplying it to others, teaching hunter education classes, initiating public information and outreach efforts, providing first responder medical assistance, being involved in search and rescue operations (e.g., finding stranded or lost hunters), and performing general administrative work (e.g., time reports). For most jurisdictions, the law enforcement focus for the position involves fish and wildlife, watercraft, and off-highway vehicle recreation. For jurisdictions where conservation officers (as wildlife managers) are assigned a full array of wildlife management duties, district activities may center on wildlife and fish surveys; data compilation and assessment; the development and presentation of hunt, harvest, and protection recommendations for the various species within the district; habitat assessment; the development and implementation of habitat improvement projects; agency representation at interagency and public meetings; nuisance wildlife abatement and capture; animal attack investigation and resolution; public information, outreach, and education; training (both receiving and administering it to others); and general administrative work (e.g., time reports).

Conservation officer and wildlife manager positions are field-oriented, and work priorities must be set so that field time and presence occur in locales situated throughout an assigned district. These efforts have to be directed toward prime public outdoor recreation times, as well as to places where the public is likely to be and, thus, where violations are likely to occur. This entails working a lot during weekends, holidays, and nights and responding to call-outs, which all mean time spent away from family and friends. To be successful, conservation officers must have supportive and involved families and be able to balance their personal life against these regular job demands.

Conservation officers must stay in contact with local landowners, community leaders, and organizations, in order to assess issues, provide two-way communications and feedback for their jurisdiction, and inform and educate the local public regarding jurisdictional directions (guidance, policies, regulations, rules, laws). Conservation officers have an ability and an opportunity to be leaders in their local areas. To be most effective, they have to seek out, organize, and use local community involvement, sportsmen, and volunteers to get their jurisdiction's mission accomplished.

Background Needed

A conservation officer's job is very diverse, and you must have many skills and abilities related to the responsibilities of the assignment. Besides demonstrating top educational performance and an ability to complete the training noted above, a background in and affinity for outdoor skills and pastimes are musts. Successful conservation officer candidates generally have personal experience in hunting, fishing, camping, hiking, backpacking, wildlife watching and handling, firearms

handling and marksmanship, and survival skills. In addition to having a variety of outdoor abilities and wildlife knowledge, you must be adept in the use of various kinds of equipment, including four-wheel-drive vehicles, watercraft, off-highway vehicles, firearms, wildlife capture equipment, survey gear, telemetry devices, optics, radios, and computers. Depending on the remoteness of an assignment, you might also have to be skilled in using horses, or even be licensed as an aircraft pilot. Those who serve on the front line of an agency as law enforcement officers often work alone, in remote areas. They must be able to present themselves well, be physically fit, be a people person, be self-driven and not require close supervision, and be able to make good, logical decisions.

During the hiring process, a successful conservation officer candidate generally is subjected to a criminal background check, an oral board interview, medical and physical fitness assessments, a polygraph examination, and psychological test-

Game wardens protect our nation's wildlife resources from land (*left* and *center right*), water (*top right*), and air (*bottom right*) (photos: Leonard Ordway)

ing, all of which are necessary to rigorously assess a person's ability to perform and survive as a law enforcement officer. The criminal background check usually determines if you have been involved in felonious activities or shown a regular pattern of any type of criminal behavior; minor infractions, such as common traffic violations, generally are overlooked. A board interview verbally examines your technical skills, knowledge, and abilities; assesses the way you present information, your reasoning ability, and your demeanor under stress; and generally evaluates how you may fit into the culture of the respective agency. In the physical and medical fitness assessments, you must be able to demonstrate that you can meet the standards for your jurisdiction. A polygraph exam checks if you have been truthful in the application you submitted and if you generally meet the law enforcement standards for your jurisdiction; any deception or lying disqualifies a candidate. A psychological test is administered to determine if you have the character and the mental reasoning capability to survive under the stresses and rigors of a law enforcement career.

A crucial requirement is that those wishing to be conservation officers must have demonstrated proper legal, ethical, and moral decision making and behaviors in their lives. It's so disheartening, as a hiring supervisor, to see young applicants, who appear to be perfect for the job, be disqualified because of a poor choice they made early in their lives, not realizing the later consequences.

Demonstrated outdoor and wildlife experiences put outstanding candidates at the top of the list. You must show that you understand and are part of the public wildlife-advocacy community. Applicants who have spent their lives participating in such experiences or cultures have a leg up for a successful career as a conservation officer. If you have not grown up in this type of environment or culture, you can certainly obtain—and enjoy—such experiences if you seek them out and participate in them. For example, while you are a student, become involved in wildlife organizations, such as a wildlife society or advocacy groups for various species. In addition, wildlife agencies across North America have educational programs and volunteer opportunities where you can gain the necessary knowledge and hands-on skills.

Most agencies provide ride-along programs, which are opportunities for interested members of the public to spend time with and learn from local district conservation officers. In doing ride-alongs with officers, prospective candidates can see if this is really a correct career choice for them. Candidates also obtain key information about the hiring agency, and, most importantly, the agency can observe the character traits of those individuals beforehand.

The hiring process to become a conservation officer is very competitive, and you must be the one to take the initiative in order to become a top candidate. In 2013, based on the annual survey done by the Law Enforcement Committee of the Western Association of Fish and Wildlife Agencies, most member agencies hired less than 10 new officers per year, for which there were more than 400 applicants.

Education Required

For entry-level conservation officer positions, wildlife conservation agencies generally require a bachelor's degree in wildlife biology, ecology, or a related field. A college degree is not necessary in some states, but those conservation officers generally have law enforcement duties only. The majority of officers do have college degrees, with majors varying from criminology to biology, although the preponderance have wildlife biology–related degrees. Also, because conservation officer jobs often are entry-level positions for career ladders within agencies, many candidates have an MS degree in wildlife biology, ecology, or a closely related field.

After being hired, conservation officers must attend and graduate from an agency-approved law enforcement academy, in order to meet either state or federal law enforcement standards. With few exceptions, conservation officers graduate from these academies with a full peace officer certification and can be expected to handle any law enforcement issue discovered by or put before them, if the need arises. Academies vary in the length of their training (from three to five months), are generally very regimented, and ensure that an officer candidate can stand up to the high stress levels, physical fitness needs, rigors, and legalities of law enforcement duty. Most agencies put their officer recruits through a post-academy training regime. This involves both classroom time, where recruits learn the agency's processes and culture, and a field training phase, where senior officers oversee and test the recruits to ensure that they can make it on their own in the field. This post-academy training lasts from six months to a year.

Once new conservation officers are placed in their respective districts, they generally have a probationary period, where they are closely managed, supervised, and evaluated by their immediate supervisor to assess their ultimate fitness for a career with their respective agencies as conservation officers. These probationary periods generally last from six months to a year, after which the officers are considered to be full-fledged district conservation officers. Nonetheless, they are still the rookies until the next class of officers comes on board. For most agencies, the entire hiring and training process—from applicant status to being a trained and assigned district conservation officer—takes about two years.

Pay Scale

An agency's field officer cadre is typically broken into classification levels, based on tenure, assignment, performance, and supervisory duties. The compensation scale is structured and tiered accordingly. Agencies that tend to have their officers primarily perform law enforcement work generally use a military-based classification (officer, sergeant, captain). Agencies that assign other activities besides law enforcement to their officers usually are more general in their classification terminology (e.g., wildlife managers I, II, and III; game warden; conservation officer). Also, jurisdictions treat their officers differently, in accordance with the Fair Labor Standards

Act, depending on whether or not the officer positions are classed as professional or nonprofessional. You must check with each agency to determine its classification levels.

Based on the annual survey done by the Law Enforcement Committee of the Western Association of Fish and Wildlife Agencies in 2013, member agencies reported starting salaries for conservation officers ranging from $31,000 to $65,000 per year, with the average being between $35,000 and $40,000 per year. Top salaries for conservation officers can reach from $47,000 to $80,000 per year. Salaries vary across jurisdictions, due to many factors, including available budgets, laws, the political atmosphere, benefit packages, and labor relations, to name a few. Top-performing field officers are promoted to administrative, managerial, and supervisor positions within an organization, and such promotions result in higher salaries. Officers with advanced job responsibilities, specialty assignments, and additional training credentials are generally awarded extra salary compensation. For most agencies, an officer's tenure generally is from 20 to 30 years, with a law enforcement or public employee pension available on retirement. Nonetheless, it is common for conservation officers, even those with a four-year degree, to earn less than other law enforcement officers in their jurisdiction, such as highway patrol officers or city police.

Conservation Officer Employment Resources

For additional information concerning game wardens / conservation officers, visit the websites of the following organizations (appendix B):

North American Wildlife Enforcement Officers' Association

National Association of Conservation Law Enforcement Chiefs

North American Game Warden Museum

Western Association of Fish and Wildlife Agencies

Association of Fish and Wildlife Agencies

PRIVATE AGENCIES

WILDLIFE CONSULTING AGENCIES
Lisa K. Harris

Agency: Private Wildlife Consultancy

Companies employing wildlife biologists are typically consulting firms that work for a wide variety of clients, including governmental agencies, utility companies, private developers, and nonprofit organizations. Work is typically available on a project-by-project basis, although in some instances, corporations supply staff to governmental agencies to work as stand-in government employees for long-term assignments. Governmental clients include federal, state, county, and municipal (cities and towns) agencies. At the federal level, many entities, particularly in the Department of Defense, Department of Agriculture, and Department of the Interior, contract out wildlife and natural resource studies to private firms. Utility clients include electrical, gas, and water companies, which develop and maintain utility corridors, such as electrical transmission lines, gas lines, water and sewer lines, and treatment centers. Private developers build infrastructure, such as residential subdivisions, industrial structures, and apartment buildings. Nonprofit organizations hire consulting firms to conduct research that is supported by a variety of funding sources, particularly grants and endowments, and to add to their staff in cases where the organization cannot support hiring such expertise internally. The latter scenario typically requires consultants to join their workforce at a discounted (and sometimes volunteer) rate.

Wildlife Consultant Biologist
Job Description

Consultancy projects usually involve alterations to the landscape, including degrading it from its natural state (building a road, dam, utility corridor, mine, housing development) or restoring it after it was developed previously (wetland restoration, prairie restoration). In most cases, activities such as undertaking ecological and wildlife management research and publishing studies to further our understanding and knowledge base of species, habitat, and the environment in general are not typically conducted by consulting firms. This type of research is generally limited to universities, nongovernmental organizations, and, to a lesser degree, governmental agencies. For example, the Audubon Society might use consultants to conduct habitat studies related to birds or to define important habitat on land that has been donated to them. The Nature Conservancy hires consulting firms to conduct environmental assessments on land that is used for conservation mitigation. Hiring a consulting firm or a third-party expert validates the findings, as neither of these two entities have a vested interest in the study's outcome.

Consulting firms sell intellectual expertise and services provided by individuals who maintain particular certifications and permits and have the education and experience to conduct specific work. For example, a strip mine company was mandated by law to make an effort to relocate Texas horned lizards, a threatened reptile in the state, from future mining areas. The mining company hired a wildlife consultant who possessed the proper permits to move the horned lizards on their behalf. Consultants advise clients on environmental policy and laws, obtain permits, adhere to environmental compliance rules, and, basically, keep their clients out of trouble from a regulatory standpoint. Biologists in consulting firms, regardless of their level within the organization, are subject matter experts. *Project deliverables* (a tangible product or products produced as a result of a project), typically consisting of technical reports, are developed by the consulting firm and submitted to the client as supporting documentation, facilitating compliance or satisfying specific permitting needs. Many times, the client submits the report to a permitting agency for review and comments. Clients pay consulting firms to produce high-quality, well-written reports that demonstrate adherence to environmental laws and policies, and they expect minimal revisions to be requested by the reviewing agency, so that the proposed project permitting process is not delayed.

Businesses are geared to make a profit, and management within publically traded companies place a higher emphasis on the percentage of profit made than smaller companies, which are not under pressure from stockholders. Consulting firms often operate within a feast or famine framework, with either too much work or not enough. Staff may feel overwhelmed by their workload or, when there is not enough work, they may worry about job security. Because staff salaries are the largest expense for consulting firms, when there is not enough work, staff can be furloughed or even terminated, depending on the circumstances.

The structure of larger consulting companies is generally based on a team model, with a program manager, project manager(s), and midlevel and junior-level biologists, working together on multiple projects and tasks. Big projects include all of these staff members, but a small one may only have a project manager and a junior-level biologist dedicated to it. Higher-level staff members (from program manager through midlevel biologists) communicate with the client and develop the scope of the work (including deliverables), the time frame for completion, and the project budget. They also negotiate the contractual terms. Internally, the program manager and project manager assign tasks to individual team members. Team members, with specific direction from higher-ups, understand how many hours are assigned to each task, what deliverables are associated with which task, and the time frame for each deliverable.

Junior-level biologists are often assigned to fieldwork and data entry. Ideally, they work as members of a cohesive project team, taking orders from higher-level staff members. They should have well-developed communication skills and be comfortable as part of a team. Junior-level biologists spend most of their time (50%–75%) in the field and are responsible for conducting fieldwork, following the methods developed when the project was first set up. After returning to the office, they are responsible for recording and archiving all documentation, including photographs, field notes, GPS coordinates, and field data. Depending on their skill level, junior-level biologists may be responsible for producing maps and data analyses. They also may be responsible for writing sections of a report or, on smaller projects or with small companies, for writing the entire report for final review by senior-level staff, who are responsible for submitting it to the client.

Background Needed

In addition to the education required, demonstrated fieldwork capabilities and good written communication skills are key to securing an entry-level job in a private firm. Field experience can be obtained through course components, summer internships, work on graduate student projects, and volunteer experiences. In evaluating candidates, consulting companies seek individuals who can demonstrate an ability to work on their own outdoors, with little supervision. Letters of references explicitly describing this quality in applicants are very helpful. Demonstrated experience should include time spent working in difficult environments (e.g., deserts, mountains, places with extreme and variable terrains and temperatures). This shows that applicants can take care of themselves outdoors, so management does not have to worry about their safety in the field. A wide variety of fieldwork experiences are ideal (e.g., plant identification, plant measurements, habitat studies, telemetry, GPS techniques, mark-and-recapture skills, census techniques). In addition, experience in environmental education (e.g., as a docent, summer camp counselor, nonprofit organization volunteer) rounds out a candidate's ability to work as team member to achieve a desired goal.

Communication experience can be demonstrated through papers and publications (scientific and popular articles, creative writing, college assignments). Published articles or short stories are preferred, but many entry-level candidates may not have had anything published yet, so they should choose a well-written student assignment in which the applicant is the sole author. Written work helps management evaluate a candidate's thought processes and grammatical expertise. Junior-level biologists are responsible for producing legible field notes, appendices, and, in some instances, portions of a project's final report. Incorrect grammar, misspellings, and wrong syntax reflect back on the entire project team as well as the company, as reports are often the sole product provided to clients.

Education Required

A bachelor's or master of arts or sciences degree in wildlife, plant biology, ecology, or a closely related field from an accredited college or university is required. Classes should cover applicable sciences, such as biology, chemistry, and physics; mathematics; English composition, including technical writing and creative writing; and specialized courses within the biological field, including wildlife management, field techniques, and botany. Other important coursework includes GIS, plant identification, and environmental policy. A sound understanding of the National Environmental Policy Act, Endangered Species Act, and Clean Water Act is essential, because many consulting jobs are a direct result of these laws.

All things being equal, a candidate with an MA or MS degree will progress further, and faster, in the business environment, so advanced degrees are encouraged.

Pay Scale

There are two employment classifications for consultants: project-based and salary-based. Project-based positions are usually temporary. Employees work on a well-defined project, which could last from a day to several field seasons. When the project is finished, the job is completed, and employees are terminated if there is no additional work available. A project-based pay scale is usually hourly, and the rate varies, both regionally and by the degree of difficulty associated with the job. Projects in urban areas with higher standards of living pay more than ones in rural areas, where the cost of living is lower. Work that involves a significant degree of difficulty, either from a technical standpoint (using specific knowledge, e.g., statistics) or with regard to safety (e.g., backpacking to remote areas) has a higher level of compensation. Employees might be eligible for overtime pay if they work more than 40 hours per week, but this depends on the project's budget and pertinent labor laws.

Hourly rates for project-based, entry-level biologists with a bachelor's degree range from $14 to $18. For those with an MA or MS degree, the hourly rate can rise to $16–$20. This job classification typically does not include benefits, such as paid vacation time, sick leave, health insurance, or a pension plan. Hourly employees most often are not eligible for quarterly, semiannual, or annual bonuses.

Salary-based positions typically are permanent. In addition to an annual wage, they include benefits such as paid vacation time, sick leave, health and dental insurance, long- and short-term disability insurance, and pension plans. The employees in this category work on several projects, either back to back or concurrently. When a project is completed, they remain with the firm and begin working on another one, as well as help with developing proposals for new work. Salaried positions are based on a 40-hour workweek (2,080 hours per year), and these employees are typically not eligible for overtime pay. Compensation varies regionally, and takes into account the degree of difficulty involved in the job and the technical knowledge required. Work that involves little supervision also garners higher pay. Salaries in urban areas with higher standards of living pay more than the same position in a rural area, where the cost of living is lower.

Salary rates for permanent entry-level biologists range from $32,000 to $62,000 per year, which is equivalent to federal government GS-5 to GS-9 levels. Candidates who have earned a bachelor's degree are on the lower end of this pay scale, and those with an MA or MS degree and significant expertise in a given subject matter (e.g., statistics) are on the higher end. A benefits package (paid vacation time and sick leave, health and dental insurance, and a pension plan) would be offered, in addition to this base pay. Also, quarterly, biannual, or annual bonuses are possible. Bonus pay scales vary among employers. They are often subjective and depend on many variables, such as the firm's profitability, the project's profitability, the company's employee retention strategy, and length of employment.

PRIVATE CORPORATIONS
Kevin McKinstry, Darren A. Miller, and Michael J. Rochelle

Agency: For-Profit Company

Large corporate landowners are commonly timber companies that use their property to grow and harvest wood fiber for lumber or paper products. For-profit corporations in the forest industry sector, whether privately owned or publicly traded, seek to sustainably and profitably manage natural resources. Private landowners generally operate under third-party forest certification systems, which use an auditing process to ascertain if they follow sustainability standards that help ensure the conservation of soil, water quality, and wildlife habitat and afford protection to unique ecological communities, as well as possibly provide opportunities for recreational leases (e.g., hunting, fishing, other outdoor activities). Collaborative empirical research often is used to gain an understanding of the effects of forest management systems on biodiversity and other ecosystem functions. Other types of corporate landowners include agricultural companies that use their land to grow crops or raise cattle, and refineries and chemical companies that own the surrounding lands and have them act as buffer zones to mitigate against industrial accidents. Hunting sometimes is layered as another business platform that produces profitable revenues, in conjunction with the core industry. Under these circumstances, land is considered to be a capital investment, and it must produce financial returns to the shareholders who have invested money in the company. Recreational hunting on corporate lands, especially in the southeastern United States, is a significant contributor to a return on investment. Not all corporations with land, however, integrate recreational hunting into their endeavors. Some conduct threatened and endangered species programs on their property, others offer conservation education programs to the surrounding community, and still others combine several of these ventures into their overall enterprise. In addition, corporations may hire either a single biologist to manage all activities on their property, or a team of biologists, each with specific duties. Therefore, depending on the company, actual job descriptions within the corporate world may have a significant overlap of job duties or none at all.

Corporate Wildlife Technician / Entry-Level Biologist
Job Description

The primary responsibility in this type of a position in the forestry industry is to assist in providing day-to-day operational support to staff engaged in growing and harvesting timber. In most cases, an entry-level wildlife technician / biologist functions under the supervision of a more senior biologist or someone in another resource manager position. Duties as an entry-level technician/biologist are likely to be highly variable, but in general they usually include activities such as conducting and managing terrestrial and aquatic surveys for threatened, endangered, and sensitive species, as part of ensuring compliance with state and federal regulations, forest practices, and best-management practices; reviewing planned harvest activities, timber sales, and land acquisitions for potential wildlife considerations or constraints; assisting operations personnel with the permitting processes associated with planned management activities; facilitating communications among operations personnel, regulatory agencies, other external stakeholders, and the public; assisting with the identification and marking of stand-level structural elements

that are to be retained, as part of the harvest-unit layout process; assisting the senior biologist with technical input needs that relate to third-party certification efforts, regulatory and policy discussions, and the development of internal policies and guidelines; providing training on wildlife-related issues to company operations personnel and contractors; collecting data in support of internal and collaborative research efforts; making GIS-based maps; and, perhaps most importantly, responding in a timely manner to site-specific wildlife issues that may arise during normal forest operations, to help minimize or avoid potential delays in and disruptions to daily operations. Most entry-level technicians/biologists working for private companies can expect to function more as species generalists, rather than plan on focusing on a single species or a small group of species. Issues and opportunities probably will arise that provide exposure to a wide range of wildlife and forest management considerations.

At the entry level, tasks may include assisting with inventory, maintaining field equipment and company vehicles, participating in fire suppression activities, entering data, providing logistical support for field tours or meetings, writing reports or other technical documents, and performing any number of other occasional and nonroutine tasks that may be assigned. These duties are not glamorous, but they are an important aspect of the job.

A typical position at this level consists of a mix of field time and office time, and the relative balance between the two is likely to vary by the season. Spring, summer, and early autumn are generally going to be focused on field-based activities, with a significant portion of time spent out of the office. Winter months tend to center on compiling field data, maintaining and repairing equipment, producing summary reports of field activities, and planning and preparing for the following field season. Thus they are heavily weighted toward office time. On an annual basis, entry-level technicians/biologists are generally going to spend 60%–70% of their time in the field and the remainder in the office. Because of the dynamic nature of commercial timberland management, you can expect a high degree of variation in your tasks, unexpected changes in schedules, potentially long hours (particularly during the field season), work taking place under challenging physical conditions, and an underlying sense of urgency in ensuring support for operations personnel that is adequate, timely, and consistently delivered, when requested. Beyond the wildlife-related duties associated with this position, other obligations exist. Depending on the size and nature of the company, all employees can expect to engage in some level of mandatory activities, including attending safety meetings and staff meetings, reading and responding to emails, returning phone calls, participating in required trainings (e.g., first aid / CPR training), completing expense reports, and taking care of a host of other routine, and sometimes mundane, but required tasks.

If you are employed by a company with significant holdings across a wide geographic area, brief overnight travel may be required, and you can expect a substantial amount of driving. In most cases, entry-level technicians/biologists with significant field responsibility will have their employer supply a vehicle for use in the field, or, at a minimum, compensate employees for the use of a personal vehicle for company-related fieldwork. The necessary equipment for the position (desktop or laptop computer, tablet, GPS unit) is also provided by the company, as is required personal protective equipment (e.g., safety glasses, hard hat). In most firms, participation in professional organizations (e.g., The Wildlife Society) is encouraged, and membership dues are often covered by the employer.

Because of the for-profit nature of corporate timberlands management, technicians/biologists at all levels within this environment are expected to be cost-sensitive, judicious in their use of company resources, and focused on helping deliver value to the company. At times, job security can be more tenuous than a position in a state or federal agency, and corporate technicians/biologists should be prepared for the possibility of changes in company ownership and, thus, in leadership, specific position requirements and expectations, corporate requirements and policies, and, perhaps, an expectation that employees will relocate.

Background Needed

In addition to formalized education, those seeking corporate entry-level technician/biologist positions should ideally possess a background in outdoor work and recreation, with strong abilities in map reading, compass use, and the operation of mobile devices for navigation. Experience with driving in wooded areas is highly desirable, and entry-level wildlife technicians / biologists should be prepared and willing to work long hours, in potentially challenging terrain, and in less than optimal weather conditions. Experience with field repairs for instrumentation and equipment is a valuable asset. An ability to work independently and, at times, with minimal direct supervision is imperative, as is a tolerance for last-minute schedule revisions and a redirection of effort in response to both changing and newly identified issues. Written and oral communication skills are critical, as interactions with a diverse range of audiences can be expected. Similarly, having a strong, team-centered orientation is a valuable skill.

Prior experience as a field biologist, either as a volunteer or in a paid position, is highly desired. This can include seasonal survey jobs, work at wildlife refuges or state and federal parks, time spent at wildlife rehabilitation centers, participation in research projects as a part of the data-collection efforts, or even time working as a member of a trail crew.

Education Required

Formal education should, at a minimum, include a bachelor's degree in wildlife biology, wildlife ecology, or conservation ecology, with a strong emphasis on forestry and wildlife in managed forest systems. Due to the relatively small number of corporate entry-level technician/biologist positions available, an MS degree can be a distinct competitive asset, particularly if you intend to advance within the organization as your career progresses. Regardless of the level of your degree

or specific area of emphasis, strong skills and demonstrated abilities with GIS, computer applications, oral and written communications, and statistics are a necessity. Exposure to business concepts is also desirable. In general, a curriculum following the requirements for Associate Biologist certification by The Wildlife Society should provide an adequate educational foundation for a position of this type.

Pay Scale

The pay scale for corporate entry-level wildlife technicians / biologists is at least competitive with and quite probably somewhat higher than entry-level technician/biologist positions with agencies or other organizations. The pay is usually hourly, with compensation for overtime work at 1.5 times the base hourly wage. Annual pay can range from $32,000 to $38,000. Corporate entry-level wildlife technicians / biologists can usually expect a reasonable benefits package that includes some component of insurance coverage, a 401(k) or other retirement plan, and the possibility of annual bonuses, based on the overall performance of the company. Raises and promotions are merit-based and do not follow specific formulas that are centered on years of service. Such advances are always accompanied by expectations for increased levels of responsibility and performance.

Corporate Wildlife Hunting Lease Manager
Job Description

Hunting lease managers (HLMs) oversee the operation of a corporation's hunting business. Specifically, HLMs market and sell available hunting leases on corporate properties; verify that federal and state laws and corporate policies are upheld by lease customers; provide technical knowledge to and answer the questions of lease customers; coordinate leasing and hunting activities with other company operations, such as timber and mineral management; prepare and interpret business status reports and budgets; and handle internal and external conflict resolution. In addition, HLMs develop, implement, and recommend game management programs (e.g., quality deer management programs [QDMs]) on corporate properties to benefit both the lease customers and the corporation. Some companies may have hunting lodges on their property, and the HLMs oversee the facility, including scheduling, maintenance, upkeep, and hospitality for lodge guests. Duties of HLMs also include writing wildlife management articles for corporate newsletters and websites.

Background Needed

This is a customer-service position, so excellent people skills are mandatory. HLMs are the liaison between lease customers and the corporation and, therefore, are always in the public eye. HLMs should always keep in mind that they represent the corporation and thus conduct themselves accordingly, including in how they answer phones, drive in public, and meet and greet potential customers. Excellent oral communication skills are paramount. HLMs need to be comfortable conversing with a variety of people who have diverse backgrounds,

Corporate wildlife hunting managers lease land to sportsmen (photo: Kevin McKinstry)

from blue collar to white collar. In addition, excellent writing skills are required, because they must compose competent business reports that are forwarded to corporate management and newsletters that represent the corporation. HLMs must be proficient with GIS and Microsoft programs (e.g., ArcView, Word, PowerPoint, Excel). A hunting background is necessary to understand customer needs and wildlife management applications. An ability to operate and maintain all-terrain vehicles and farm equipment is a plus. In addition, a basic understanding of other corporate property interests, including silviculture, minerals, and livestock management, is very beneficial. Other helpful skill sets include team collaboration, business experience, and photography.

Education Required

The minimum educational requirement is a bachelor's degree in wildlife management or wildlife science, with a minor in forestry or business, from an accredited college or university. An MS degree in wildlife management or in business is preferable, however. An ideal candidate would have a bachelor's degree in wildlife management and a master of business administration (MBA) degree. Certification as an Associate Wildlife Biologist through The Wildlife Society is beneficial.

Pay Scale

The starting salary for a corporate HLM is in the range of $35,000 to $40,000 per year, plus corporate benefits. Pay increases typically are merit-based and are dependent on the profit margin of the corporation for that year. Advancement opportunities are not frequent, but they can occur within a business group or across other businesses within the corporation.

Corporate Wildlife Biologist
Job Description

Corporate wildlife biologists provide technical assistance to a corporate landowner that must consider wildlife management implications as part of the firm's overall land management

program. This position is frequently supported by a recreational business platform that is also based on that company's land ownership (commercial hunting, hunting leases). Corporate biologists generally spend about 50% of their time in the office and 50% conducting fieldwork. Companies that own large amounts of property and manage this land for resources such as agriculture, forestry, and minerals often have staff positions for wildlife biologists. Wildlife biologists typically are actively involved in monitoring various species of wildlife and provide input on management strategies. In addition, these wildlife professionals could be the business managers for various forms of outdoor recreation that are offered on the company's lands, including hunting leases, sport fishing leases, and commercial hunting lodges. From a technical standpoint, corporate wildlife biologists are expected to have a full understanding of the biological needs and management applications for species of interest that inhabit the property. Depending on the core business of the company, this could include typical game species, such as white-tailed deer, waterfowl, wild turkeys, and bobwhite quail. These species often are essential to running a recreation-based business. Therefore, management practices need to ensure that they flourish. In addition, nongame species or species of special concern, such as bald eagles and red-cockaded woodpeckers that also inhabit the property, would be part of management considerations. Corporate landowners want to ensure that ongoing management activities do not violate industry regulations and laws, such as the Endangered Species Act or the Clean Water Act. Corporate wildlife biologists sometimes assist with educational and training efforts about game and nongame wildlife management to customers or company associates by giving professional presentations, writing popular articles, or maintaining websites or social media accounts. The reality of most days would be talking on the phone, returning emails, and attending meetings. Working with harvest data and implementing land management plans do not take place as often.

From a business standpoint, corporate wildlife biologists often are expected to actively manage business ventures that are a good fit with the company's core industry. They have annual profit and revenue goals to meet and need to be able to prepare business plans and budgets, while also leading a team that carries out daily operations. The skill sets needed for managing a business include marketing and sales, strategic planning and written reports on the results, leadership, conflict management, and customer service skills.

Background Needed

Typically, a hunting background is necessary to understand both the customers' experiences and practical wildlife management applications. Most individuals that are in the corporate hunting business developed a passion for the sport and a taste for land management practices at a young age. It is essential to understand not only good management practices for the resource, but also the needs of hunting customers, as well as to communicate effectively with them. Therefore, good interpersonal skills are crucial.

The majority of corporate wildlife biologists need good communication skills. They deal with customers on the telephone and in meetings. Verbal communication skills are taken for granted sometimes, but it is clearly a challenge for young biologists who have been acclimated to communicating by text messages or emails. Young biologists must keep in mind that a good public image is very important to corporations, so they must always be cognizant of how to treat customers with respect. Good writing and presentation skills also are valuable. As part of maintaining positive public relations, most companies publish newsletters, have an active social media campaign, and require that articles be published in popular media. Presentation and public speaking skills are necessary for customer education programs, seminars, and marketing.

No matter what the resource used by a corporation, knowledge of the management practices for that resource is essential (e.g., oil management for refineries, silvicultural practices for timber management). For example, most large private corporations that own sizeable amounts of property are in the timber-growing business, so an understanding of forest management practices is important. Daily dialogs with coworkers and customers center on management of the land and the resources. Wildlife biologists are expected not only to understand how relevant practices are applied, but also what strategies would best benefit species' diversity and quality. The majority of private timberland companies are managed under a forest certification program, which offers assurance that the products produced on the property are from a sustainable resource and an environmentally sound source. These firms' wildlife biologists typically are involved in the certification programs and must be experts in certification requirements and silvicultural practices.

Education Required

The minimum educational requirement is a bachelor's degree in wildlife management or wildlife science. Strong candidates also would have a minor in forestry or in business. The majority of applicants have an MS degree, but actual job experience is sometimes more valuable than two more years of college. An ideal educational combination would be a wildlife-based undergraduate degree plus an MBA, which would open many more opportunities for employment in the corporate workplace. There are now abundant opportunities for college students to gain temporary employment as summer interns with corporate landowners. These intern positions are critical, as students can gain practical experience, plus find out if they would like a career in the corporate world. Companies use internships to evaluate prospects for future permanent positions.

Pay Scale

A typical starting salary would be in the range of $40,000 to $50,000 per year, plus benefits. Corporate positions typically are salaried, which means that the work can frequently exceed 40 hours per week. Benefit packages include paid vacations, and most have a year-end bonus, based on how well the

company does that year. Pay raises usually are annual, but, like bonuses, they normally are tied to both the firm's performance and employee merit. These positions are like any other job in the business world: they exist to support a profit center and could be eliminated if the company is suffering financially. Advancement opportunities are not frequent, but they can occur within a specific business group or across other businesses. For example, internal advancement opportunities could include upper-level management, as well as timber production, land acquisition, and sales positions. To facilitate future advancement, many companies provide cross-training and opportunities for professional development.

Corporate Wildlife Scientist
Job Description

The primary responsibility of corporate wildlife scientists is developing and directing a scientifically rigorous research and technical support program that seeks to understand wildlife/habitat relationships on commercially managed forest lands. This includes providing technical support for threatened and endangered species programs; developing approaches that integrate the management of wildlife habitat with intensive forestry; ensuring that sustainable forestry certification requirements are met; establishing liaisons with universities, conservation and forestry organizations, and governmental agencies; and serving as a focal point for technical, biodiversity-related communications internally and externally. One example of this integrated approach was the development of a hunting club cooperative (HCC) on 4,452 ha of Weyerhaeuser Company land in Kemper County, Mississippi. The goal was to promote the concept of QDM for white-tailed deer by establishing common harvest criteria across six different hunting leases and improving habitat conditions for white-tailed deer and associated species in the area. One of Weyerhaeuser's interests was to determine if recreational leaseholders would be willing to pay a higher fee for lands managed in this way, which not only helps the company make money, but also provides a more cost-effective approach for implementing habitat management practices that benefit game and nongame species. Habitat management recommendations and the ensuing benefits for biodiversity were based on the results of a long-term study investigating the biodiversity response to prescribed fires and selective herbicide applications in midrotation pine plantations, and from other studies of white-tailed deer habitat management on industrial timberlands. I [DM] worked with individual hunting clubs; biologists and law enforcement officers from the Mississippi Department of Wildlife, Fisheries, and Parks; Mississippi State University; the National Wild Turkey Federation; our area foresters; and other partners to make the HCC a reality. I conducted the program for eight years, which included some related human dimensions research. In the end, we learned a great deal about the efficacy of habitat management practices on industrial forests and how to productively implement QDM programs with multiple partners, and we better under-

stood the expectations of our recreational lease customers. I was able to publish the results of this experience in a variety of popular and technical publications, in order to share what was learned with others.

This job generally provides flexibility for scientists to develop applied research programs. Research efforts, however, need to align with the informational needs of the business. For example, in the timber industry, much of this research is focused around understanding the biodiversity implications of intensive forest management, including short rotations, silvicultural treatments (e.g., fertilization, herbicides, forest thinning, clear-cut harvesting); implementing forestry best-management practices or regulatory requirements; documenting the contributions of intensively managed landscapes to the conservation of biological diversity; and meeting regional conservation objectives. The expectation is for scientists to develop a high level of credibility, both among stakeholder groups and within the company. Companies often work with universities to develop and conduct research collaboratively. This also means that the results of this work—whether positive or negative, relative to forest management—can be objectively presented in the scientific literature, which helps establish and maintain scientific credibility.

Incoming scientists are expected to spend considerable time forming professional relationships, understanding the silvicultural systems used by the company, developing expertise in managing wildlife residing in industrial forests, and becoming familiar with the various businesses and the leadership structure of a large corporation. Acquiring this knowledge and these skills requires spending time on the company's managed forest lands, collecting data, scoping potential research projects, understanding landscapes, meeting with company land managers, and working with potential cooperators. As scientists advance within the company and receive additional responsibilities, they generally spend less time in the field conducting research and more time engaging with people within and outside the company, in order to communicate research findings and participate in program planning with business leaders and decision makers. The actual percentage of time someone spends on each activity depends on the person and the needs of the organization.

A career in the corporate world differs from an academic or government career, in that company scientists work within a system designed to return value to the shareholders or owners. Thus conservation and management actions must make good business sense. Because some corporations own millions of hectares of land, however, there are great opportunities to positively influence wildlife policy and management across a broad spatial scale and to engage with a diversity of stakeholders on a variety of natural resource management issues. For example, the establishment of habitat conservation plans for federally protected species benefits the species and reduces risks to landowners from having a protected species on their acreage. For example, in Louisiana, Weyerhaeuser Company has entered into a safe-harbor agreement for federally endan-

gered red-cockaded woodpeckers. The company has agreed to maintain a certain number of breeding pairs of this species on a specific portion of its land, which includes intensive habitat management (e.g., prescribed fires, selective herbicide applications, nesting-cavity inserts, longer rotations). In return, Weyerhaeuser can practice intensive silviculture on the rest of the land it owns within that region and be protected from the potential incidental take (i.e., direct harm or habitat destruction) of red-cockaded woodpeckers.

Company scientists generally address management and conservation topics across a wide range of issues and taxa, which change over time. This provides an opportunity for company scientists to broaden their perspective on natural resource management and interact with many different organizations. For example, when I [DM] began working for Weyerhaeuser Company in the late 1990s, it became apparent that we scientists in the firm understood very little about bat communities within the southern lands the company owned. Therefore, I initiated a research program to investigate the conservation value of our managed forests for bats. This led to multiple research projects with a diversity of partners, including Mississippi State University, the University of Georgia, the University of North Carolina at Greensboro, the National Council for Air and Stream Improvement, Bat Conservation International, the Joseph W. Jones Ecological Research Center, and others. I also became very involved in the circles of those interested in bats, serving in leadership roles locally, regionally, and nationally. In the end, we learned a great deal about bat communities on intensively managed landscapes and contributed to an understanding of bat conservation in forests.

In general, scientists are encouraged to be highly engaged with professional conservation organizations, which also can be very rewarding. With experience and advancement, scientists can assume increasing responsibilities for research program management, interact with policy makers and governmental affairs personnel, and provide key input to business leaders within the company. Budgets for corporate environmental research are not unlimited, but internal funds for traveling to conferences and meetings, conducting fieldwork, purchasing equipment, and engaging in cooperative research often are available. Corporate scientists are expected to be cost sensitive, to efficiently use company resources and to show benefits—from a business perspective—for investments in research.

Although there is less corporate emphasis on obtaining grant funding than there is in academic settings, company scientists do face other challenges, particularly with shifts in the structure of a business. It is not unusual for firms to be bought or sold, and programs must adjust to changes in leadership. As a consequence, job stability may be more tenuous than with an academic or governmental agency appointment. Therefore, company employees must be able to succeed in an atmosphere of potential, unexpected job changes. Nonetheless, there are many examples of employees who have been with the Weyerhaeuser Company for at least 20 years and

The primary responsibility of a corporate wildlife scientist is to develop and direct a wildlife research program that supports both wildlife and the mission of commercially managed property (photo: Darren Miller)

have made substantial contributions to wildlife conservation. Secondly, because companies necessarily are driven by financial goals, the implementation of conservation measures can be challenging and, in most cases, must fulfill a clear business need or benefit. This may include such activities as improving habitat conditions, entering into formal conservation agreements with agencies (e.g., Habitat Conservation Plans and Incidental Take Permits with the US Fish and Wildlife Service), or working proactively with hunting leaseholders to manage wildlife populations and habitat.

Background Needed

Incoming wildlife scientists are required to have a demonstrable research capability, similar to what is expected by a land grant university, including a clear record of peer-reviewed publications and presentations to professional and lay audiences. They are expected to develop, conduct, and interpret original research to address current and emerging issues. They must be able to work in teams across disciplines and collaborate effectively with others in the private and public sectors. Preferred skills include an understanding of landscape ecology and management, experience in developing strategies to integrate habitat management into working forests, knowledge of biometrics and habitat modeling, research skills relative to wildlife-habitat relationships, familiarity with working forests, and a close involvement with hunting and the recreational use of forests. Most wildlife scientists have little knowledge of or experience with integrating economic and conservation goals, but a successful company wildlife scientist will have developed these skills within the for-profit environment.

Although experience working within or with a variety of conservation organizations and governmental agencies is desirable, a career as a corporate wildlife scientist can begin following a terminal degree (typically a PhD), with additional capabilities being developed on the job. Excellent written and verbal skills are critical. Company scientists must be able to communicate with diverse audiences, from loggers to other scientists to key decision makers. They need to understand and translate science into applications on the ground, as well as effectively convey this science to land managers and business leaders. Familiarity with the larger context of sustainable forestry and environmental issues (water quality, carbon, soil productivity) is also of benefit, as these issues are intermingled with the conservation of biological diversity and are often addressed within a single research platform.

Education Required

A PhD degree in wildlife ecology, landscape ecology, or conservation biology, with an emphasis on forestry and wildlife management in forested systems, is required. A curriculum following the requirements for certification by The Wildlife Society should provide a firm educational base for a career in the forest industry.

Pay Scale

Starting salaries and benefits are competitive, in the range of those for faculty positions in land grant universities. For example, the salary for a new hire with minimal (less than three years) experience would be commensurate with that for an assistant professor (ca. $65,000). Promotions generally are merit-based, rather than on a specified timetable, and promotions are accompanied by increased levels of responsibility. Annual raises are also merit-based.

NUISANCE WILDLIFE CONTROL OPERATOR
Stephen M. Vantassel

Agency: Private Wildlife Damage Control Business

Wildlife damage management (WDM), also known as wildlife control, specializes in the control of vertebrate pests whose activities conflict with human interests. The field of WDM is divided into four areas: product manufacturing and supply, research development and education, governmental wildlife control agents, and private wildlife control operators (WCO). Position titles, however, are not standardized in this industry at the present time. Wildlife control operators are known by a variety of names, including nuisance wildlife control operators (NWCOs, pronounced "new-coes"), animal control operators (ACOs), nuisance wildlife trappers (Georgia Department of Natural Resources, Law Enforcement Division 2016), animal damage controllers (ADC), problem animal control agents (PACs), vertebrate pest controllers or operators, wildlife control agents, and wildlife damage control agents. The diversity of titles is due partly to the development of the industry in different states when wildlife control became privatized in the 1980s and 1990s. For convenience, I use WCOs for the owners / operators of wildlife damage management

companies and wildlife control technicians (WCTs) for the employees of a WCO.

Wildlife Control Owner/Operator
Job Description

Presently, the majority of wildlife control companies are operated by single-owner WCOs. Only a small minority are large enough to hire workers (WCTs). Opportunities for WCTs are expected to improve as the industry matures and as more pest control companies add wildlife control to their list of services.

Consider several issues before deciding to start your own company. First, determine if your personality can handle the stress of self-employment (Gerber 2001). Second, assess the potential demand for your services. You need enough customers willing to pay your fees to make your business solvent. Service areas with populations of less than 50,000 often lack the needed customer base to operate full time. As a rule, your take-home income is 50% of your service fee. The other 50% covers gas, overhead, taxes, and equipment. Third, fulfill the licensing requirements and business regulations issued by your state and local governments. Fourth, gather the necessary materials to start your business. You need approximately $5,000 for equipment, phone lines, stationary, licenses, and business insurance. In addition, you must have a service vehicle capable of carrying ladders. The National Wildlife Control Operators Association, the Internet Center for Wildlife Damage Management, and the US Small Business Administration (appendix B), as well as private consultants, can provide assistance to you in evaluating the likelihood of the success of your startup and point out pitfalls to avoid. Because the majority of wildlife control companies are operated by the owner, the information provided about WCTs also apply to WCOs.

Wildlife Control Technician
Job Description

The diversity of abilities and skills demanded of WCTs cannot be overstated. Wildlife control technicians locate, identify, remove, manage, and exclude a variety of nuisance wildlife for municipalities, businesses, and property owners. They inspect structures and landscapes and interview clients to identify the source of the problem, determine its severity, and suggest potential solutions that fall within legal restrictions, since WCTs use both lethal and nonlethal methods to resolve human / wildlife conflicts. For example, you may be asked to evaluate holes in a structure and determine whether they were caused by insects or woodpeckers. If woodpeckers are the culprits, you are restricted to nonlethal control measures until the client obtains a federal depredation permit. WCTs work directly with customers to design effective control strategies that directly meet individual needs. They also advise customers about proper sanitation, habitat modification, and structural maintenance, in order to minimize the likelihood of future human / wildlife conflicts. For example, you may install

Nuisance wildlife control operators set traps to remove nuisance animals from residential properties (photo: Scott Henke)

trench screens to prevent skunks from denning under a deck or place spikes on ledges to deter birds from loafing there.

While the exact numbers of species managed by WCTs vary by region, most of the calls involve the control of gray squirrels, fox squirrels, red squirrels, flying squirrels, raccoons, house mice, deer mice, voles, moles, rabbits, woodchucks, chipmunks, thirteen-lined ground squirrels, pocket gophers, bats, striped skunks, beavers, opossums, armadillos, snakes, and various birds, particularly pigeons, house sparrows, starlings, and woodpeckers.

Wildlife control work varies significantly, according to the habitat and construction practices in the service area. For example, two-story buildings require a great deal of high ladder work, while single-story, ranch-style houses do not. The description below seeks to be comprehensive and cover the skills required anywhere in the continental United States.

Physical requirements: Wildlife control work is physically demanding. Technicians must be of reasonable weight and mobile enough to handle the labor and environmental conditions involved. If WCTs cannot fit into standard-sized attic and crawl space entrances (e.g., 57 × 76 cm), then they cannot perform many job functions. Technicians need to be capable of loading, carrying, positioning, and removing a 10-meter long ladder rated for their body weight, plus the weight of any equipment. For example, a Type 1–rated ladder weighs 29 kg and can support 113 kg. In contrast, a Type 1A–rated ladder weighs over 32 kg and can hold up to 136 kg. WCTs should expect to have to move ladders to and from their trucks multiple times a day, often over uneven terrain.

Technicians must have a valid driver's license and hearing good enough to understand conversations on the phone and in person. Endurance is another key quality, as WCTs often have to work in extreme conditions, including high humidity, cold or freezing rain, and wind, as well as stand on a ladder for an hour at a time. Because bathroom facilities are not always readily available at job sites, WCTs need to have normal bowel and bladder control (not needing to use a restroom more than four times in eight hours). Technicians may be exposed to pesticides and insect stings. Thus they cannot have severe sensitivities or suffer anaphylactic reactions when these circumstances are encountered.

Psychological requirements: The diversity of conditions encountered by WCTs requires technicians to have a certain mental toughness. They must be able to control phobias regarding heights, confined spaces, dark smelly rooms, and dangerous insects and other animals. Technicians may have to euthanize both ill and healthy animals. A positive attitude is important for success in this field, as the work cannot always be completed within normal 0800–1700 working hours. For example, you may need to check traps located in high-visibility areas before dawn, to prevent unwanted public scrutiny. Technicians should expect to work some weekends and to be out of town for two weeks or less. They should be willing and able to follow directions, such as company policies, and learn from their mistakes, so errors can be avoided in the future. WCTs also must display exceptional customer service skills, including being able to help customers relax and feel confident that their pest problem will be resolved.

Background Needed

Because most work is done away from direct supervision, technicians must be trustworthy, reliable, and honest. Most employers require WCTs to have a clean record, with no criminal convictions, or at least to have committed no offense more serious than a misdemeanor. Crimes involving theft, threatening behavior, or physical assault are likely grounds for rejection. Insurance requirements typically demand higher standards, such as having an excellent driving record. Technicians must be able to drive a truck or van, typically with an automatic transmission, and be able to back it up safely, using only side view mirrors. Some companies require potential employees to pass a drug-screening test.

WCTs must have good communication skills, and fluency in Spanish is a plus. They need to write legible, clear, and accurate reports, describing the findings from inspections and service calls. Technicians must balance work efficiency with customer service. Multitasking and prioritizing the responsibilities of the job are essential skills. For example, a technician may be required to talk on the phone with the office while setting a trap in a customer's backyard. WCTs should be comfortable with using a computer, vehicle GPS system, tablet, and smart phone. It is critical that technicians be able to solve problems without assistance. Though most of the job duties are done independently, WCTs should be willing and able to work in teams to accomplish tasks. They must follow driving laws and be able to recognize and respond to difficult driving conditions, including excessive traffic.

Education Required

Wildlife control work does not require an extensive education. Employers want technicians that can read, write coherently and legibly, follow directions, and communicate effectively with coworkers and customers. Most companies train desirable job applicants. In fact, some companies prefer to hire employees who have not done this type of work previously, so they can be trained properly. In general, most companies are willing to hire candidates with a high school diploma or a GED. Some prefer those with a college degree, particularly in the wildlife-related sciences, as this training demonstrates the needed comfort level required to work with wildlife. Candidates with skills in the trades (e.g., carpentry, roofing, chimney sweeping, electrical work) are highly valued, as they demonstrate mechanical ability and a willingness to perform manual labor.

Pay Scale

Starting pay can range from minimum wage to $15 per hour. Hourly rates typically increase after WCTs have successfully completed a probationary period, which can last for 90 days or less. Benefits vary widely, from none to assistance with health care costs, as well as paid sick leave and vacation time. Additional income may be earned by meeting and exceeding production goals, selling add-on services, or receiving bounties for certain types of animal control. First-year technicians should expect to earn at least $30,000. Efficient and sales-savvy technicians can earn $50,000 or more. If the wildlife control industry continues to grow at its present rate, positions for midlevel managers should become available within 10 years.

ENVIRONMENTAL PUBLIC RELATIONS COMPANY
Edgar Rudberg and David J. Case

Agency: Private Public Relations and Communications Company

All organizations have a need for communications, outreach, and public relations (PR) work. The extent of these efforts is dictated by the size and the budget of the organization. You can find opportunities for employment in communications in state and federal governments, nonprofit organizations, and the private sector.

There are many state and federal agencies with communications positions that deal with a wide range of topics. Depending on the agency, your work might focus on outreach and communications regarding wildlife, fisheries, agriculture, environmental protection, or pollution prevention. Governmental agencies include state departments of natural resources, fish and wildlife, agriculture, or environmental protection. Examples of federal agencies are the US Fish and Wildlife Service, the US Geological Survey, the National Oceanic and Atmospheric Administration, the Environmental Protection Agency, the Bureau of Land Management, and the US Department of Agriculture. If you are restricted in your ability to relocate, it is important to identify where these state or federal agencies are situated, to ensure that either a headquarters or a field station is within commuting distance. In particular, if you hope to be employed by a federal agency, keep in mind that career paths often lead to or through Washington, DC.

The goals of the communications function at governmental agencies can vary widely. Some positions are mostly informational in nature, telling the agency's story to the public using examples of its current work and initiatives. Other positions are more educational, seeking to provide information in an effort to influence the public about a specific issue or issues and potentially change people's behaviors, such as promoting the use of natural areas, planting buffer strips to minimize erosion, or reducing litter.

The degree to which nonprofits employ communicators depends largely on the size and the scope of the organization. They range from very large entities (e.g., The Nature Conservancy, Sierra Club) to small, local clubs. Governmental communications and PR positions are often focused on specifics, while nonprofit organizations have the ability to seek broader social change through advocacy. Communications work with nonprofits often includes member relations, volunteer recruitment, media relations, and fundraising.

Communications and PR positions that focus on the conservation of natural resources are more limited in the private sector than they are in governmental agencies and nonprofit organizations, although opportunities do exist. Large, general-purpose public relations firms are far more prevalent. They may have staff that work on conservation issues, but these projects tend to be just one part of a larger mix of other activities that might have no connection with natural resources. Public relations firms, however, can provide a foundation in PR skills, so you could later specialize in a communications role more closely aligned with wildlife and natural resource conservation.

Communication Specialist
Job Description

Communications specialists work on a broad range of tasks, as communications and public relations can be multistep processes. Duties range from assessing the communications landscape, to planning and executing a strategy, to evaluating a campaign.

Communications work starts with research, to provide an organization with data on the current state of affairs. This can take the form of conducting primary research through surveys, focus groups, and in-depth interviews or by tracking media stories on a certain topic. Based on that research, specialists develop targeted messages and create a communications strategy using outreach materials, websites, news releases and interviews, and social media campaigns. As part of this plan, specialists look back and evaluate the effectiveness of their strategy, in order to shape future efforts in this area by the organization. Were the objectives met? Why or why not? What can the organization do differently to better meet those objectives?

Thus the specific responsibilities of communications specialists are diverse. For example, specialists, guided by an organization's strategic plan, might work internally with leaders to create materials and conduct outreach activities, implement external communications, build the organization's online presence, support the development of coalitions, and act as a point of contact for media inquiries. In this instance, a percentage breakdown of responsibilities could include coordinating outreach activities, including exhibits, events, written materials, and presentations (25%); managing successful communications with and the delivery of consistent messages to stakeholders, including the use of electronic means (25%); maintaining positive communications with internal staff, volunteers, and other partners (15%); initiating and implement a social media strategy, including proactively pitching media stories and answering all media-related inquiries (15%); recommending improvements to communication and outreach efforts in support of the organization (5%); documenting the work on and progress in developing outreach events (5%); maintaining consistent and accurate records (5%); and assisting the organization's leadership in developing long-term strategies and tactics for education and outreach (5%).

Communications work is not for everyone. Unlike many natural resource professions, the bulk of these activities are largely done indoors, either on a computer or over the phone. Site visits out in the field may be common, however. These can be trips to coordinate media events, collect videos or photographs, or research specific stories.

While time spent outdoors may be limited in communications and PR positions, the breadth and depth of the subjects encountered is often high, providing great interest and job diversity. In times of organizational crisis, communications and PR staff are often at the forefront of the action. This could include being on-site to work with the media covering a natural disaster, such as a flood, or a human-caused event, such as an oil spill. These staff members play a key role in addressing every contentious wildlife and environmental issue, from an overpopulation of white-tailed deer in an urban setting to relaying the anticipated impacts of a rising sea level on coastal natural resources.

Background Needed

Compared with wildlife biologists, ecologists, or other wildlife-related careers, entry into the communications field can take many forms. Some come to it straight out of college, with a degree in communications, public relations, or journalism. Others start out in the biological sciences but see their career paths migrate into communications as they become more experienced with the subject matter and realize that they have good communicative abilities. The best practitioners in this line of work are experts with both written and verbal communications. Not every position requires talents in public speaking and in engaging with peers internally and externally, but these skills are highly recommended and can mean the difference between success and failure for a communications effort. In general, specialists need an ability to understand and convey scientific and technical information to a diverse, nontechnical audience; have excellent verbal and written skills; be self-motivated and able to work independently; complete tasks on schedule; be proficient in the use of electronic communications tools, including emails, listserves, collaborative workspaces, social media vehicles, and, in most governmental agencies, Microsoft Office programs; and have a valid driver's license.

Education Required

To be most competitive for communications and PR positions in the fields of wildlife, natural resources, and the environment, it is important to be able to understand scientific and technical information and have an ability to convey that information to a broader, more general audience. Ideal candidates would have formal training in communications and in some type of natural resource–related science. Most practitioners, however, have formal training in just one discipline and then receive cross-training, continuing education, and on-the-job experience in the other. State and federal agencies often have specific course or credit requirements in particular areas of study. Therefore, if employment in one of these sectors is of interest to you, it is important to check each agency's class-work requirements well in advance.

For entry-level positions in communications, outreach, or public relations, you must have a bachelor's degree either in communications, public relations, or journalism, or in the natural resource sciences, such as forestry, wildlife, environmental studies, or biology. If your BA is in communications, public relations, or journalism, then you need to have a strong interest in and aptitude for natural resources, wildlife, and the environment. If you hold a bachelor's degree in the natural resource sciences, then you need a strong interest in and aptitude for communications theory and practice.

Pay Scale

The expected earnings for entry-level positions vary, depending on whether they are in the private, governmental, or non-profit sector. These positions typically start at around $30,000 per year. Advancing to a leadership position generally requires at least five years of experience and an MA or MS degree. A midlevel career employee might expect an annual salary of up to $50,000. Director-level posts in communications are less common, requiring at least 10 years of experience and, most likely, an MS or MA degree. Individuals in these positions could earn from $75,000 to more than $100,000. Above and beyond years of experience and graduate degrees, success in this field is determined by savvy for communicating effectively and a passion for conservation.

WILDLIFE PHOTOGRAPHER / WRITER
Wyman Meinzer

Agency: Self-Employed

The Yukon Territory can be an intimidating environment for both man and animals, especially in the mountainous terrain that defines so much of the landscape in this Canadian province. I was on a photo shoot for *Sports Afield* magazine, and my goal was to stay 10 days atop Rose Mountain, along the Pelly River, with guides and hunters, documenting the rigors of hunting wild Stone's sheep, a very elusive trophy that inhabits the most remote reaches of this rugged system of mountains. My pack weighed 23 kg, inclusive of camera gear and a change of clothes, and the ascent up the steep slopes would take about 17 hours of grueling labor. I had my reservations after some eight hours of climbing, but once on top, the exhilarating view and potential photo opportunities were cause for celebration, both from overcoming the physical obstacles and for the opportunities that were presented, which would conclude in a successful photo shoot for one of the premier outdoor magazines in America.

Chances for in-house editorial employment in the field of outdoor photography and writing are few and far between in today's job market. As a successful freelancer for more than three decades, I know of only a handful of instances where individuals have achieved positions as staff photographers and writers. In most cases, the person who was hired had actually engaged in freelance work for many years prior to being offered a full-time position, thus having acquired a marvelous portfolio to submit in job interviews. On a personal note, freelancing is a very tough business, but once your work garners a high level of respect from prospective clients, self-employment is immensely satisfying and can sometimes be quite lucrative.

Potential job opportunities include photo editors, art directors, staff writers, staff photographers, editors, associate editors, and photo archivists. Although the list appears impressive, these are positions that come up rarely, and competition for them is intense. Unlike the very early years in this field (at least four decades ago), publications now depend more on freelance work for photography and writing, instead of hiring a full-time employee.

Wildlife Photographer / Outdoor Writer
Job Description

Outdoor photography, editorial work, and book authorship can be very gratifying endeavors, especially with the focus on environmental and preservation issues that exists in the nation today. Notions of spending significant time afield, immersing yourself in fact-finding missions, or perhaps creating imagery for editorial venues or book publishers abound. Outdoor communicators, whether wielding a camera or an iPad, project a lifestyle that is quite attractive to anyone harboring an appreciation for experiences in the outdoors while, at the same time, wanting to disseminate facts of educational value to selected reading audiences.

Despite its alluring qualities, a career in outdoor communications can be a hefty challenge, especially with the advent of over-the-counter cameras technically capable of elevating mediocre photographers to levels of excellence that might otherwise have been unattainable. With self-motivation, the proper educational background, and a solid marketing strategy that incorporates professional standards, however, you can possibly achieve regional, national, and even international fame, despite the competition that currently exists in this profession.

Background Needed

Having plenty of initiative, superior work ethics, a viable marketing strategy, and a passion for the art of effective communications are determining factors for success in the areas of outdoor photography and writing in today's competitive marketplace. With a strong will to succeed, despite the odds often confronting neophytes in this exciting field of work, a rewarding future awaits those who are willing to stay the course, regardless of the adversities that often lie ahead.

Wildlife writers/photographers spend long but satisfying hours in the field to capture their story (photo: Darrel Franks)

Self-motivation: Over the past three decades, I have observed too many talented individuals fall short of their intended goals, due to a lack of internal drive. This is not a stumbling block specific only to communicators, but is frequently an obstacle for those engaged in a career goal that is largely (but not always) dependent on self-employment.

Gainful employment or maintaining a viable presence as a contributing freelancer in the outdoor photography / writing arena is a difficult achievement in today's marketplace. A very significant factor that greatly enhances your marketability is expanding your communication skills as both a photographer and a skilled writer. When clients realize that a single individual can effectively cover both imagery and the accompanying written text, assignments from prospective clients will increase proportionally. Once you are established as a productive and reliable photographer/writer, you will attain a new level of professional respect.

In the few examples I am aware of, an essential criterion for any employer or editor seeking to fill an in-house job or editorial or creative needs is to have a substantial background in publishing. Whether your goal is full-time employment or successful freelancing, employers/editors are looking for an impressive portfolio, combined with a thorough knowledge of how to edit and post-process digital images, including the use of Photoshop, Lightroom, and other software applications for cropping and finalizing color balance. Acquiring these skills, especially from the standpoint of being self-employed, is an exercise in discipline and motivation, two elements that cannot be taught at any level of academia. One such example is a willingness to rise at 0200 and drive four hours in the snow to be on location for a sunrise image.

You need to set personal goals and follow rigid, self-disciplined guidelines that are directed toward achieving a high level of success. With few editorial venues today retaining full-time staff photographers and writers, self-motivation is of the utmost importance if your goal is to be a marketable freelancer. A career as an outdoor writer/photographer is extremely unlikely without a willingness to succeed through personal determination.

Marketing: Successfully marketing your skills is highly dependent on the level of proficiency you exhibit in your field or profession. In photography and writing, a strong portfolio of tangible works ups your chances of marketing success. An old adage that pertains to those aspiring to be writers and outdoor photographers is, "Creating a concept and effectively conveying the finished product to the readership is essential to success." I would add, "Creating a good single photo/article intermittently does not make for a marketable career." A suggestion for beginners is to select topics that are of great interest to you and the intended readers. Research in locating past editorial coverage minimizes the possibility of expending time on a subject that has already been treated, thus increasing the chances of your concept reaching the editorial desk. Consistency and productivity are keys to subsequent marketing success.

An effective marketing strategy is essential for almost all genres of successful writers or photographers, and it is vitally necessary for the longevity of your freelance business and your gross income. Of course, no amount of marketing is effective if you do not project professionalism. Thorough research on target subjects, the submission of sound concepts, and punctuality in meeting deadlines are top priorities for photographers and writers who aspire to levels of excellence beyond the norm.

Good marketing is not a short-lived event. It must be engaged in regularly, using venues such as social media outlets (e.g., Facebook, Twitter, blogs). Securing a large following or readership on social media, with optimum numbers of engaged readers and contributors in the thousands, is also an essential factor in determining the success of your marketing strategy. My social media venues have more than 10,000 readers, which is a highly viable base of potential customers to whom services can be offered. To be successful as an author, you need to establish a credible reputation for the reliability and accuracy of your writing style. Consistently providing erroneous or misleading ideas or concepts significantly limits your audience.

Projecting professional standards in your marketing approach is essential in securing and maintaining the interest of those who subscribe to your media outlets. A constant presence on Facebook, Twitter, and blogs, sharing your expertise or anecdotal experiences to the reading and viewing audience, substantially expands your base on these sites. I have found that submitting comments covering a variety of subjects (history, natural history, photography, geology, archeology, aviation) keeps your reading base broad and the interest level high. The extent and depth of your contributing posts are directly proportional to your own educational background, hands-on experience, and personal preferences. For example, I have effectively used written accounts describing the natural history of various wildlife species or historical events that pique the interest of readers throughout my state, and even nationwide. Experimentation in concept selection is a good way to gauge the likes and dislikes of your target audience. Find out which blogs or posts are well received and maintain production in that genre.

Style: Perhaps one of the most cherished compliments to the ears of any communicator is, "You have a distinct style, and I know your work without seeing the byline." For marketing purposes, a personal signature or style should be the goal of anyone aspiring to engage in an outdoor photography and writing career.

Personal style can come in many forms, not least being an author's sentence structure or, in the case of a photographer, an effective application of light on the subject matter. Both represent a brand that is unique to a particular individual, and your work should have a style that resonates with editors and audiences. Without a characteristic signature to your writing or photographs, they become redundant creations, with little value to editorial markets or the general public.

Education Required

Having the proper educational foundation is important if your goal is creating a niche in the outdoor communications market. Although several individuals without an academic degree have successfully achieved a certain, if not great, reputation as wildlife or outdoor photographers and writers over the last three decades, that is rare. Without question, an educational background in wildlife behavior, zoology, biology, range management, and basic English (including some creative writing and journalism courses) greatly enhances your viability in this field of work. As for photography classes, it is my experience that unless an academic institution offers a course in actual outdoor photography in their communications or wildlife biology curriculum, little is gained through studies in a university's College of Arts. Although photography classes are applicable to careers in other disciplines, my 12 years of experience as a photo instructor at Texas Tech University have shown me that an arts curriculum does not effectively prepare students for a career in outdoor communications.

Pay Scale

Salary levels for outdoor photographer and writer jobs are dependent on a magazine's circulation numbers and whether the publication is a state, federal, or private enterprise. The annual income for an employee may range from $30,000 to $100,000 per year, with the average being about $50,000 per year, depending on the job title. Very successful and market-savvy freelancers, on the other hand, can make more than $100,000 a year, especially if they have established a broad income base through writing, photography, educational workshops, and speaking engagements. Earning a substantial income from workshops and speaking tours requires a level of knowledge and fame that encourages private, social, and professional groups to seek your advice. For example, I am frequently asked to conduct workshops on photography techniques and give PowerPoint presentations on inspirational career choices, historical events, or simply the beauty and diversity that define my home state of Texas.

WILDLIFE RANCHING ENTERPRISE
Don A. Draeger

Agency: Private Wildlife Ranches

Wildlife ranching enterprises arose from the livestock production industry. Incorporating wildlife ranching and hunting opportunities into a traditional livestock production operation initially offered landowners a new way to get some extra income, especially when the livestock market fluctuated. Within the past 30 years, however, wildlife ranching has grown substantially, and in most of these types of ranches today, wildlife biologists have supplanted cattlemen for jobs as ranch managers. Ranches can be small (less than 40 ha) to quite large (more than 40,000 ha) and can be managed for a few wildlife species or a multitude of species. Many wildlife enterprises offer hunts for native wildlife, such as white-tailed deer, bobwhite quail, and bison, and others also incorporate hunts for exotic species, such as fallow deer, sika deer, axis deer, blackbuck antelope, and many more. Because exotic species have domestic animal status in some states, they offer year-round hunting opportunities for sportsmen and potential year-round income for ranches, while hunts for native species are restricted to state-mandated hunting seasons. Some wildlife enterprises raise animals for a variety of products, including meat and fur, as well supplying stock for zoos, hunting preserves, research programs, and even for restocking natural populations. Many wildlife enterprises also offer ecotourism packages, where visitors view, photograph, and enjoy the aesthetic beauty of free-ranging wildlife. Whatever the goals, ranch owners hire wildlife technicians, interns, and managers to operate their wildlife ranching enterprises.

Ranch Technician
Job Description

Unfortunately, it is nearly impossible to get a job as a wildlife ranch owner. You either have to earn a substantial amount of money in some business venture to purchase a ranch, or you need to inherit one. Wildlife ranching enterprises sometimes are operated by the owners, but more often than not, owners hire a ranch manager to conduct daily business and oversee the wildlife operations. For many wildlife students, managing a wildlife ranch is their dream job. Rarely, however, can they obtain such a position without first having to pay their dues and demonstrate their competence.

I have been in the wildlife business since I was 15 years old. I was lucky enough to have a cousin who was earning his MS degree in the wildlife field, and he let me tag along with him. We worked very hard, and my only compensation was that he let me stay on his couch and fed me, but I was happy. Through the web of wildlife connections I made then, during my later high school and college days, I was able to convert that experience into valuable employment opportunities each spring break and summer, working for various ranches in South Texas. None of the jobs paid much, and some did not pay at all, but each one added to my experience and, more importantly, they all gave me networking connections.

I didn't know it at that time, but I was a ranch technician. I worked cheaply, for long hours, and did the jobs no one else on the ranch wanted to do. I never complained, and I was truly happy to be there. Today I am the wildlife manager for four large ranches in Texas, totaling 52,610 ha, that are owned by a single individual. I credit my current employment to those early days.

Ranch technicians form the backbone of a ranching enterprise, as they are the ones who perform the majority of the labor required to keep a ranch operational. Ranches may hire full-time, year-round technicians, but more often than not they hire interns, who are college-age wildlife majors who wish to gain hands-on experience in their field. Each ranch technician or internship opportunity can be different, so it is difficult to describe everything you might be asked to do. Nonetheless, from my 14 years as a ranch manager, I can give you details on what I expect technicians and interns to do. The duties for our ranch technicians or interns are a mixture of wildlife and ranch work. On any given day, they could be maintaining feeders and filling them with protein feed, corn, or cottonseed, as well as taking care of ranch equipment and vehicles (including mechanical work) and water sources (e.g., windmills, submersible pumps, cattle and wildlife troughs). In addition, technicians and interns construct and set up hunting blinds, build cattle- and hog-proof fences around feeders, and map natural resources using GIS and GPS technology. They assist with guest services, including washing vehicles; setting up and maintaining vehicle ice chests; cleaning and organizing the contents of wildlife-related buildings, such as the skinning shed, intern housing, the hunt room, and other guest areas; and providing hunting guide services for guests at the ranch. Post-harvest duties for deer collected by ranch guests include skinning, cape preparation, and meat processing. When guests are not present, technicians and interns spend much of their time conducting habitat modifications for the ranch, including brush clearing with heavy equipment (e.g., dozer, excavator, tractors) and chemical applications. Ranch technicians and interns also use game and video cameras to locate and find the pattern in deer movements; scout for deer for future

hunts; conduct predator control activities, including trapping and hunting feral hogs, coyotes, and raccoons; and assist the wildlife manager during wildlife surveys, including helicopter surveys, spotlight surveys, and call counts (determining wildlife numbers by the sounds they make).

Our interns spend approximately 80% of their time outdoors. Daily routines vary greatly, depending on the seasons and weather conditions. During nonhunting seasons, work typically takes place Monday through Friday, from 0800 to 1700, with occasional weekend work required. During the hunting season (September–March), we work seven days a week and only have an occasional day off.

It is surprisingly easy to be a superstar technician or intern, but what puzzles me is that finding one is rare. Why is that? I believe people lose their focus on what is important for the long-term aspects of their career. We live in a society that tends not to think very far into the future, and the side effects can bleed over into work ethics. I have had interns who believe they are too good to do certain jobs, such as washing ranch vehicles, picking up trash, and cleaning flower beds. This is shortsighted, in my opinion. The issue is not about being overqualified for menial labor, but about being willing to do anything to earn your spot in the wildlife industry. There are jobs that are not fun but still have to be done, and they exist for everyone employed on a ranch. For interns or technicians, it may be filling protein feeders during the summer from bags weighing 23 kg. For upper management, it might be keeping accounting spreadsheets. I have a saying, "I am paid to do all the mundane stuff, but I do the wildlife work for free." In other words, all jobs at every level have negative aspects. That's why they pay you! But if you work hard and don't complain, that will be noticed, and you will be rewarded over time.

Internships are like ships—they come in all shapes and sizes. Some are pretty, and others are utilitarian. Some ships sail, and some sink. So how do you know which internship is right for you? First, make sure that you're working in some aspect of the field in which you're interested. Find out if the internship has the species and a location that interest you, and if the employer conducts data-gathering techniques and has a management style that is compatible with how you operate. Next, interview the interviewer. Ask detailed questions about your daily workload (including the number of hours required and if your weekends would generally be free), vehicle use, housing, things you will be billed for (e.g., utilities), and the intent of the internship. Your goal is to avoid surprises about any aspect if you are hired. Once you accept a position, dedicate yourself to completing the internship, even if it's not what you expected. If you leave early or flat-out quit, it reflects badly on you and shows an inability to complete projects. Each one of your future employers has had an unpleasant work experience, and they will respect you more if you make lemonade out of a lemon of an internship.

Another one of my sayings is, "When you think you have paid all your dues in wildlife, you are halfway there." Advancement most likely will not happen in the time frame you hoped for or dreamed about, but it can and will occur if you stay

Ranch technicians fill feeders to entice white-tailed deer to an area for wildlife viewing (photo: Scott Henke)

true to your work ethic and maintain a good attitude. One of the great things about the wildlife field is that we truly enjoy helping others succeed. My peers and I want to see you move forward in your career when you're ready to do so. Where the friction may occur is on a matter of opinion: when *you* think you're ready, and when *your boss* thinks you're ready for the next step. My suggestion for not getting frustrated is to talk to your boss and others in management positions about how long it took for them to move into more-responsible positions on the ranch. Such communication offers perspective about where you are on your journey.

Background Needed

It is preferable for technicians or interns to have prior experience with wildlife, ranching, or farm operations. Students who grew up working on farms and ranches have an added advantage for our internships. In general, they have an impressive work ethic. They also tend to know the value of maintaining equipment. Students who are raised in a city, with no exposure to rural life, and get a wildlife degree have a steeper learning curve in terms of what happens on a ranch. Among other things, they tend to break more equipment early in their careers (which drives ranch managers crazy—and not in a good way!). City-raised students can certainly succeed in the wildlife field, especially if they recognize that they have more to learn and take a humble but determined approach to climbing this mountain of knowledge. The other group of students who usually impress me are those who served our country in one of the branches of the military. They have been trained to accept orders and take care of equipment. Their work ethic is strong, and honesty and loyalty are ingrained in them.

During the past 14 years, I have employed about 300 interns. I have seen the full spectrum of employees, ranging from superstars to those who are frightening to be around.

I hired some of the standouts. Former interns are now the on-site managers for three of the four properties I oversee. I employ people I know and trust, and these end up being the interns who have proven their worth. Some of the superstars I did not hire are now managing and working for other ranches across the state. When you have a great intern or technician but do not have any positions open at that time, you feel an obligation to that person, and to our industry, to try and place your former employee in a job that will be rewarding. The truth of the matter is, the really good ones come along so rarely that, when it happens, you bend over backward to make sure they advance in this field.

The wildlife ranching industry is extremely tight knit. We share information, which includes the quality of interns and technicians. My peer group of biologists, ranch managers, and professors have a connected history. Most of us have worked together early in our careers. Through the years, we have slowly paid our dues and are now in management positions. This peer group reaches across states and into universities and agencies. Because of this network of shared information, interns and technicians who are good employees get known by others and are able to advance in their careers. Often an individual who interns for one ranch will receive a full-time job with another ranch, because that person is highly recommended by someone in the peer group. Keep in mind, however, that the reverse is also true. If an intern does a poor job for one of us, it will be difficult to get a position with any of the other members of this group.

Over the past five years, my peers and I have seen a dramatic decrease in the number of applicants for the internships we offer. In my opinion, the positions have not changed, but the general attitude toward them has. My peer group and I have discussed this in length. It seems as though the average wildlife student now wants a full-time, salaried position, with benefits, immediately after graduating with a bachelor's degree, even if that person has little to no true wildlife experience. As nice as that would be, it is not the real world. Employers are more comfortable hiring from within, or at least offering a job to someone who is recommended by a trusted source. Internships are a good way to establish your trustworthiness. That's why I urge each wildlife student to try out several internships while still an undergraduate, even if that means taking a semester or a year off from school to do so, and gain the experience. This can be a hard sell, however, to parents who are paying for your college education. You, and they, must understand that, by itself, a wildlife degree (a bachelor's degree or an MS) in no way guarantees you a job in the field. Today's prospective employees have a much greater chance of being hired if they have the educational background *and* the level of experience desired by employers.

I believe that the secret to a successful career in any field is a combination of a positive demeanor and connections. If you have a great attitude and know the right people, you will go far in your selected field. An internship is your pathway to network connections, but your pattern of behavior is a choice you make daily. Most work skills can be taught, and employers are generally more than willing to supply the training. Few employers, however, are willing to help you change your attitude. So it's up to you to be pleasant, friendly, outgoing, honest, loyal, self-motivated, and not be a complainer.

My peers told me about the work ethics they look for in interns and technicians. They desire people who always show up for work early; have good hygiene, a clean-cut appearance, and a positive attitude; ask questions, especially if they don't know how to do something (i.e., ask before you break it); volunteer when the opportunity arises; work during weekends, and especially after 1700, without complaining about it; own up to a mistake (others will respect you for that); and do not text or email the ranch manager about a personal request, but instead ask in person. Remember not to quibble over the pay. You are not working as an intern to get rich, but to gain experience and make connections that will result in a full-time job in the future. Realize that every day is an interview, so act like it.

I strongly believe internships are a valuable asset in your early wildlife career, since the field is extremely competitive. In my opinion, internships give your resume a decided advantage over those for applicants who just have a college degree. Ask yourself whom you would hire from the two following candidates. Candidate A has a 3.7 GPA from a well-known wildlife university and gives the names of three professors as his references, who attest to how good a student candidate A was. Candidate B has a slightly lower GPA (3.2) from the same university, but he completed three internships during his undergraduate career. He lists his former employers and a professor as references, who praise how good an employee candidate B was. You know and respect one of his former employers. Who gets the job? Now look at your resume. Do you resemble candidate A or candidate B? If you said candidate A, then go to the job boards today and start looking for a wildlife internship to improve your marketability for a full-time ranch job.

Education Required

Applicants for internships range from freshmen wildlife undergraduates to post–MS degree students looking for full-time employment. Although education is important and necessary if you wish to advance to a position as a ranch manager later, a solid work ethic and a good attitude are more valuable if you want to initially be hired as an intern or a technician. Applicants who have had coursework that emphasized hands-on components, such as classes in wildlife techniques, surveys, and GIS applications, find that they are more competitive for internship or technician positions.

Pay Scale

The pay for technicians and interns varies widely, but I can provide some general guidelines. In 2015, ranch technicians and interns in Texas were paid from $800 to $1,600 per month, as contract labor. The salary difference usually depends on the fringe benefits that come with the job. Some common benefits include free housing (in shared living quarters), free utilities, and free phone or satellite TV service. A work truck is included, as are hunting opportunities. Not every internship position, however, offers all of these benefits. Pay rates also fluctuate, depending on the time of year. For example, we pay more for our hunting season internships, because they have more-demanding hours and a greater workload, than we do for our summer internships. We also have fewer applicants for internships during hunting season than for summer ones. Full-time, yearlong technicians typically are paid at the higher end of the salary range listed above.

Wildlife Manager
Job Description

Wildlife managers typically move up the hierarchy from an intern to a ranch hand or wildlife technician position before they are given an opportunity to work as a wildlife manager. They most likely have spent many years at each of the previous jobs before becoming a manager. Because wildlife managers have worked their way up the job ladder, probably there is no job on the ranch that they cannot or will not do. These include but are not limited to developing work schedules; hiring employees, including interns; advertising ranch products, such as hunts and animal sales; developing management plans for livestock and wildlife and seeing to it that these plans are carried out properly; being in charge of hospitality for ranch guests; maintaining ranch records; regularly meeting with and updating the ranch owner concerning ranch operations; acquiring and maintaining permits and other documents, such as vaccination records for livestock, hunting permits, and disease records; overseeing daily ranch operations; ordering supplies as needed; and developing and maintaining the ranch budget, including payroll. There is no typical ranch workday, and emergencies often occur that demand immediate attention. All of the listed duties must be done on a regular basis. Wildlife managers need to prioritize each task and make sure everything gets done on an as-needed basis, according to its priority.

Background Needed

Previous work on a ranch is required for wildlife managers. Owners rarely, if ever, consider someone who has no prior experience. A strong work ethic and a willingness to do anything the ranch needs, no matter how menial, are expected traits. Wildlife managers need to have a positive attitude, and they must do what the ranch owner wants. Always remember, it's the owner's ranch; therefore, the owner is the boss. Wildlife managers and landowners should work diligently at creating and maintaining an open, honest, and candid dialogue. This creates an environment where wildlife managers can offer suggestions based on their background and expertise but still allow the landowner to make the final management decision. Owners want managers to carry out the goals of the ranch, and those goals ought to be determined by a cooperative effort between the ranch owners and their wildlife managers.

In addition, wildlife managers need excellent people skills. They must be able to converse and positively interact with employees, ranch owners, and ranch clients. Often this means quite a large array of people from very diverse backgrounds, ranging from blue collar to white collar, and from laborers to the mega-rich. Remember, this is a service-oriented industry and, as a wildlife manager, you are there to serve the needs of the ranch owner and the guests.

Education Required

A bachelor's degree in wildlife management, ecology, biology, or a closely related field is typically required. An MS degree in a similar discipline is preferred, however. Owners demand managers who are competent with both livestock and a variety of wildlife species. Knowledge of grazing management, large mammal ecology, and avian ecology is needed to develop management plans. Familiarity with wildlife laws, especially concerning the permitting process and state regulations, also is necessary. An ability to identify species of plants and animals and discuss their benefits to ecosystems is impressive to ranch clients. Therefore, a strong candidate is well versed in mammalogy, ornithology, herpetology, and plant taxonomy. Accounting skills and a working knowledge of GIS mapping are beneficial. A minor, or at least coursework in the hospitality disciplines (often associated with restaurant and hotel management departments) and meat processing (associated with animal science departments) is useful.

Pay Scale

Pay can vary greatly for wildlife managers, and it is usually correlated with the amount of land for which a manager is responsible (more land = more pay). Thus salaries could range from $30,000 to $200,000. The jobs available at the upper end of the scale are extremely rare. There are few very large landholdings, and jobs on those properties do not have a high turnover rate, because managers usually remain in their positions until they retire. Therefore, the average pay rate for a wildlife manager is probably closer to the $45,000–$65,000 per year range.

WILDLIFE EDUCATION

HIGH SCHOOL
Jessica Reeves Bogart

Agency: Education, High School

High school teachers help prepare students for life after graduation. They teach academic lessons and various skill sets that students need to function well in college and to successfully enter the job market. Agricultural science teachers (ASTs) are more specialized in what they teach than general high school teachers. ASTs are in charge of classes in the agriculture, food, and natural resources cluster within the high school curriculum and typically are advisors for an active chapter of the Future Farmers of America (now referred to as *FFA: Agricultural Education*).

Agricultural Science Teacher
Job Description

In general, ASTs educate young people about the history of agriculture and teach them to value the future of the industry. This is done both in and outside of a traditional classroom setting. More importantly, ASTs help turn students into life-long learners who possess skills such as leadership, good communications and dedication.

Within agricultural education, there are seven pathways that can be taught in the schools, including (1) animal systems; (2) agribusiness systems; (3) power, structural, and technical systems; (4) plant systems; (5) food products and processing systems; (6) environment service systems; and (7) natural resource systems. A school district can require ASTs to teach in any of the abovementioned pathways, but an AST whose interests lie in natural resource systems can seek out a school that offers this pathway.

Currently, numerous courses offered in the natural resource systems pathway at the secondary school level can be taught by a certified AST. These include principles of agriculture, food, and natural resources; wildlife, fisheries, and ecological management; range ecology and management; forestry and woodland ecosystems; energy and natural resource technology; advanced environmental technology; and practicums in agriculture, food, and natural resources. ASTs also may be asked to teach additional courses, such as professional standards in agribusiness, agribusiness management and marketing, mathematical applications in agriculture and natural resources, equine science, livestock production, small animal management, veterinary medical applications, advanced animal science, food technology and safety, food processing, agricultural mechanics and metal technologies, agricultural power systems, agricultural facilities design and fabrication, principles and elements of floral design, landscape design and turf management, horticultural science, and advanced plant and soil science, depending on the school district's goals and student-driven interests.

Along with teaching, ASTs usually serve as the school's FFA advisor. These responsibilities include but are not limited to (1) building and maintaining a professional relationship with parents, fellow teachers, and the school's administrators; (2) overseeing student-supervised agricultural experience projects (SAEs); (3) maintaining school, student, and FFA records and reports; (4) teaching students how to keep good records; (5) coaching fall and spring leadership career development events (LCDEs), career development events (CDEs), and public speaking contests; (6) overseeing parent support organizations (e.g., Young Farmers, Booster Club, Livestock Show Syndicate); (7) supervising the student FFA organization; and (8) supervising school facilities (e.g., farm, greenhouses, labs).

Student SAE projects can take many forms. For example, SAEs can be entrepreneurships, where a student plans, implements, runs, and accepts financial risks in an agribusiness, such as raising quail, growing plants or trees for sale, or operating a guide or lawn care service. They can be placement projects, where a student becomes part of in an agriculturally based business, lab, or community service area outside of normal school hours. This can be a paid or a volunteer position. Examples of placement SAEs can be in game ranches, veterinary clinics, animal shelters, farms, livestock ranches, and floral shops. SAEs also can be research projects, where a student uses the scientific process to conduct experiments or develop plans, such as to compare and contrast the effects of high fences on wildlife movements, or to develop a ranch management plan that incorporates livestock and wildlife enterprises. SAEs can be exploratory projects, where a student investigates various aspects of agriculture, such as shadowing a wildlife biologist, interviewing a fish farmer, assisting a game warden, or raising livestock for a project. SAEs can be improvement projects, where a student adds to the value of a community building or home. Such projects might involve restocking a pond, landscaping, building fences, or restoring a piece of equipment. Lastly, SAEs can be supplementary, where a student engages in an activity that enhances his or her SAE (e.g., performing quail counts, capturing deer, giving injections, assisting in castrations).

For ASTs who are passionate about natural resources, there are several spring FFA career development events that specifically focus on natural resources, including wildlife and recreation management, range, range plant identification, land, forestry, and natural resource use. Training for fall leadership development teams on such topics as agricultural issues, FFA broadcasting, job interviews, and agricultural skills demonstrations can be focused on wildlife and natural resource topics. In addition, for those ASTs and students who love wildlife and natural resources, prepared public talks, soil stewardship, and even agriscience fair projects can be geared toward this passion of the outdoors.

Teaching agricultural science at the secondary school level is definitely not the typical 0800–1500 teaching position that boasts of weekends and summers off. It is not uncommon for ASTs to work 12 to 14 hours per day. An AST's daily schedule could consist of the following:

0630–0730: Train students for a fall leadership career development event (LCDE), a spring career development event (CDE) team, or public speaking (takes place before the school day starts)

0730–1430: Teach six to seven different hour-long classes

1430–1600: Train another team or speaker

1600–1800: Check livestock projects

1800–2000: Supervise an FFA meeting, participate in a parent-group meeting, or serve on a school district planning committee

Somewhere in this daily schedule, ASTs must create lesson plans, purchase and lay out lab materials, make copies, grade papers, plan FFA activities, write scripts for LCDE teams or speakers, create practice tests or contests for CDE teams, answer emails, and make phone calls to parents. Weekends are usually filled with competitions, livestock shows, community service projects, and other school or community-related events. For student/teacher travel, ASTs are required to gain school district approval, find hotel rooms, reserve rental vehicles, pay contest entry fees, get parental permissions for each student, arrange quality substitute teacher plans, obtain money for meals, and so much more. This all takes large amounts of time, organization, and money. Professional behavior and the constant supervision of minors are of the upmost importance when attending off-campus and overnight activities. There are not a fixed number of competitions or activities that ASTs must attend, but each community and school district has its own set of expectations ASTs should strive to meet. Traditionally, the more activities students attend, the more well-rounded and successful they usually are.

Camps, conventions, workshops, curriculum writing, and project supervision quickly fill the entire summer. Also, a certain number of professional development hours are required for ASTs to maintain their teaching certificates. For example, the Vocational Agriculture Teacher's Association of Texas (VATAT) hosts a week-long conference each summer that allows teachers to learn new skills from seasoned professionals while earning these necessary hours.

Background Needed

First and foremost, people considering going into the agricultural teaching field should have an open mind, an eagerness to learn, and a willingness to ask for and accept help. Every community, school district, and group of students is different, so being open-minded about their wants and needs is important. Do not be afraid to teach a course or train a team just because you have not done it before. Use the networks available to you. Because the amount of required knowledge is so vast, there is no way one person can know everything, but there is a strong network of ASTs across the country who are successful in individual areas and willing to help those who ask.

It is usually helpful for future ASTs to have taken some agricultural science courses while in high school and to have been involved in their local FFA chapter, but many extremely successful ASTs do not come from agricultural and FFA backgrounds. Not every school offers agricultural education courses or has a successful FFA chapter. Furthermore, some great ASTs are former agricultural students who watched those in other schools have opportunities that were never offered by their local FFA chapter. Future ASTs lacking high school opportunities should spend extra hours observing successful teachers and chapters. Volunteering to judge competitions, helping to train teams, assisting with local project shows, taking advantage of internships, and lending a hand at local school functions while attending college help young teachers gain useful experiences and make contacts that could be valuable when trying to secure an agricultural teaching position. Due to the fact that ASTs should be diversified in their agricultural knowledge, have a strong background in educational theory and application, and have sound student teaching skills, please pay close attention to the listed educational requirements.

Education Required

Secondary school agricultural science teachers must have a bachelor's degree and a current teaching certificate. Along with completing the required number of observation hours (watching a state-certified teacher in a classroom setting) and the student teaching component, college seniors must take and pass the state's Agricultural Content and the Pedagogy and Professional Responsibilities (PPR) tests to earn their teaching certificates. Most ASTs pursue a bachelor's degree or an MS in agricultural science. It is not uncommon, however, for students to obtain a degree in another closely related agricultural field and continue to double major, minor, or even obtain an MS that satisfies the required education courses and student teaching period. Undergraduates should take classes in all of the diversified agricultural fields: animal science, wildlife science, plant and soil science, business and economics, and agricultural mechanics. Education courses also should be taken, in order to better understand how to deal with student learning styles, classroom management, and lesson development.

Many states currently offer an alternative educational program, where anyone holding a bachelor's degree can quickly and easily obtain a teaching certificate. These programs, however, usually consist of very limited training inside and outside the classroom. I do not recommend that alternative educational programs be used to take the place of postsecondary education programs and university-supervised student teaching. Student teaching allows future educators to spend several months observing and assisting an experienced teacher. Classroom management, time management, parent/teacher communications, administrative expectations, community involvement, and professional networking are just a few of the key concepts developed in student teaching that drastically impact the success and effectiveness of first-year ASTs.

Pay Scale

Due to differing numbers of contract days and pay scales, salaries vary for agricultural science teachers. Most ASTs earn a base pay of $30,000–$70,000 per year. A school district may employ an AST on a 10-month, 11-month, or 12-month contract. Those with 10- or 11-month contracts may have additional contract days to compensate for weekend and holiday responsibilities, and these are usually paid at a daily rate. Other ASTs are given a stipend to cover any extra duties (training teams, supervising projects, travel) above and beyond that of a traditional teacher. Stipends vary from district to district.

Secondary school teachers are offered benefits, such as health, dental, vision, and cancer insurance, and they are given five state and several local paid personal/sick days. State teacher retirement benefits are obtained through years of service, and some school districts pay extra money into your retirement account as an added benefit.

MUSEUM
John M. Bates

Agency: Education, Museum

Museum collections exist in all types of places. There are federal, state, nonprofit, and private museums. They can be associated with state or federal agencies or protected areas or affiliated with public or private universities and colleges. Thus museums can be freestanding entities or part of a much larger agency or organization. Sizes of museum collections vary greatly. The Smithsonian Institution is the country's largest federal museum. Some of the bigger nonprofit museums include the American Museum of Natural History (in New York), the Field Museum (in Chicago), and the California Academy of Sciences (in San Francisco). There are also sizeable museums associated with major universities, such as the Museum of Vertebrate Zoology (at the University of California, Berkeley), the Museum of Comparative Zoology (at Harvard University), and the Museum of Natural Science (at Louisiana State University). Smaller universities and colleges maintain collections that have fewer but still important holdings (e.g., Occidental College, in Los Angeles) or well-used teaching collections.

Museum Curator
Job Description

Collections can consist of many different kinds of things, ranging from fossils and meteorites to all types of biological and cultural diversity. Because museums can be public or private, large or small, freestanding or part of a school or governmental agency, the duties of a curator vary widely. Many curatorial positions are not full time, meaning that curators may oversee collections as only a part of their overall job description. This makes it a little difficult to generalize about these positions. The common thread in all curatorial posts is that curators must demonstrate expertise in a scientific field and have an understanding of and passion for collections and how they can be used.

The idea of taking care of collections of dead animals or plants may seem dreary, but developing and maintaining collections involve much more than just putting dried or preserved specimens into cabinets. Collections have value because of their associated data. Animal collections, for instance, offer useful insights into wildlife and wildlife management, because the objects and their associated data can be used for long-term monitoring of wildlife populations and observations on how they change over time. Botanical collections (in herbaria) provide data on shifts in the plant communities on which animals depend. Modern research done by curators combines traditional approaches and new technologies to collect data from specimens. These data can address critical scientific questions associated with evolution, ecology, and wildlife management. Gathering information from animal specimens can include measuring morphological (form and structure) characteristics of specimens; collecting reproductive data;

recording basic natural history data; performing elemental analyses (e.g., of stable isotopes) and genetic analyses; utilizing spectrophotometry (measuring the properties of materials as a function of the wavelengths of light) and scanning devices; analyzing geospatial data; and modeling environmental niches. New analytical tools allow curators to use collections in increasingly different ways. Curators frequently have the freedom to develop research and educational programs on their own, and they go into the field to collect and prepare specimens. An opportunity for fieldwork is a primary reason why many curators became interested in pursuing this career.

Specimens also can be added to collections through salvage (animals that have died and are brought to the museum). Once specimens are collected, they must be archived and maintained. An understanding of data management is essential for curatorial positions. It is important to recognize that every series of specimens in a collection present data points that can, if the items are properly cared for through time, be used to answer questions that we have not even realized need to be asked yet. Thus a curator's work is not just about safely maintaining information that was collected in the past. It is also about continuing to build a baseline of data for use far into the future. Curators provide a service to other wildlife professionals and the public, because they are experts on the identification of wildlife and plants and at responding to questions related to the collections in their care.

What do museum curators provide to the broader community? One of the best recent examples is the Grinnell Project, which has been carried out by curators in the Museum of Vertebrate Zoology at the University of California, Berkeley. In this project, contemporary curators are using museum collections to assess the effects of climate changes in the mountains of the western United States. This research is anchored by detailed elevational transects made by the museum's visionary curator, Joseph Grinnell, in the early 1900s, where small mammals (and other vertebrates) were collected in a systematic way throughout the mountains of California (including Yosemite National Park) and the Great Basin. These data, in the form of specimens and field notes describing trap lines and capture rates, are housed at the museum. Not only have they permitted a reassessment of the basic patterns associated with changes in the elevational range of species across a 100-year time span, but they also allow further investigations to be carried out. Grinnell's data, combined with modern surveys, enable researchers to investigate whether changes in the elevational ranges of small mammals are simply due to a shift in those ranges, or whether these movements are associated with modifications in the genotype (genetic makeup) and phenotype (observed properties) of the animals, indicating the possibility of natural selection affecting the populations. Researchers also can compare how adjustments are occurring in areas with varying levels of climate changes or geological differences, in order to investigate responses across regional spatial scales that could be the initial stages involved in population divergence (differences among groups that can lead to the formation of new species).

By studying collection data like Grinnell's, wildlife managers can understand how plants and animals are responding to a changing world.

Lastly, museum collections offer special opportunities for educating the public about wildlife. Teaching future generations to appreciate the value of wildlife and the natural world is one of the more rewarding aspects of being a museum curator. Educational opportunities for curators range widely, from guiding graduate students doing cutting-edge research on wildlife in the untamed parts of the world to teaching an undergraduate course about biogeography (the geographic and geological distribution of species). Curators can work with museum exhibit developers to put together a new display about the wonder of ants, birds, or lichens, or they can lead a tour for a group of young children and show them specimens of wildlife they have never seen before, as well as helping them understand that it is up to them to preserve this wildlife as they become adults.

A museum curator is someone who takes care of a collection, but there is much more to a career in this field. Far too often, collections are underappreciated by higher-ups, so curators must interact with administrators to make sure that the value of these incredible resources is not lost on short-sighted decision makers. Museums are extraordinary resources for professionals, students, and the public, and curators are responsible for interacting with all of these audiences on behalf of the museum.

Background Needed

Because museums can represent a variety of organizations and sciences, the job requirements and skills needed to be a curator can vary widely. Museum curators generally have a love for natural history and wildlife and usually have acquired taxonomic expertise in a particular area. Most curators first became interested in museum work by visiting a museum while they were young. The best way to learn about curation and collections is to seek out opportunities in active museums. Museums frequently rely on volunteers, but there also may be paid internships.

Education Required

A degree in biology (often a PhD), with a focus on systematics and taxonomy, is highly desirable for curatorial positions. Courses in organismal biology (e.g., mammalogy, ornithology, herpetology) and botany supply important basic knowledge, and labs for these courses can provide introductions to preparing and working with specimens. A few higher educational institutions offer specific coursework in museum science. Universities with museums are likely to provide the best opportunities for learning about collections and working with them. It takes time to master the techniques for preparing specimens, so do not expect to pick everything up easily or quickly. Top curatorial candidates generally have experience working in a museum collection, including training in specimen preparation and data management, as well as skills in the use of specimens and their data for research and education.

Museum collections have specimens for scientific research, as well as for display. Well-prepared, mounted specimens provide tremendous opportunities for people to see and learn about wildlife up close (photo: John Bates).

Pay Scale

Typical starting salaries for curators range from $30,000 to $70,000 per year. Pay scales vary widely, depending on the size of the museum and the scope of the position. At major research universities, curatorial posts are often set up jointly with a professorship, and the pay is similar to that for other faculty. Top private museums (e.g., Field Museum, American Museum of Natural History) strive to keep their salary levels comparable with these institutions, but they may lag behind. Other curatorial positions are often lower paying, particularly when collections are not seen as being as important as they should be by administrators and agencies. Promotions are possible, but these happen only a few times during a curatorial career, unless you go into museum administration.

UNIVERSITY
Scott E. Henke

Agency: Education, University

The goal of a wildlife professor is to educate the next generation of wildlife biologists in ecological theory, species management, and the North American Model of Wildlife Conservation, as well as to prepare students to be marketable for employment as wildlife biologists. There are more than 400 universities in the United States and Canada that have wildlife programs at the bachelor's degree level, with an additional 200 higher educational institutions providing two- to three-year programs (typically an associate's degree) that relate to wild-

life. Many of these universities also offer MS and PhD degrees in wildlife science and have an officially chartered student chapter of The Wildlife Society (appendix B).

University Professor
Job Description

On first glance, a university professor may seem to be just a teacher of college-age students. It would be wrong to assume that the only duties of a professor are to give lectures to students, assign and grade homework, grade exams, and sit in an office thinking of things that matter to few people. Some folks believe that if professors are not teaching in the classroom, then they are not working. This is actually far from the truth, as they have many other duties that they must fulfill. In addition to teaching, they conduct research, advise students about courses and career choices, mentor students within their profession, assist the university with administrative duties through participation in various committees, conduct pro bono community service, and are active members within professional societies. The amount of time professors spend on each activity depends on their specific positions. The types of appointments do vary and often are allocated into 25% allotments. For example, a faculty member with a 50:50 appointment is expected to teach a half-time load and conduct research for the other half. Another professor may have a 25% teaching, 25% research, and 50% extension appointment, teaching a quarter-time load, conducting research 25% of the time, and assisting landowners with natural resource issues 50% of the time. Most universities require all professors to spend at least some percentage of their time in each of the listed activities. The difference is the amount of time they are expected to devote to the various components of the job. A typical workweek for university professors consists of about 55–60 hours. Rarely is it a 0800–1700 job. Often you work in the evenings and on weekends, either at home, in the office, or out in the field with animals or habitat.

Teaching loads vary among universities, with full-time loads ranging from one to four classes per semester. Some professors teach the same class (e.g., an introductory natural resource class) at multiple time periods (e.g., at 1000, 1300, and 1500) during the same semester, to handle a large volume of students. Each period counts as one class toward their teaching load. Other faculty members teach several different topics during a semester. Although professors are hired for their specialized training in a certain area of wildlife science (e.g., predator ecology, habitat restoration), they may be asked to offer classes on a variety of wildlife-related topics. For example, I was asked to teach a course in the management of North American big game during my very first semester as a new faculty member. I had not previously given lectures on that topic, nor had I ever conducted research on it. The largest animals I had experience with at that time were coyotes. I read books and papers at night to learn enough about big game management to teach these undergraduates. I tried to stay about two weeks ahead of what I was telling the students so

I could feel prepared for class. That situation reinforced my conviction that we always need to learn more.

Class sizes can vary, depending on how large the university is and the physical constraints of the room. Whether 10 or 500 students take a particular course, it typically counts as one class toward a professor's teaching load. Most universities also mandate a minimum class size. Where I teach now, the minimum for an undergraduate-level course is 10 students. Therefore, the course will be cancelled if less than 10 students enroll in it. Maximum class size typically depends on the available facilities. For example, when I was an undergraduate student, my introductory biology class was held in a concert hall. I was one of 500 students taught by a single professor!

Teaching styles need to vary, dependent on class size. I find that most students retain information best if new knowledge is coupled with a hands-on activity. This works if class sizes are limited, but it is difficult to do so with 50 students or more. In large classes, it takes a lot of time to have every student complete a particular activity, and they become bored while waiting for their turn. Therefore, large classes typically are taught in the more-traditional lecture style.

Professors are expected to maintain office hours, which are set days and times when they will be available in their offices for students to stop by and ask questions. These typically are centered around classwork material, upcoming courses to take, or career options, but, sometimes, are about life in general. Basically, be prepared for anything. Although it can be unrealistic, students often expect you to drop whatever you're doing and help them with their problem or need. This can be frustrating, especially when you have other priorities and deadlines, but be ready for the situation to happen. It eventually will.

Research can occupy much of a professor's time, because it includes developing a question to investigate; designing an experiment to answer the question; writing and submitting grant proposals to funding sources to pay for student stipends, travel, equipment, supplies, and any other associated expenses; conducting the experiment; analyzing the results; writing papers about the research findings, for publication in scientific journals; and mentoring the students working with you throughout the entire process. A typical professor has numerous undergraduate and graduate students involved in several research projects at once.

This research component is often considered the fun part of a wildlife professor's job. It gets you out of the office and into the field with your students to collect data on animals and their habitat. This is often hands-on work, such as capturing animals, obtaining blood or tissue samples, placing telemetry transmitters on the animals, and recording GPS locations. While these activities may seem exciting, remember that an important aspect of research entails writing scientific papers about what you have discovered. A mentor of mine is noted for having said, "Research without publication is recreation." Being out in the field and working directly with animals is very pleasurable, but unless you take the time to write about what was done and get it published, then the entire experience was nothing more than that—just a lot of fun.

Research is an essential component of the job. Most universities require professors to obtain a specific level of grant money each year to conduct research and expect them to have a certain number of publications per year in scientific journals. As a new professor, you should ask your department head or dean what the expectations are concerning research, so you can fulfill them adequately. Another aspect of research involves presenting your results at scientific conferences. This is expected and required by most universities, but it is up to individual professors to cover their conference expenses: the registration fee, travel, and hotel and food costs. Universities rarely pay for their faculty members' participation in conferences, but funds within research grants are typically allocated for such purposes.

Universities also ask professors to assist with administrative duties and governance. Be prepared to work on a variety of committees to help your department, college, and university. Some committee assignments may only require a little of your time, and the task can be accomplished within one or two meetings (e.g., selecting a commencement speaker). Some committees, however, such as a university's Institutional Animal Care and Use Committee may require weekly meetings, lasting for several hours each. Such intensive assignments can take up much of your workweek, so your evenings and weekends need to have time devoted to catching up on other job duties. Do not be afraid to tell your immediate superior that you are serving on too many committees, however, when they drain inordinate amounts of your time. As chair of my department, I have attempted to slowly introduce new faculty to committee assignments. I want to give them a chance to develop their class presentations and research programs first, before they get inundated with committee work.

Sometimes professors are asked to be the academic advisor for a student organization. This can be as simple as offering advice to students on the best ways to organize and conduct business and then attending their meetings. In other instances, advisors can become highly involved and assist students with the running of the organization. It really depends on a professor's attitude, enthusiasm, and willingness to serve. Most wildlife departments have a student chapter of The Wildlife Society, but larger universities may also have student chapters of Quail Forever, Ducks Unlimited, and other conservation groups.

University professors often assist landowners with wildlife-related issues and problems on their property. Professors conduct workshops and seminars to teach landowners current theories and methodologies concerning wildlife and habitat, and they often receive phone calls and emails with questions from landowners. In addition, wildlife professors are members of professional societies (e.g., The Wildlife Society). Involvement in these groups includes giving presentations at scientific meetings, editing other members' papers, and helping to advance that society and the profession as a whole.

All in all, professors have significant demands on their time, but the job also can be very rewarding. Faculty positions offer many freedoms that are not associated with other wildlife jobs, such as an ability to choose your own research program and, sometimes, a chance to set your own working hours. For example, I was hired initially to conduct predator research, but I have had the freedom to delve into other areas that intrigued me, such as invasive species and wildlife disease–oriented research. Thus I have been able to work with a variety of species, from horned lizards to alligators, and from mice to exotic ungulates. I have found that conducting research on a number of different topics has kept my interest at a high level, rather than my becoming bored by doing the same type of work every day. Also, as long as I complete my job duties, I have had the freedom to set my own work schedule. I do not have to punch a time clock or be in my office from 0800 to 1700, Monday through Friday. This can be especially important if you have a family. I can attend my kids' school events, whether they take place during the day or at night. If needed, I can take a day off in the middle of the week. Don't get me wrong. I am expected to work at least 40 hours each week, but I have the freedom to choose when (within reason) my work hours will be. Many jobs do not offer such flexibility.

Background Needed

Besides the educational requirements, you need to show previous teaching experience. It is best to accomplish this in undergraduate-level courses while you are obtaining your PhD degree. Some universities require PhD candidates to teach a class, or at least to assist in conducting the laboratory portion of a course. The more instructional experience you can obtain, however, the more marketable you will be for a career in academia. I consider myself lucky. I had a teaching assistant position when I was a PhD candidate. I was the lecturer for two courses: introductory wildlife management and natural resource conservation. When I finished my degree, I had taught these classes six times each. Although it required much of my time when I was a PhD student, I believe the teaching experience I gained was why I was hired as an assistant professor right after graduation. My time as a lecturer while I was a student was a good investment in my career. Excellent written and oral communication skills are also needed. You should present evidence of having written and published papers in scientific journals, as well as articles in popular magazines and similar outlets. Oral skills can be demonstrated by having given presentations at scientific meetings and to lay groups and organizations (e.g., Rotary Club, elementary and high school groups). Previous grant-writing experience is a plus, especially when your grant proposals have been funded. When you are working toward your MS and PhD degrees, seek out grant opportunities to help support your research projects and write those proposals, no matter how small the funding source is. The experience is invaluable, and it pays off in enhanced employment opportunities. Lastly, community involvement, especially in workshops and seminars for the general public, can be an important way in which to make yourself marketable for a job as a university professor.

Education Required

A PhD degree in wildlife ecology or a closely related field from an accredited college or university is required. This must include classes in the sciences (biology, chemistry, physics), mathematics (algebra, trigonometry, calculus), communications (English composition, technical writing, speech), and specialized courses within the wildlife field (wildlife management, field techniques, GIS). Most university accreditation organizations require an institution's professors to have taken a minimum number of graduate-level classes within a specific discipline to be able to teach courses in that discipline. For example, at my present university, a professor must have a minimum of 20 credit hours of graduate-level coursework in wildlife ecology to teach wildlife-related classes.

Pay Scale

There are typically three levels of professorial positions: assistant professor, associate professor, and full professor. New entrants into academia would generally be hired at the assistant professor level, with a salary of about $55,000–$65,000 for a nine-month appointment. Six to seven years of satisfactory performance are required to be eligible for promotion to the associate professor level, and another five to seven years of acceptable work before being promoted to a full professor level. Pay raises occur at each promotion, but the amounts vary greatly among universities. Professors receive a salary, rather than an hourly wage, so there is no overtime pay. Many positions are nine-month appointments, where salaries cover the academic year but not the summer months (even though the research component of the job occurs throughout all 12 months). Professors can supplement their nine-month income by teaching during the university's summer sessions or by including an amount equivalent to a portion (usually three months' worth) of their salary in a research grant proposal, and thus be paid (assuming the grant is funded) for 12 months. Keep in mind that with a nine-month appointment, you would need to continually write grants to ensure sufficient funds for those otherwise unpaid months each year or teach summer school to receive financial support during the summer months. By doing so, however, you can increase your annual salary by about $20,000.

ZOOS
Cindy Pinger, Clayton D. Hilton, and Heather Richardson

Agency: Education, Zoos and Aquariums

The Association of Zoos and Aquariums (AZA) is a nonprofit organization dedicated to the advancement of zoos and aquariums in the areas of conservation, education, science, and recreation (appendix B). As of September 2016, there were 232 zoos and aquariums that are accredited by the AZA. (For an updated and complete list of accredited zoos and aquariums, see appendix B). Zoos can be small, housing a few animals each from several popular species and covering a few hectares of land, such as the Cosley Zoo in Illinois. Or they can be extensive, encompassing several square kilometers and maintaining thousands of animals that represent hundreds of species, such as the San Diego Zoo Safari Park in California. Zoos may be owned privately; owned and managed by municipal, county, or state governments; or owned and run by a zoological society or private nonprofit corporation (e.g., a public/private partnership). The functions of zoos are to provide education, outreach, and entertainment for the public and to conserve animals and their habitats through propagation programs and in-situ and ex-situ conservation projects. Approximately 200 million people visit zoos annually.

Depending on their level of education (bachelor's degree, MS, or PhD) and range of experience, there are many ways for wildlife biologists to engage in conservation biology while employed by a zoo or aquarium. Zoo positions include zookeepers, curators, conservation biologists, and directors of conservation programs. Zoos classically are arranged taxonomically, zoogeographically (according to the past and present geographic distribution of species), or both. Zookeepers typically are the people in charge of a taxonomic group of species within a zoogeographic region of the zoo. Curators are in charge of the zookeepers for that region. For example, a zookeeper would oversee the health and maintenance of birds within an Australia exhibit, while the curator would be responsible for both all the animals (mammals, birds, reptiles, amphibians, and invertebrates) and the zookeepers within the various Australia exhibits. Conservation biologists design and conduct research for a zoo, while directors oversee a zoo's research programs or other major projects. Most zoo positions are posted under the jobs tab within the AZA website (appendix B). In addition, employment opportunities are posted on the website of the hiring zoo and in job-specific journals, such as *Animal Reproduction Science* or *Conservation Biology*, as well as on the respective websites of these journals.

Zookeeper
Job Description

Zookeepers provide daily care to many animals that are threatened or endangered in countries in their native range. This goes a long way toward conservation of these species, especially if they may be repatriated to their range, as has been the case for black-footed ferrets, red wolves, Karner blue butterflies, Wyoming toads, California condors, golden lion tamarins, and Arabian oryx. Zookeepers at an AZA-accredited zoo can be members of a taxon advisory group, a Species Survival Plan®, or a scientific advisory group. They also may serve on conservation-based AZA committees or be studbook keepers. Many zoos provide intramural and extramural grants (appendix B) that sponsor in-situ and ex-situ conservation research programs. Grantees are given paid time off from their

regular duties to conduct their own research or to be part of a larger, multi-institutional research initiative, such as the Marianas Avifauna Conservation Project or Elephants for Africa (appendix B).

Zookeepers start their workday by conducting a live check of all animals in their care. This entails stopping by all the exhibits and holding areas to ensure that the animals are healthy and that the exhibits have not sustained any damage overnight. Animal abnormalities are reported to the veterinary staff and the area supervisor. Problems with the exhibits are conveyed to the area supervisor and the zoo's facilities staff. For exhibits that housed animals overnight, the animals are shifted into their holding areas, the exhibits are cleaned, and enrichment items are placed in the exhibit. Animals often receive breakfast, usually prepared by the zoo nutritionist and kitchen staff, while in their holding areas. After eating breakfast, the animals are returned to their exhibits. All of this is accomplished before the first guest enters the zoo for the day. Next, the holding areas are cleaned and made ready for the next time the animals are placed there. Throughout the day, zookeepers interact with the public, monitor the animals in their care, attend area-wide or zoo-wide meetings, make notations in the records for each animal, and engage in training the animals for medical husbandry behaviors. Zookeepers participate in veterinary medical procedures and refurbish exhibits, as needed. They also may take part in training and drills to prepare for animal escapes or human intrusions into exhibits.

Background Needed

Past experience with animals is always useful, from owning pets to working in veterinary clinics, pet stores, grooming facilities, farms, and vivariums (enclosures simulating a specific habitat, such as an aquarium). In addition, animal handling experience, especially the live trapping and and tranquilizing of wildlife species, are beneficial when pursuing a position as

Zoo biologists check the health status of each animal daily (photo: Scott Henke)

a zookeeper. Demonstrated experience operating a jabstick, blowgun, air-pump tranquilizer gun, and CO_2 tranquilizer gun is helpful. Because zookeepers often assist in the maintenance and repair of facilities, previous mechanical, electrical, carpentry, and plumbing experience is valuable, as is a knowledge of horticultural practices.

Education Required

A bachelor's degree in wildlife biology / ecology / management or in animal science is required at most zoos. Additionally, courses in animal behavior, nutrition, husbandry, horticulture, and genetics are beneficial.

Pay Scale

An entry-level zookeeper with minimal prior experience can expect an annual salary of approximately $25,000, with increases as you advance in your career. A zookeeper position includes a typical benefits package of health and dental insurance and an allotment of paid sick leave and vacation time. Some zoos have a retirement benefit plan.

Zoo Curator

Job Description

Zoo curators need to be able to fulfill many duties. The most important is to develop collection plans, in order to decide which animals best fit the needs of the zoo. Factors to consider are the region's climate, the resources of the zoo, and the level of staff expertise with certain animals. Examples of staff expertise are successful reproduction for a species and years of demonstrated success in exhibiting species. Another important duty is to supervise the day-to-day activities in the animal areas. Curators generally recruit and hire animal handling staff members, deal with personnel issues, and create work schedules and routines. They collaborate with the veterinary staff to identify health problems, develop diets specifically tailored to an animal's needs, and keep the animals as healthy as possible. Curators also supervise the exhibit buildings and the maintenance of the animal areas. For example, if you are managing an aquarium or other type of water exhibit, it is essential to have a good working knowledge of aquatic life-support systems and water quality. About half of the curators' time is spent handling issues related to employees, and the other half in dealing with animal issues. The latter may involve collection planning, exhibit planning, and the development of or participation in conservation efforts. The people issues may center around constructing work schedules, setting goals with staff, and training keepers. Other duties are developing budgets and ensuring that animal records and required permits are current and accurate. Zoos may need to have specific documents, such as a US Department of Agriculture Class C Exhibitor Permit, a US Fish and Wildlife Service Captive-Bred Wildlife Permit, and a US Fish and Wildlife Service Eagle Exhibition Permit. Arrangements for transporting animals from zoo to zoo, whether across town or across the world, are part of a curator's responsibility. In most cases, these transports are to keep captive populations as genetically healthy as possible. In short, curators deal with the management and administration of animal areas. In general, 50% of the job is talking to animal keepers and being out in the collections, and 50% is working behind a desk, although this time allocation depends on the needs and philosophies of each zoo. When applying for a job as a curator, first make sure that the zoo's focus matches your own. Some zoos expect a curator to actively be performing husbandry duties on a daily basis, and other zoos have strict guidelines against it. This is usually controlled by whether a zoo is a member of a labor union. If a zoo's employees are union members, then curators may be restricted in the amount of hands-on animal duties they can perform. Smaller zoos and nonunion zoos usually have fewer constraints on job duties. As a curator, most likely you will be required to serve on AZA committees and be active in Species Survival Plans® and animal programs. These demand some computer skills (e.g., Word, Excel, payroll software system) and an ability to effectively communicate with curators at other zoos. A typical day for zoo curators includes arriving at work and checking emails; reading daily reports; going to each area under your supervision and assessing staffing levels; checking animals for any health or behavioral issues; seeing if there are any pressing maintenance issues, such as a downed tree or heating and cooling problems; touching base with the veterinary staff and communicating with them about any animal health issues; and then returning to your desk and handling paperwork, including animal care manuals, breeding plans or budgets, animal transaction forms, purchase orders, work orders, and payroll. During the day, there are staff meetings with other curators and meetings with other staff members about new exhibits, new animals, or how to deal with animal issues. In the afternoon, curators check on the animal areas again, to make sure there are no issues at the end of the day.

Curators are expected to develop conservation plans and goals for the areas they supervise. These may involve conservation at local, national, and international levels. For example, the conservation goals for Birmingham Zoo's bird department for 2014 were to send two animal-care staff members to participate in the Great Lakes Piping Plover Recovery Effort and another two staffers to take part in assisting with the rearing of whooping cranes for Operation Migration (appendix B).

Background Needed

Education is only part of what it takes to be successful in this job. Experience as a zookeeper goes a long way in helping you develop into a good curator. In fact, it would be rare to obtain a zoo curator position without first being a zookeeper. Most zoos have volunteer programs, and many zoos have intern programs that provide useful experiences in the animal care field. Human relations skills also are valuable, because curators are, in large part, managers of humans. On a daily basis, curators must interact with staff, help solve problems, and be in contact with zoo visitors. Be sure to take a course in personnel management. This advice is often shocking to new curators, because most likely they did not get into this field to manage humans. If you do not have the ability to be a good people manager, it will be impossible to succeed as a curator. Writing and public speaking skills also are essential. Most curators need to be able to give presentations at professional conferences and to local groups. A typical curator has such speaking engagements at least once a month. An ability to write protocols and guidelines to help employees do their jobs is essential. To be effective as a curator, you must be a fearless problem solver. In this field, there often is no clear path to the right decision. You are tested on a daily basis to think outside of the box, inside the box, and in places where there are no boxes. Keep in mind that zoo curators are expected to work on weekends, holidays, and often at night to make sure the animal areas are covered. In addition, they often work outside, in the elements.

Education Required

A minimum of a bachelor's degree is required, but a graduate degree is typically preferred. Biology and environmental science are the most pertinent fields of study. Other degrees may be acceptable if candidates have the right combination of experience and education.

Pay Scale

This depends on the zoo and on your level of education and experience, but expect a salary range of $30,000–$60,000 per year for zoo curators. Curator positions, like those for zookeepers, include a typical benefits package of health and dental insurance and an allotment of paid sick leave and vacation time. Some zoos have a retirement benefit plan.

Zoo Educator
Job Description

Zoo educators are responsible for many duties each day. Although there may be similarities, no two days are usually the same. Zoo educators develop, implement, and evaluate age-appropriate curricula and lesson designs for all aspects of the educational programming at a zoo. They plan and implement program instructions that adhere to the zoo's mission, philosophy, goals, and objectives and write curricula for many educational programs. Excellent communication skills and attention to detail are required to appropriately accomplish these tasks and ensure that all lessons, activities, and programs are the best they can be. Quality outcomes are the result of using many different instructional methods (e.g., lectures, hands-on activities, role playing) and teaching in many different settings, which are sometimes unconventional. Many days may be spent in a classroom at a school, a business, or the zoo. Educational activities, however, also may require you to teach in other places within the zoo, in a retirement home, or even in a park. The ages of your audiences will vary, so it is important to be open to interacting with a variety of age groups. For example, zoo educators may present a program at a local elementary school during the morning, conduct a

demonstration within the zoo to high school students in the afternoon, and participate in a fundraising event for zoo supporters that evening. Zoo educators are responsible for the safety and care of the students in their charge, as well as for any animals used for demonstration purposes. Therefore, they must be familiar with all zoo safety protocols and procedures. Many zoos require their educators to handle animals during their presentations. Being comfortable around numerous animal species is a must, and training is usually provided by the zoo staff. Knowledge of animals is an important part of being a zoo educator. Knowledge about conservation and environmental education, however, is just as important when planning an effective lesson. Workshops, trainings, and conferences are available specifically for people who work directly or in conjunction with animals. Most zoos take advantage of such continuing educational opportunities and have dedicated a portion of their budgets to do so. Customer service skills are used daily when working with zoo guests. An unpleasant guest experience can redirect people's focus from the educational program and make the time spent unproductive for everyone involved. Because volunteers play a vital role at most zoos, zoo educators should be able to lead, plan, and prescribe purposeful assignments for volunteers. Many events take place at zoos (e.g., community outreach events, fundraising galas), and educators are often asked to take part. Work schedules are altered to accommodate these events, so flexibility with your work time should be expected.

Background Needed

A teaching background in science education, especially environmental education, is becoming the norm at most zoos, aquariums, and other educational sites when hiring a zoo educator. Experience in conducting informal educational programs typically is not required, but this certainly helps you stand out in stack of resumes from applicants with formal educational backgrounds only. A portfolio of previous educational presentations, including photographs, is beneficial. Animal handling experience (at a farm, wildlife rehabilitation center, veterinary clinic, another zoo) is a plus. Being a zoo educator requires a strong ability to work in a team-oriented atmosphere, so demonstrated experience in such a setting is needed.

Education Required

A zoo educator should be a certified teacher, or at least be able to obtain teaching certification. A bachelor's degree in education, conservation of natural resources, wildlife science, or animal science, in addition to teaching certification, is beneficial. Coursework in mammalogy, ornithology, herpetology, ichthyology, biogeography, ecology, and wildlife conservation can be useful in providing a solid knowledge base.

Pay Scale

Zoo educators can expect to earn $25,000–$45,000 per year, depending on experience and the zoo's location, and receive a typical benefits package of health insurances and paid sick leave and vacation time.

Additional Information

Similar positions can exist within state and local governments and with private organizations that own property. Positions may be called *naturalist, conservation educator,* or *environmental educator,* but the key aspects of the job are basically the same—develop and provide educational opportunities to various age groups within the general public. Employment opportunities often are available in parks departments and nature areas. Examples of educational programs typically include presentations about the local flora and fauna, the natural history of the area, ecosystems, weather, and the stars in the night sky. Work hours and days typically vary, to coincide with times when it is convenient for the general public to attend presentations. Persons in such positions should be physically fit, as the job often entails hiking. Additional background needed, educational requirements, and pay scales are similar as those described for a zoo educator.

Zoo Conservation Biologist
Job Description

Wildlife biologists with advanced degrees may become conservation biologists or directors of conservation programs. These are positions that do not directly involve care of the animals in the zoo's collection. Rather, they support the conservation mission of the zoo and the AZA's strategic plan (appendix B). The AZA very strongly encourages accredited zoos to spend 3% of their annual budget on conservation projects, which can be used to maintain a zoo conservation biologist on staff or to fund conservation projects designed by them. Zoo conservation biologists fashion their own research programs or incorporate their research into a larger project, such as the Cheetah Conservation Fund, the Blue Iguana Recovery Program, or the Marine Mammal Center (appendix B). Such projects often have been initiated at a different institution or are institutions unto themselves. Zoo conservation biologists tend to work on one project at a time, which may take place solely in the field, only in the laboratory, or both. They may or may not write their own grant proposals to receive funding for their projects.

A typical workday for zoo conservation biologists can start anywhere in the world, at any time of the day or night. If they are in the field, there may be mist nets or camera traps to check, observational or biological data to be collected or processed, reports to write, and gear to be maintained. Field staff may need to be organized and their work scheduled. Videoconferencing often takes place. If zoo conservation biologists are in the laboratory, biological samples may require additional processing or analyses, data may need to be compiled, reports and manuscripts have be written and submitted, expenses should be tracked, and communication with the director of conservation programs must take place. It is expected that zoo conservation biologists present their findings at scientific meetings and in scientific journals.

Background Needed

In addition to the necessary education, anyone involved in conservation biology at a zoo needs to have excellent interpersonal skills; an ability to talk with people from all walks of life; a passion for their job; and abundant patience, since animals never do what you want them to do every time, and fieldwork rarely goes as planned. Previous research experience is very beneficial when seeking a position in conservation biology. One of the best ways to do so is to get involved in research projects while earning an undergraduate degree. There often are paid and volunteer positions as part of research projects, and these provide many opportunities to learn field techniques, animal handling techniques, laboratory tests, and data input. Coursework in statistical analysis, as well as evidence of having prepared written reports and journal articles and given presentations at scientific meetings, provide a tremendous benefit for applicants. Many zoos have wildlife rehabilitation programs. If you cannot join a research project, you can still participate in these programs to learn valuable animal handling and animal behavior skills that increase your chances of securing a job at a zoo.

Education Required

At a minimum, zoo employees who engage in conservation biology need a bachelor's degree in wildlife biology, wildlife management, or range and wildlife science. Extra coursework in animal nutrition, (reproductive) physiology, animal behavior, or a related field increases your chances of getting a job and being successful at it. An MS degree in the same or a related field adds to your marketability, and a PhD makes you very attractive as a candidate.

Pay Scale

Zoo conservation biologists earn $35,000–$60,000 per year. A benefits package includes health and dental insurance and an allotment of paid sick leave and vacation time. Some zoos have a retirement benefit plan.

Director of Conservation Programs
Job Description

Directors of conservation programs oversee all conservation and research projects within the institution, although they may or may not be directly involved in research activities. Their primary responsibilities are to ensure compliance with the protocols of their Institutional Animal Care and Use Committee; coordinate the schedules of multiple research projects and personnel; write grant proposals; manage budgets; see that reports are submitted to funding agencies on a timely basis; confirm that information gained from the projects is disseminated through the popular press, such as in zoo newsletters, articles published in scientific journals, and by presentations at conferences; coordinate efforts with state and federal agencies; and secure scientific collection permits or other permits that are required by individual projects. Directors of conservation programs often serve on various committees and advisory groups, such as a taxon advisory group, a Species Survival Plan®, a scientific advisory group, and conservation-based AZA committees.

For directors of conservation programs, their workday often starts by checking emails from the zoo's conservation biologists and from laboratory staff that report to the director. Managerial duties, including managing budgets, writing and submitting reports, attending in-house meetings, and video-conferencing tend to occupy much of the remainder of the workday. Regular participation in donor events and scientific meetings is expected of directors of conservation programs, as is the continued publication of articles in scientific journals.

Background Needed

Directors of conservation programs need to have excellent organizational and time management skills and be proficient with word processing programs and spreadsheets. Since project funding can come from zoo members or zoo donors, they have to have the interpersonal skills to convey their passion about a project in a manner that entices a donor into providing financial or logistical support for projects.

Education Required

An MS degree in wildlife biology / ecology / management is typically required. A PhD degree in a conservation-related field is often required.

Pay Scale

Directors of conservation programs can expect to earn $70,000–$100,000 per year, depending on experience. A benefits package includes health and dental insurance and an allotment of paid sick leave and vacation time. Some zoos have a retirement benefit plan.

NONPROFIT ORGANIZATIONS

WILDLIFE/CONSERVATION ADVOCATE
Greg Schildwachter

Agency: Nongovernmental Organization or Independent

A nongovernmental organization (NGO) is typically understood to be a private charity that, like government, provides a public service. In wildlife conservation, many NGOs focus on a species or community, a habitat type, or a certain kind of conservation work. The Boone and Crockett Club promotes "the conservation and management of wildlife, especially big game, and its habitat," which is a broad mission. Ducks Unlimited "conserves, restores, and manages wetlands and associated habitats for North America's waterfowl." The Nature Conservancy began by pioneering a real estate transaction called a *conservation easement*, which secures conservation uses on private lands. The Nature Conservancy now also holds and manages land for conservation.

Most NGOs are exempt from federal and state taxes because of their charitable public function, which is defined by federal law at 26 US Code 501(c)(3). They are often referred to as *501(c)(3) organizations* or *nonprofits*, because "no part of the net earnings [may benefit] any private shareholder or individual."

There are two closely related types of organizations that employ advocates to do similar work. Operating foundations use endowments to run conservation programs, and they also issue grants. Associations are organizations whose members are other organizations. These can be a 501(c)(3) charity or another type of organization that is defined in paragraph (c) of 26 US Code 501, such as a trade or political association listed under 501(c)(4), (5), or (6). Though not always strictly an NGO, some associations are a close equivalent, such as the Association of Fish and Wildlife Agencies, which includes all state wildlife agencies.

Advocate/Lobbyist
Job Description

Titles for entry-level advocates include associates, research assistants, or program managers. Midlevel job titles are often directors or vice presidents. Senior positions can be executive vice presidents or chief conservation officers. Certain titles specify the type of advocacy (e.g., director of grassroots organizing). Some jobs involve *lobbying*, which is a particular type of advocacy that promotes support for or opposition to a policy decision by contacting an official decision maker, such as a member of Congress. Lobbying is defined and regulated by federal and state laws, including the federal Lobby Disclosure Act of 1995. Both lobbying and other advocacy positions also involve many additional responsibilities.

Advocates must understand and act on issues, in a particular sense of that word. *Issues* are the real-life situations that advocates hope to change, in order to advance a mission. Such situations can present problems or opportunities. For example, a common problem is a declining wildlife population. An opportunity could be a slump in corn prices that prompts farmers to let some of their land lie fallow, thus helping advocates recruit more farmers to enroll in the Conservation Reserve Program under the Farm Bill.

Every issue has a life cycle that advocates must follow. In fact, they and their employers talk about their work in exactly that way, asking, "Who's following this CRP thing?" or "Can you track down this problem with species X?" Issues begin with a moment of recognition, followed by devising solutions, proposing the adoption of a particular solution, and debating and deciding whether to move forward on that proposal. Then the cycle repeats.

Throughout that issue's cycle, as an advocate you must track its real-life details, policies, coverage in the media, debate by decision makers, as well as its costs and what its relevance is to you and your employer. You may be responsible for only one aspect of one issue, but usually you deal with all aspects to some degree, as they are interdependent. In an entry-level job, maintaining awareness and reporting is often all you are asked to do. Higher-level positions involve influence: deepening the understanding of an issue, expanding possible solutions or initiatives, changing others' opinions, and getting a favored solution adopted.

Advocates work with documents and relationships. You must know people close to an issue to get real-life details, and you need to test their views against your research. An elk hunter may complain that the restoration of gray wolves has decimated an elk herd, but that herd may have simply moved beyond the hunter's area. A farmer may talk about crop-planting rates and prices, but the US Department of Agriculture publishes official statistics on these factors. If laws or regulations already govern your issue, you can study these online in the US Code or the Code of Federal Regulations, but sometimes the most relevant policy is in a 10-year-old agency memo that can be tracked down only by knowing the organization's staff attorney, who has worked there for 30 years.

Public debate, as reported by the media, is available to all, but you may be conducting surveys and measuring the success of your communications (e.g., how many people click on your social media posts, attend events, make personal contact with decision makers). *Policy debate* is different. Much of it occurs in private meetings among staff and decision makers. Many advocates try to extract information on these debates by Freedom of Information Act requests for documents, but the best advocates have the trusting personal relationships it takes to get answers from phone calls, lunches, or official meetings.

Money is often the most regular and powerful factor regarding an issue. Budgets are set annually and adjusted during the year, but the life cycle of an issue can run for decades. Budget makers are the most active policy makers. The Endangered Species Act has not been amended since 1988 and no longer needs official authorization to be implemented, but every year the congressional appropriations committees still write a budget to carry it out; that's the policy. Congress and the relevant agencies fund both listing decisions and recovery programs for endangered species, but decisions are cheaper than recovery. Therefore, like it or not, Endangered Species Act policy is to list species but not necessarily to recover them.

You and your employer decide which issues are the most important ones to work on. Part of your job is to keep abreast of facts and use sound reasoning to justify priorities, weighing each issue for its relevance to your organization's mission, the cost of the effort, and the likelihood of its success. Advocates often talk about juggling issues, because of the constant flux among their organization's intentions for its highest priorities, urgencies on lower priorities, and new issues needing evaluation. This is called *agenda setting*. It budgets your time, but formulating that agenda also takes time. Because time is the coin of the realm in policy work, there is a market of sorts for it, which forces some decisions on how you spend what's available to you.

For example, if your issue is high enough on the national agenda to be debated on the floor of the US House of Representatives or Senate, professionals call this *floor time*. You have to devote long days and nights to being on the phone and in the hallways of congressional offices, as well as contacting members of your organization and the general public to make your case for what you want the decision on that issue to be. You had better have completed your research at this point and be able to call up pieces of it from your files as needed, or the decision may be made before you can catch up.

Far more often, you in your office, spending about half your day in meetings, on the phone, or emailing, or in researching and writing up your top one or two issues. The rest of the day fills up with opportunistic or urgent tasks on other issues: a new study is published or an event occurs that must be investigated, or an adversary makes a move that could destroy your efforts, or someone may bring you a new pressing issue.

The situation is not complete chaos, however. Agendas are manageable, and you and your organization each need have one. The organization's complete list is its *formal agenda*. Your list is your *portfolio* or *issue set*. You and your colleagues should focus on the priorities and reserve (or be ready to sacrifice) some time to keep your files on each issue ready to use at a moment's notice.

Time needed for travelling depends on how tied your job is to decision makers or to the public. For decision makers, you stay close to a state's capital or to Washington, DC. For the public, you may travel constantly among communities to give presentations to clubs and local organizations, or attend public meetings or rallies. Mid- and senior-level advocates travel more frequently, in order to build relationships with allies, board members, funders, and rank-and-file members.

There is an excruciating reality in policy work: issues abound, but time does not. Many worthy issues go nowhere for lack of time. This can be painful for you and impossible to explain to people affected by the issues. You eventually develop a professional detachment, but you must guard against apathy. If you lose your passion for wildlife and compassion for people, your confident focus on a vital few issues may blind you to the right time to shuffle priorities.

Background Needed

Entry-level advocates are judged more on their enthusiasm and their speaking, writing, and reading skills than on their backgrounds. Among equally strong communicators, the edge goes to a background in real-life issues, even if they are not exactly aligned with the job's issue set. If you grew up tending livestock, you can seek a job focusing on wildlife concerns. If you grew up in a city, you can get a job dealing with farm policy. Being yourself is extremely important, because genuineness is key to your main task of building a network of relationships that help you identify, understand, and act on issues. If your background includes a lot of previously developed relationships, such as through volunteering and serving in clubs, you have a head start. Advocacy is a people-oriented job, so solid interpersonal skills are mandatory. You must be equally comfortable in speaking to laborers as you are to government dignitaries.

Table 5.2. Pay scale levels for advocates in the United States, based on 2016 salary ranges

	Most places	High-value location
Entry-level	$40,000–$70,000	$50,000–$80,000
Midlevel	$80,000–$120,000	$100,000–$170,000
Senior-level	$100,000–$150,000	≥$200,000

Education Required

Entry-level positions usually require a college degree, but the subject area of that degree is less important. Many political science, economics, pre-law, and history majors go into advocacy jobs right out of college. In competing among these candidates, you can gain an advantage by taking one or more policy courses, especially policy analysis. Policy courses typically review major laws in state and federal land, wildlife, and water conservation. Here you also acquire the basics of formal government: the organization and operations of the agencies, Congress, state legislatures, and the courts. In policy analysis, you learn to understand issues, using concepts from law, economics, math, political science, and sociology. On wildlife issues, a student who has expertise in this topic area has an advantage over someone who majored in one of these other subjects (or even double-majored in two of them). A wildlife degree is also helpful because, by knowing the subject matter, a wildlife student can be the most insightful analyst in devising solutions that transform the policy debate.

Pay Scale

Advocacy jobs often are located in high-value markets, such as state capitals. Jobs in these locales tend to have higher salaries, to cover the greater costs of living there, than pay levels in smaller cities and towns, where you may work for a local organization. Within each scale, the salary range depends on the size of the organization and negotiability (table 5.2).

ARCHBOLD BIOLOGICAL STATION
Hilary M. Swain

Agency: Nonprofit Organization, Archbold Biological Station

A biological station is a scientific resource for field studies of the natural environment. The Archbold Biological Station, in southcentral Florida, is one of the most renowned in the world. Founded in 1941 as an independent not-for-profit organization, Archbold's mission is to build and share the scientific knowledge needed to protect the life, lands, and waters in the heart of Florida and beyond. Its strategic priorities are to understand and conserve biodiversity and ecosystem processes; to provide solutions to threats by addressing invasive species, guiding land management, and anticipating climate changes; and to share knowledge and solutions by offering learning opportunities to many audiences, fostering conservation stewardship, and engaging with issues on the national and global stage.

Archbold's cutting-edge research has created real conservation solutions across diverse ecosystems, ranging from the globally threatened Florida scrub ecosystem of the Lake Wales Ridge, which is home to threatened and endangered plants and animals found nowhere else on Earth, to the waters, wetlands, and wildlife of the vast open lands of cattle ranches lying within the headwaters of the Everglades. Archbold manages 8,094 ha in this region, including a pristine scrub preserve, a working cattle ranch, and a large-scale restoration site. The station is a wellspring of knowledge in this region, and its scientists stand at the forefront in the fight to build a sustainable future for Florida.

Research Assistant
Job Description

Research assistants are staff members working in the various Archbold research programs under the direction of a PhD scientist, usually a research biologist, program director, or postdoctoral fellow. Research assistantships range from temporary seasonal (three to six months) jobs to longer-term positions. A seasonal research assistant II usually helps with fieldwork and data entry during a breeding or flowering season. Research assistants II–IV are often grant-supported, term-limited, and typically dedicated to one or more research projects for one to two years. Longer-term career positions, typically research assistants IV and V, are in charge of at least one research project, including the design of the project and its related experiments, grant writing, fieldwork, data analyses, lead authorship on publications, and supervision of students and junior staff.

Annually, Archbold supports an average of four seasonal research assistants and eight other research assistants. Positions are advertised with colleagues nationwide or appear online, typically on the Texas A&M University Job Board and on Ecolog-L (appendix B). Advancement depends on educational achievements and increasing skills. Because Archbold is a small nonprofit organization, there is a limited career ladder, but research interns are often promoted to seasonal research assistants, and seasonal assistants to longer-term positions. Some research assistants take jobs that serve as two- to five-year, resume-building stepping stones to a future career in academia, an agency, or another nonprofit organization. Others, however, have established longer-term careers at Archbold.

Research assistants work in Archbold's programs in plant ecology, avian ecology, herpetology, agroecology, restoration ecology, and entomology, or for a project directed by a visiting scientist in another discipline. Responsibilities differ, depending on the taxa or ecological process being studied, as well as on the level of the position (research assistants II–V) to which they are appointed.

A large portion of the research assistants' duties involve

Research assistants help biologists with every aspect of fieldwork in collecting research data (photo: Dustin Angell)

fieldwork, which includes observing, identifying, and conducting a census of plant or animal populations; marking and tracking individual plants or animals; collecting samples or tissues for analyses; taking physical measurements of environmental conditions; and deploying and managing field experiments. Some assistants handle field instrumentation, ranging from using simple GPS devices or downloading sensor data to programming, installing, and operating advanced equipment. Fieldwork typically takes place daily. On average, about 20 hours per week, mostly from early morning until early afternoon, is spent conducting fieldwork. Employees should expect to work variable and unusual schedules, however, such as long hours during peak breeding seasons or overnight stays. Off-site travel sometimes makes for long, arduous days, but everyone usually goes home at night. Fieldwork can be tough physically, requiring extensive walking in deep sand, in wetlands, or through thick vegetation, with repeated kneeling, bending, and lifting. It requires a lot of endurance, especially due to Florida's heat and humidity. Research assistants may help with management and research protocols, ranging from setting prescribed fires, manipulating hydrology, installing nest cavities, being involved with cattle grazing, and applying herbicides.

After fieldwork, most of the research assistants' time is spent in meticulous data management. Archbold does not operate advanced laboratories, but research assistants are engaged in processing samples, including plant material; analyzing soils and nutrients; taking tissue samples and blood work; and readying such samples for shipment to laboratories. Research assistants IV–V may oversee equipment, such as microscopes and −62° C freezers. Some assistants have curatorial responsibilities in on-site, multitaxon specimen collections. Senior research assistants serve as the lead author or a coauthor on scientific papers and give presentations at scientific conferences.

Research assistants III–V oversee fieldwork, schedule teams, organize transportation, and maintain equipment. They may recruit and supervise junior staff, train postbaccalaureate interns, and oversee high school students and volunteers. With experience, research assistants IV–V have responsibilities for conceptually developing new research; writing grant proposals; managing grants, reports, and permits; and participating in collaborative research.

General administrative responsibilities are minimal, although they include staff meetings or occasional donor events. Research assistants may serve in official positions and on committees in academic societies, as well as be engaged in professional activities, including proposal reviews for agencies. Archbold links research to conservation and land management needs, so research assistants become familiar with sites regionally and, to a lesser extent, are engaged in meetings, reviews, and workshops with nonprofit organizations, agencies, and landowners. They occasionally present research findings to diverse audiences, from elementary school to college students, and from the general public to conservation professionals and decision makers. This happens through workshops, interpretive materials, talks, tours, and online venues. Research assistants with skills as talented photographers, writers, and illustrators may be asked to contribute their work for print publications and online.

Background Needed

Archbold has a great reputation, and the competition for research assistant positions is fierce. Assistantships are exciting learning opportunities, with participants engaged in cutting-edge science with globally threatened species and ecosystems, working in a lively scientific atmosphere with top-quality scientists. Applicants for research assistantships should submit a resume and cover letter to the relevant program director. These should demonstrate significant achievements, consisting of a relevant degree in the natural sciences, prior experience, and proven abilities that are directly related to position responsibilities. Applicants who are intellectually bright and curious, able to withstand the demands of fieldwork, display evidence of being in good physical shape to cope with demanding environmental conditions, are self-sufficient outdoorspersons possessing common sense and mechanical know-how, and are competent naturalists receive a request for an interview. Applicants who have the fortitude for never-ending, accurate data management; are reliable; pay meticulous attention to details; and have shown themselves to be skilled science writers, mathematical modelers, or statisticians receive serious consideration to join the Archbold team of research assistants. We desire people who have references that attest to their great attitude, people skills, and inspiring leadership. Never bluff or embellish your resume, or do so during an interview. There is nothing more immediately obvious and frustrating to employers than an applicant who has exaggerated. Recent graduates with insufficient experience should first try for a research internship at Archbold (see appendix B for websites with a written description and a video), or at another field station appearing on the website of the Organization of Biological Field Stations (appendix B).

Research assistants need to have ample practical skills, in order to operate four-wheel-drive vehicles competently in difficult terrain; follow safety protocols diligently; be alert for approaching inclement weather, wildfires, or other threatening situations; and be responsible around cattle, equipment, and other land management activities. Mechanical competence is a plus. Archbold requires research assistants to have prior computer experience (e.g., Excel, Access). Research assistants II–III are given advanced training in working with ArcGIS, processing streaming data, or using R, SAS, and SPSS. Prior working knowledge of some of this software, however, may be required for research assistants IV–V. They also take the lead in data analyses for many projects, and applicants with advanced programming, analytical, and modeling skills are greatly valued. In addition, curatorial experience (specimen preparation and cataloging of arthropods, bird and mammal skins, herptiles, fishes, and plants) is a plus and may even be a requirement for some assistantships. Applicants may need to have already given evidence of excellent organizational skills and flexible leadership qualities. Miscellaneous attributes, such as being talented photographers, writers, and illustrators, are an added plus.

If selected for an interview, applicants should be prepared to demonstrate that they have knowledge of the sciences used at Archbold, such as regional biodiversity, climate, fire, hydrology, and long-term studies of taxa (e.g., Florida scrub-jays, scrub plants). Applicants must enjoy working in collaborative teams. They need to display a generosity of spirit and friendly engagement in a close-knit community, one ranging from scientists to cowboys, from cooks to accountants, and from students to wealthy donors. We seek applicants who would relish the privilege of working at Archbold and recognize that the setting accelerates scientific productivity. In addition, we desire people who clearly share our organization's passion for coupling science, conservation, and education, as well as those who are content living in a rural area with restricted social opportunities.

Education Required

Minimum educational standards for research assistant II or III (lower-level) positions require a bachelor's degree in ecology,

wildlife, or related biological sciences, as well as demonstrated field experience, such as summer jobs, internships, or post-baccalaureate research. Research assistant III–V positions require an MS degree in ecology, environmental science, wildlife, or a related biological discipline, as well as extensive field experience, such as that gained during an MS degree and part-time or other jobs. A bachelor's degree with extensive experience may substitute for an MS degree for research assistant III–IV positions. Some assistantships require an educational background in and experience with programming and installing advanced instrumentation. Research assistants with these skill sets are highly sought after and command higher salaries.

Pay Scale

Seasonal research assistants typically earn $18,000–$25,000 per year, which includes other benefits, such as temporary housing in shared cottage or trailer-type accommodations. Research assistants II and III generally earn $22,000–$33,000 per year, plus benefits. Research assistants IV usually earn $30,000–$40,000 per year, plus benefits, while research assistants V normally earn at least $40,000 per year, plus benefits. Benefits for full-time staff include health, dental, and life insurance. Beginning after the first year of employment, Archbold offers a retirement contribution of 5% of an individual's salary into a 403(b) plan. There is no long-term housing available.

NATIONAL AUDUBON SOCIETY
Cendy Hernandez

Agency: Nonprofit Organization, National Audubon Society

The mission of the Audubon Society is to conserve and restore natural ecosystems, focusing on birds, other wildlife, and their habitats for the benefit of humanity and the Earth's biological diversity. For more than a century, the Audubon Society has built a legacy of conservation success by mobilizing the strength of its network of members, chapters, Audubon centers, state offices, and dedicated professional staff to connect people with nature and the sources of power to protect it. A potent mixture of science, education, and policy expertise are combined in efforts ranging from the protection and restoration of local habitats to the implementation of policies that safeguard birds, other wildlife, and the resources that sustain us all across the Americas.

The Audubon Society employs a number of people with different types of expertise. Jobs include conservation biologists, GIS analysts, assistant land stewards, and field technicians.

Conservation Biologist
Job Description

Conservation biologists promote wildlife conservation practices as a tool for protecting, enhancing, and creating quality breeding habitat for regional-priority bird species. They work directly with private landowners, public land managers, foresters, loggers, and communities, and establish close liaisons with the Natural Resources Conservation Service to implement the Working Lands for Wildlife program. The primary function of conservation biologists is to engage landowners and the primary partners within priority important bird areas (IBAs) and forest blocks to implement bird-friendly management practices. The work requires actively identifying and connecting with landowners and managers to engage them in the program. This includes networking with partner agencies; becoming familiar with a variety of tools for conservation and management, such as the Working Lands for Wildlife program; supervising a set of landowners, so they complete habitat assessments; making management recommendations, including scheduling visits; writing management plans; utilizing ArcGIS software to prepare maps; maintaining data sets of landowners, acreages, and other factors within the program; working with partners on implementation projects for selected properties; participating in workshops and presentations; assisting in the development and implementation of fundraising strategies for our Putting Working Lands to Work for Birds program by working with development staff to identify potential major donors and participate in their subsequent cultivation, as well as in solicitation and stewardship strategies; representing the Audubon Society at workshops, conferences, and seminars, including developing and delivering presentations; reviewing and staying current on research about forest practices and their effects on birds; and maintaining ongoing communications with participating landowners and foresters by preparing occasional email updates.

Background Needed

Excellent written and verbal communication skills are a must. Skill in working with a diverse community of landowners and managers is essential, as is an ability to independently manage a flexible schedule that may require evening and weekend time, as well as additional hours during field seasons. Conservation biologists must be physically fit, including being able to walk in steep terrain throughout all seasons, bushwhack, and work in challenging outdoor habitats and terrain, with or without sleeping or other accommodations.

Education Required

A bachelor's degree in natural resources, forestry, or a related field is required. A graduate degree (MS or PhD), however, is strongly preferred. Familiarity with best-management guidance and basic silvicultural and habitat management practices and knowledge, bird identification skills (sight and sound), and an understanding of habitat and bird relationships is essential. Therefore, ornithological knowledge and experience is required. Excellent written and verbal communication skills are a must.

Pay Scale

The salary range for conservation biologists is from $25,000 to $35,000 annually.

Geographic Information Systems Analyst
Job Description

GIS analysts apply advanced skills in data management, geoprocessing, landscape analysis, and map design to wildlife conservation issues of national and international significance. GIS analysts primarily use spatial data to create project layers and maps for renewable energy and shorebird conservation projects within the context of important bird areas. Their work involves contact with agencies to secure data on spatial layers, and collaboration with Audubon Society staff. These data are then analyzed, using GIS tools to identify and evaluate potential impacts of renewable energy projects on IBAs, critical habitats, and other such designations. Analysts create printable renewable energy maps that can be used in meetings with consultants for project reviews and mitigation programs. GIS analysts also create an online, user-friendly volunteer coastal stewardship map that incorporates critical habitat, IBAs, and volunteer opportunities. Analysts assist with biological data collection, data sharing, and data management; perform spatial analyses using ESRI tools, including Spatial Analyst and ArcGIS online; provide support through cartographic design, which may include using web-based applications, such as Google Maps; and work with a team of conservation professionals within the Audubon Society, as well as with outside organizations, including other NGOs and state and federal agencies.

Background Needed

Advanced computer experience is paramount. GIS analysts must be skilled in the use of ArcGIS 10, Spatial Analyst, and Microsoft Office.

Education Required

A bachelor's degree (or at least three years of completed coursework) in environmental studies, ecology, biology, geography or a related field, and at least one year of GIS coursework or equivalent experience are required. Knowledge of and experience in cartography (map design), spatial analysis (geoprocessing), and ArcGIS are preferred. Completion of an applied conservation GIS project, as well as interest and education in wildlife ecology, are helpful.

Pay Scale

The salary range for GIS analysts is from $30,000 to $40,000 annually.

Assistant Land Steward
Job Description

Assistant land stewards provide support to the land manager and help oversee the ecological restoration of the entire property. They monitor the diverse natural systems throughout the grounds, including controlling invasive plant populations; monitoring wildlife populations; removing hazardous trees near structures and trails; cultivating, planting, pruning, and watering native species; applying herbicides as needed; and maintaining a daily invasive plant management log. Assistant land stewards also work on other property improvement projects, including timber stand improvements; research restoration methods; use GIS and GPS programs; lead volunteer groups; assist in training interns; safely operate power tools and heavy equipment; and maintain and repair restoration equipment.

Background Needed

Good communication and interpersonal skills are needed. Assistant land stewards must have the ability to perform tasks requiring physical exertion, in the outdoors, in all weather conditions and on difficult terrain, with or without sleeping or other accommodations. They must have the ability to lift and carry 9–23 kg regularly. This position requires a valid driver's license. Assistant land stewards must be able to work unusual hours and maintain flexible work schedules.

Education Required

A bachelor's degree in the natural sciences or education is required. A basic knowledge of ecology and restoration principles and a desire to learn about land stewardship and restoration are necessary. Simple mechanical skills are helpful in being selected for the position.

Pay Scale

Assistant land stewards are paid $9–$11per hour.

Field Technician
Job Description

Field technicians assist the Audubon Society's conservation department on projects that advance the organization's mission to conserve and restore natural ecosystems, which focuses on birds, their habitats, and other wildlife. Specifically, they provide support for habitat restoration and enhancement projects; assist others in biological surveys and field trips involving school-age children; help with habitat restoration, trail maintenance, and landscape care; performs basic office procedures, including printing, copying, and time tracking; pick up and deliver equipment and supplies; and maintain and organize equipment and maintenance areas. Evening and weekend work and travel are often required.

Background Needed

Field technicians must be comfortable working with school-age children. Other requirements are a willingness to work in the field during various types of weather conditions and to work outdoors for extended periods of time, with or without sleeping or other accommodations. Technicians must be physically fit and able to lift at least 23 kg. A valid driver's license is required.

Education Required

A bachelor's degree in biology, wildlife management, environmental science, or a related field, or equivalent education and work experiences are required. Field technicians must have knowledge of and experience in habitat management, invasive species control, and related fieldwork.

Pay Scale

The pay range for field technicians is $9–$12 per hour.

BOONE AND CROCKETT CLUB
Tony A. Schoonen and Paul R. Krausman

Agency: Nonprofit Organization, Boone and Crockett Club

The Boone and Crockett Club (B&C), founded in 1887, is the oldest wildlife conservation organization in North America. Its mission is "to promote the conservation and management of wildlife, especially big game, and its habitat, to preserve and encourage hunting, and to maintain the highest ethical standards of fair chase and sportsmanship in North America." The B&C was founded by Theodore Roosevelt and George Bird Grinnell, among others, and is named after Daniel Boone and Davy Crockett, the heroic archetypal explorers and adventurers of the nineteenth century. Teddy Roosevelt and other early leaders of the American conservation movement who were members of the B&C witnessed the decimation of wildlife populations and the nation's natural resources and sought various means to protect them.

The B&C's efforts to establish a foundation and framework for conservation in America include what is known as the North American Model for Wildlife Conservation and the user-funded concept. B&C members contributed to the development of such legislation as the Lacey Act, National Wildlife Refuge System Act, Migratory Bird Conservation Act, Federal Aid in Wildlife Restoration Act, and what is commonly referred to as the Farm Bill. In addition to Yellowstone National Park, the B&C was instrumental in the establishing Glacier National Park, Mount McKinley [now Denali] National Park, Pelican Island National Wildlife Refuge, the National Key Deer Refuge, the Holt Collier National Wildlife Refuge, the National Bison Refuge, and the Theodore Roosevelt National Wildlife Refuge. The B&C published the first *Records of North American Big Game* in 1932, and it now has the most popular scoring system for big game in the world, a system built on science and fair-chase principles. The B&C also established the Boone and Crockett Professors Program at several land grant universities throughout the United States.

The B&C's visions for wildlife and conservation are for a future where wildlife and its habitat, in all their natural diversity, are managed and conserved throughout North America; hunting continues to be enjoyed under the rules of fair chase, sportsmanship, and ethical respect for the land and wildlife; all users of natural resources respect the rights of others, in the spirit of sharing; the value and conservation of private land habitat is respected and supported; North Americans are committed to the principle that their uses of resources must be sustainable for themselves and future generations; and a future in which hunting opportunities exist for all desiring to participate in it.

Regular and professional members accomplish most of the work of the B&C. There are several jobs that need to be filled by full-time staff, however, for the B&C to continue to be a successful conservation organization. The B&C's small staff is housed at its headquarters in Missoula, Montana. There is very little turnover in personnel because of the desirability of the jobs, the history of the B&C, and the satisfaction that comes from the work, but openings do occur.

Assistant Director of Big Game Records
Job Description

The assistant director of big game records oversees the data collection and dissemination of information on all entries received. That individual inserts all scoring data; manages the scoring database; and promotes the dataset to gain support and strength for the club and its missions. The latter is accomplished by working with outside authors to produce popular articles concerning the scoring data and advocating record keeping for conservation purposes. The assistant director also assists with event planning and the coordination of the Triennial Awards program.

Background Needed

Familiarity with hunting practices and regulations throughout all of North America is critical. Although not mandatory, previous hunting experience is useful, as is prior experience with conflict resolution. Excellent writing skills are essential, because much of the work for this position involves formal correspondence by letter. In addition, computer skills and a strong working knowledge of software, including Microsoft Office, database reporting programs, and ArcGIS, are needed.

Education Required

A bachelor's degree in wildlife management, biology, or a closely related field from an accredited college or university is required. Certification by The Wildlife Society as an Associate Wildlife Biologist or at a higher level is desirable. Coursework with a strong emphasis in big game management is essential.

Pay Scale

The salary for the assistant director of big game records is competitive with similar professional positions.

Director of Big Game Records
Job Description

Daily responsibilities for the director of big game records include supervising the assistant director, processing trophy entries, responding to emails and telephone calls, and corresponding with trophy owners. The director coordinates the records department program and its activities with club officers and members, staff, official measurers, outdoor-oriented media, trophy owners, hunters, and other wildlife and conservation organizations. This individual assists the publications department and other staff with writing and proofreading books, fliers, film scripts, magazine articles, and ads that are produced by the B&C. The director is also the assistant editor for the B&C's *Fair Chase* magazine, contributing a column entitled "Trophy Talk" quarterly and editing all copy for each issue. The director works with the records committee chairman to plan committee meetings; writes and disseminates the minutes; and prepares the budget, work plan, annual report, and performance evaluations. Two out of every three years, the director conducts three official-measurer-training workshops to appoint approximately 125 new official measurers. Biennially, the director renews and updates the B&C's volunteer cadre of official measurers. Triennially, the director is responsible for all aspects of the Big Game Awards Program banquet and its related activities, including site and hotel selection; trophy invitations; and the receipt, uncrating, photographing, hanging, crating, and shipment of trophies back to owners. In addition, the director is responsible for the recruitment of volunteers, the selection and activities of the judge's panel, the preparation of awards, the trophy display, the awards program photo brochure, the awards banquet, and all aspects of the awards program book. The director also is a member of the planning team for the awards program.

Background Needed

Three years of experience working in the wildlife biology or wildlife management field is required. A demonstrated interest in North America's big game and in hunting and related outdoor recreation is essential. Excellent oral and written communication skills and an ability to interact with people with vastly divergent backgrounds are required, because the director communicates with B&C members, the general public, outdoor-oriented media, and agents in state, federal, and private wildlife conservation organizations. A demonstrated working knowledge of Microsoft Office programs, Photoshop, and database management software is essential. Expertise in writing scientific and popular articles is needed, and past work in publishing is necessary. Demonstrated supervisory skills are required. Previous experience as a B&C official measurer is useful, but if not, the candidate must be willing to become one.

Education Required

An MS degree in wildlife biology or wildlife management from an accredited college or university is required. Certification by The Wildlife Society as an Associate Wildlife Biologist or at a higher level is desirable. Coursework with a strong emphasis on big game management is essential.

Pay Scale

The salary for the director of big game records is competitive with similar professional positions.

Conservation Programs Manager
Job Description

The conservation programs manager develops and conducts conservation education programs at the Rasmuson Wildlife Conservation Center (RWCC) and associated facilities on the B&C's Theodore Roosevelt Memorial Ranch (TRMR) in Dupuyer, Montana. This individual oversees all groups that visit and all programs and activities that are conducted on the ranch or at its facilities. These include but are not limited to outdoor adventure camps (residential youth summer camps), a women's hunter education program, and the Boy Scouts of America–accredited Montana high adventure base. Each event parallels the vision, mission, and objectives of the B&C's conservation education program and strategic plan. The conservation programs manager is responsible for scheduling, accommodating, and invoicing all groups that stay at the RWCC. Additional duties include marketing and advertising efforts, as well as management of the B&C's social media outlets, in order to establish and maintain business with new and previous clientele. Public relations work involves meeting the needs of all who visit the RWCC and the TRMR and maintaining a positive presence at B&C events, outdoor classrooms, youth expos, and industry shows. PR also involves outreach and business development, in order to enhance the quality of the programs and increase the guest numbers at the RWCC facilities and the TRMR. The conservation programs manager assists the TRMR ranch manager, maintains the facilities (when available), and performs other duties as needed. Other duties involve supervising the conservation educator(s) and any staff, contractors, or volunteers associated with the conduct of programs and the management of groups or activities at the education center or associated housing facilities. The conservation programs manager must be willing to serve as staff liaison for the conservation education committee and the B&C's conservation division.

Background Needed

A minimum of three years of program development and facilities management experience is required. Excellent public relation skills are necessary, especially in dealing with a diversity of people from a variety of backgrounds and ages. Computer skills and a basic knowledge of budgeting and business management are needed. Previous experience in outdoor education and program development, coordination, and implementation is useful.

Education Required

A bachelor's degree in an applicable area of scientific study, such as wildlife ecology, and natural resources, is required, along with a minimum of three years working in outdoor program development and instruction. Coursework that provides hands-on experience with common wildlife and vegetative sampling methods is extremely useful. Experience as a teach-

Learning to properly score a white-tailed deer is an important component to master if you wish to work with Boone and Crockett big game records (photo: Julie Tripp)

ing assistant in a classroom or outdoor laboratory is beneficial, as is a willingness to serve as staff liaison for the conservation education committee and the B&C's conservation division.

Pay Scale

The salary for the conservation programs manager is competitive with similar professional positions.

Additional Information

The Boone and Crockett Club maintains two other permanent positions: the director of publications and the director of marketing. The director of publications is involved with all facets of book publishing, from title acquisitions to editing, design, production, marketing, inventory management, business planning, and financial management. The director of marketing is in charge of products, messages, and brand development; the marketing of products; sales communications; and sales. In neither case is a wildlife biologist hired for these positions. Typically, the director of publications has a degree in applied arts, with at least five years of experience working in the book publishing industry, while the director of marketing has a business degree with an emphasis in marketing. The B&C also hires a conservation educator to direct conservation education as needed, but this position is not permanent.

DUCKS UNLIMITED, INC.
J. Dale James

Agency: Nonprofit Organization, Ducks Unlimited

Ducks Unlimited (DU) was founded in 1937 by a group of concerned sportsmen as a result of dwindling continental waterfowl populations, due to the effects of a long-term drought. Ducks Unlimited is now the world's largest and most effective private waterfowl and wetlands conservation organization. The mission of DU is to conserve, restore, and manage wetlands and associated habitats for North America's waterfowl, with the understanding that these habitats also benefit other wildlife, as well as people. DU is often considered to be a volunteer-based organization, due to its membership and its fundraising activities. Successful accomplishment of its habitat conservation mission, however, is achieved through science-based, solution-oriented conservation activities delivered by professional staff from various disciplines, working out of four regional offices. Ducks Unlimited's national headquarters is located in Memphis, Tennessee. Today, DU is comprised of nearly 700,000 members and has conserved more than 5.2 million ha of habitat for the benefit of waterfowl, other wetland-dependent species, and the people who enjoy them and reap advantages from the ecological services provided by these habitats.

DU's conservation staff has historically been grounded in strong, traditional wildlife (waterfowl) science and management practices. The applied nature of DUs *conservation delivery* business (monitoring, evaluating, and constantly refining its habitat programs), however, should also appeal to those trained in other disciplines, such as conservation biology, landscape ecology, wetland ecology, and conservation policy. Conservation positions within DU are varied, ranging from entry-level conservation specialists and project biologists to more-experienced regional biologists. Other conservation-oriented posts include conservation planners, geographic information systems and remote sensing (GIS/RS) analysts, conservation policy representatives, and land protection specialists. Beginning salaries for all positions within DU can vary, based on the location of the office and the applicants' qualifications. Therefore, the "Pay Scale" sections provide only a general estimate of starting salaries for each job described. Promotions are typically based on performance, an exhibited progression of skills, and increased responsibilities. These jobs often evolve into higher-level administrative management positions with regional or national responsibilities, including managers and directors of conservation programs.

Conservation Specialist
Job Description

Conservation specialists generally receive day-to-day supervision and assist other staff members with habitat or other conservation programs. Depending on an individual's skill set or the particular biological staff that person works with, these positions can be purely office jobs (e.g., GIS/RS specialist) or involve more field time if the specialists are assisting with program or project delivery. There is also some level of interaction with the public, usually consisting of phone or email communications and coordination with other conservation partners.

Background Needed

Conservation specialists usually have skill sets that are in their formative years. These abilities could include a working knowledge of general habitat management principles, GIS, and remote sensing.

Education Required

At a minimum, conservation specialists must have some technical training, such as a technical certificate or an associate's degree, and a knowledge of wildlife habitat conservation. A bachelor's degree is often preferred. Specialists usually have limited experience (less than five years).

Pay Scale

The beginning salary for a conservation specialist is $30,000 per year, plus benefits.

Biologist
Job Description

Biologists at DU focus on the delivery of one conservation program or on waterfowl habitat within one geographic area. They may be involved in project monitoring and evaluation, such as identifying potential conservation projects for development and working with staff and partners to ensure adequate progress toward the completion of a project, as well as compliance with its requirements. Biologists may work with funding partners, landowners, and permitting agencies to help keep projects on time and have them be cost effective. Some grant and report writing is required. The job may involve participation in field activities with volunteers and donors.

Background Needed

This position typically requires some level of acquired field experience beyond classwork. This could include but not be limited to any practical accomplishments in the areas of wildlife or habitat management. A working knowledge of biological principles and practices is expected. Project biologists must have excellent oral and written communication skills, be able to travel occasionally, and work effectively off site, including performing fieldwork in adverse conditions.

Education Required

A bachelor's degree in wildlife or a related conservation field, with knowledge of biological principles and practices, is required. Biologists typically need to have some level of acquired field experience.

Pay Scale

The beginning salary for a biologist is $40,000 per year, plus benefits.

Regional Biologist
Job Description

Regional biologists are directly responsible for the delivery of conservation projects on public and private lands. These projects must fulfill DU's habitat conservation objectives, meet grant requirements and partner expectations, and satisfy fiscal obligations for program fulfillment and sustainability. Duties and responsibilities include the initiation, coordination, and implementation of DU's conservation objectives for a particular geographic or conservation region. Regional biologists are responsible for developing and maintaining relationships with agencies, private landowners, corporations, and municipalities to achieve the conservation missions of DU and its partners. Management responsibilities include the performance and coordination of tasks such as project accounting, scoping, planning, permitting, oversight of design objectives, implementation, and reporting. Project development duties include grant writing and the submission of proposals to federal and state funding agencies, foundations, and other sources to secure monies for projects, as well as assisting with related fundraising activities focused on philanthropic donors. Working hours are balanced between the office and the field, and the approximate allocation of time is dependent on the region and the specific location of the position. In general, time spent in the field includes performing a biological or bioengineering reconnaissance of proposed project sites, meeting with partners regarding current and potential projects, and attending regional and state partner meetings, among other activities. Office duties include managing project budgets, developing agreements, coordinating project development and delivery, writing proposals and overseeing grants, assisting with partner programs, and various other tasks as they arise (e.g., grassroots fundraising events, state conventions). Some time is also spent responding to phone calls from private landowners interested in technical assistance or cost-share opportunities.

Background Needed

Regional biologists often have three to five years of field experience and a strong knowledge of biological principles and practices. Field experience with waterfowl, as well as wetland ecology and management, is an important consideration for this position. Regional biologists must have strong oral and written communication skills, as facility in positively interacting with private landowners, DU volunteers and donors, and state and federal resource managers is critical. An ability to prepare reports, grant proposals, and project budgets is necessary. Experience in the use of GIS applications or CAD is beneficial, as is prior operation all-terrain vehicles and GPS equipment.

Education Required

A bachelor's degree in wildlife or a related conservation field is expected, and an MS is preferred. Regional biologists often have three to five years of field experience, with a strong knowledge of biological principles and practices. Applicants' academic training should qualify them for certification as an Associate or Certified Wildlife Biologist from The Wildlife Society or for a comparable certification.

Pay Scale

The beginning salary for a regional biologist is $50,000 per year, plus benefits.

A Ducks Unlimited biologist discusses a wetland project design with an engineer (photo: Dale James)

Conservation Planner
Job Description

Conservation planners generally have regional responsibilities that include strategic planning, operational planning, and research efforts that provide a biological and scientific basis for conservation programs, including policy issues. They routinely work with the science staff for migratory bird joint ventures and landscape conservation cooperatives with regard to landscape-level conservation design, monitoring, and evaluation, and they supervise regional GIS/RS staff. While this is primarily an office position (80%–90%), there is occasional travel for meetings, conferences, and fieldwork. Approximately 45% of conservation planners' time is spent developing, coordinating, and implementing research; synthesizing and evaluating scientific information; and refining priority landscapes to help support the delivery of conservation programs. In addition, 20% is spent supervising and coordinating projects with GIS/RS staff; 15% in working with other conservation scientists, including joint venture technical committees; 10% in assisting with strategic or operational plan development; and 10% in helping with the promotion of conservation fundraising initiatives and conservation land program development.

Background Needed

Conservation planners generally are required to have detailed knowledge of and familiarity with the scientific literature, particularly in regard to waterfowl and wetland ecology. This includes a solid grasp of national and international waterfowl conservation issues and a strong working knowledge of the ecology and management of breeding, migrating, and wintering waterfowl and their habitats. A comprehensive understanding of wetland values, functions, and current management issues is also needed. A general knowledge of geographic information systems, remote sensing, landscape conservation design, and monitoring and evaluation processes is beneficial. Conservation planners must have excellent interpersonal and oral and written communication skills, as well as computer expertise.

Education Required

At a minimum, successful applicants must have an MS degree (a PhD is preferred) in wildlife or a related conservation field. At least five years of field experience is required, and applicants should qualify for certification as an Associate or Certified Wildlife Biologist by The Wildlife Society or another appropriate certification.

Pay Scale

The beginning salary for a conservation planner is $60,000 per year, plus benefits.

GIS/RS Specialist
Job Description

GIS/RS specialists work with other DU staff and conservation partners to develop spatial data, databases, maps, and graphic displays. These data and databases are used to help DU keep accurate records of conservation projects and provide the basis for spatial analyses for conservation planning purposes. Maps and other graphic mediums are used to disseminate spatial information to DU supporters and the general public, as well as to support reports and grants. GIS/RS specialists are in the office almost all of the time, with 25%–30% spent in collaboration with DU staff and external conservation partners regarding ongoing conservation projects. These specialists rarely deal with the public.

Background Needed

GIS/RS specialists should have experience with GIS and remote sensing technology, including a knowledge of ESRI products (e.g., ArcMap), as well as with the development of relational spatial databases. Specialists must have excellent communication skills, as they work with other staff to provide basic GIS services for a wide variety of DU conservation and fundraising programs. Experience in general cartographic production, data acquisition, the development and maintenance of GIS databases, metadata development, digitizing, and documentation is essential.

Education Required

GIS/RS specialists must have a bachelor's degree in GIS, geography, natural resources, or a related field, plus at least five years of experience.

Pay Scale

The beginning salary for a GIS/RS specialist is $30,000 per year, plus benefits.

GIS/RS Analyst
Job Description

The basic duties of GIS/RS analysts include ensuring that all GIS resources (SQL server, ArcGIS for the server, GIS applications) are working properly and that data products are readily available; providing mapping support internally and to partners; creating and running spatial models for data analyses and quality control; and writing desktop and web application or scripts to help streamline processes. Normally, 90% of GIS/RS analysts' time is spent in the office, with occasional out-of-office conferences and fieldwork. Approximately 30% involves reviewing or writing documentation for models, programs, or publications. In-person meetings and responding to questions by phone and emails typically utilize 15%. The remaining office time is spent on project-related activities, including landscape monitoring and evaluation for conservation planning. GIS/RS analysts typically have limited interactions with the public.

Background Needed

GIS/RS analysts are required to have a demonstrated ability with ESRI products, remote sensing, geodatabases, and web mapping. Good organizational and interpersonal abilities, basic project management skills, and a general background in wetlands and wildlife ecology are also important. GIS/RS analysts must be self-motivated and capable of working independently, with minimal supervision, as well as interacting closely with other staff within DU and with conservation partners. A strong background in GIS and a depth of experience with web mapping and geodatabases are critical. Two or more years of demonstrated experience with ArcMap, ArcSDE, and ArcServer / ArcGIS online is required. Other useful skills include computer troubleshooting, cartographic design, and basic programming knowledge.

Education Required

GIS/RS analysts are required to have an MS in GIS, geography, natural resources, or a related field and at least two years of spatial experience.

Pay Scale

The beginning salary for a GIS/RS analyst is $40,000 per year, plus benefits.

Conservation Lands / Real Estate Specialist
Job Description

Conservation lands and real estate specialists are responsible for land conservation programs, including conservation easements, fee-title land acquisition, mitigation lands, and revolving lands (acquisition, restoration, and resale). They also assist state and federal conservation partners with land acquisition and protection projects. Conservation lands and real estate specialists spend 60% of their time in the office and 40% in the field, including meeting and communicating with private landowners, as well as with other nonprofit and public conservation agency personnel.

Background Needed

Conservation lands and real estate specialists are required to be well versed in wildlife and habitat wetland management and in land protection options, with a special emphasis on wetland habitats. An ability to conduct biological field evaluations of properties is important. These specialists need to be able to manage numerous projects simultaneously, be highly motivated, and possess good written and oral communication skills. They should be computer literate and able to use Excel, PowerPoint, and word processing software. Knowledge of ArcMap or related GIS software is desirable. Conservation lands and real estate specialists also must be able to assimilate all of DU's land protection guidelines and remain current on changing laws and regulations relative to land transaction and protection options.

Education Required

A bachelor's degree in wildlife, business, agricultural economics, or a related field is required. Conservation lands and real estate specialists must understand biological principles and practices, especially those related to agricultural and conservation lands. A knowledge of land transaction and appraisal practices is desired.

Pay Scale

The beginning salary for conservation lands and real estate specialists is $50,000 per year, plus benefits.

Conservation Policy or Governmental Affairs Representative
Job Description

Specific duties include working closely with DU staff and volunteers at the local, regional, and state level as a public policy representative on conservation policies that impact waterfowl or their habitats. Conservation policy and governmental affairs representatives also provide leadership and representation for DU's conservation goals and interests with state policy makers in diverse settings, and they represent the interests of DU and its members in state legislative and regulatory proceedings. In addition, they support DU's federal policy goals by working with other DU staff members, volunteers, and in-state congressional staff. Conservation policy and governmental affairs representatives spend 70% of their time in the office and 30% at off-site meetings. Their job duties involve significant public interaction.

Background Needed

Conservation policy and governmental affairs staff need to have a working knowledge of various aspects of conservation policy and law, a general knowledge of waterfowl ecology (related to breeding, migrating, or wintering waterfowl, depending on where the position is located) and habitat management, and a general familiarity with natural resource–based industries in the region (e.g., agriculture, ranching, forestry, water-based businesses). Basic expertise with computers and excellent communication skills are required, including an ability to prepare technical information for and present it to nontechnical audiences. Conservation policy and governmental affairs representatives must possess strong interpersonal skills, in order to effectively communicate with policy makers, DU staff and volunteers, and DU's general membership. In addition, they need to be capable of thinking analytically, creatively, and strategically, in order to develop and implement current and new conservation policy in agricultural and energy-producing landscapes. The job requires working with traditional conservation partners as well as new, nontraditional partners to build support for waterfowl and wetland-friendly conservation policies.

Education Required

Conservation policy–oriented positions require a minimum of a bachelor's degree or an MS in natural resource conservation, agriculture, wildlife management, or a closely related natural resource field and at least five years of experience. Alternatively, a bachelor's degree or an MS or MA in public affairs, political science, or public policy and at least five years of experience would fulfill the requirements.

Pay Scale

The beginning salary for conservation policy and governmental affairs representatives is $55,000 per year, plus benefits.

Manager of Conservation Programs
Job Description

Conservation program managers generally are in charge of a team of biologists who deliver on-the-ground conservation programs and projects at regional and national or international levels. Regionally, these activities include land protection and wetland restoration and enhancement. Managers generally exercise considerable responsibility in completing project and program objectives, coordinating regularly and frequently with the director of conservation for their respective programs. They must also work closely with policy and fundraising staff, and oversee budget preparation and administration for a region or a conservation program. Managers develop and nurture relationships with DU volunteers, public agencies, private landowners, funding entities, and other conservation partners. They generally spend 80% of their time in the office and 20% in the field or in out-of-office meetings.

Background Needed

A waterfowl biology and wetland ecology emphasis is highly desirable, including knowledge of wetland restoration and wildlife habitat management. For its conservation program managers, DU generally seeks individuals with proven abilities and skill sets in team leadership, the implementation of strategic and business plans, budget and personnel administration, fundraising for specific programs, partnership development and stewardship, and public policy. Managers must have excellent communication skills and an ability to work both independently and as a member of a diverse team.

Education Required

At a minimum, a bachelor's degree in wildlife or a related conservation field of study is required, and an MS is preferred. Conservation program managers often have at least five years of field experience and specialized or advanced training in an area related to conservation.

Pay Scale

The beginning salary for a manager of conservation programs is $60,000 per year, plus benefits.

Director of Conservation Programs
Job Description

These individuals are in charge of conservation programs in one or more states or for a program or group of programs at the national or international level. Directors of conservation programs are responsible for substantial planning, fundraising, development, and execution for a broad array of programs that fulfill DU's conservation objectives. They exercise broad discretion in delivering these results in conjunction with partners, volunteers, and other staff. State-level directors are supervised by the regional director of conservation operations, and they may work closely with individuals in similar positions and with fundraising staff. Interactions with state, regional, and national DU volunteers are also essential components in accomplishing the job, and directors of conservation programs may serve as members of regional or national conservation leadership teams. In addition, they might represent DU on the boards of joint ventures and landscape conservation cooperatives. Directors also have a level of responsibility that affects DU's national budget and the accomplishment of its objectives. Generally, 60% of their time is spent in the office and 40% is dedicated to out-of-office meetings, conferences, and fundraising events.

Background Needed

Directors of conservation programs are typically expected to have at least 10 years of progressively more responsible work experience in DU or other organizations, and advanced training or education in conservation, engineering, or another specialized area related to conservation. This position requires excellent administrative, communication, negotiation, and supervisory skills, as well as an ability to work with a diverse set of partners to accomplish results that meet conservation-related goals.

Education Required

At a minimum, a bachelor's degree in wildlife or a related conservation field is required. An MS or a PhD is strongly preferred, however. Directors of conservation programs often have at least 10 years of field experience and specialized or advanced training in an area related to conservation.

Pay Scale

The beginning salary for a director of conservation programs is $75,000 per year, plus benefits.

NATIONAL WILD TURKEY FEDERATION
Ryan A. Boyer, Clint Carpenter, Kristen Giger, and Tyler Brown

Agency: Nonprofit Organization, National Wild Turkey Federation

The National Wild Turkey Federation (NWTF) is dedicated to the conservation of wild turkeys and the preservation of our hunting heritage. The NWTF was founded in 1973, and its headquarters is located in Edgefield, South Carolina. The NWTF is an upland habitat conservation organization that, since its inception, has worked with state and federal agencies to assist in the restoration and management of wild turkey populations and their habitats throughout North America. In 1973, there were about 1.5 million wild turkeys in North America. After decades of dedicated work by the NWTF, that number hit a historic high of almost 7 million turkeys. The foundation for that accomplishment was standing behind science-based conservation and hunters' rights. The NWTF is considered to be a grassroots organization, supported by members and volunteers who assist in fundraising to generate the dollars that provide support for the organization's goals and habitat conservation mission. Thanks to the tremendous efforts of our dedicated volunteers, professional staff, and committed partners, the NWTF has had many successes that have advanced our mission. Together, we have facilitated the investment of $488 million in wildlife conservation and in the preservation of North America's hunting heritage. The returns have included improving more than 7 million ha of wildlife habitat and introducing 100,000 people to the outdoors each year. The NWTF has staff members at the headquarters office and others that work remotely throughout the country to deliver conservation programs.

Regional Biologist
Job Description

The work of regional biologists with the NWTF varies greatly from that of those in state and federal agency positions with similar titles. In many cases, NWTF regional biologists are responsible for delivering conservation programs across large regions, which can encompass multiple states. Regional biologists work in conjunction with staff from governmental agencies, private companies, landowners, and other NGOs to further the conservation of upland habitats throughout their specific region.

Approximately 30% of regional biologists' time is spent developing and administering grant proposals to assist in funding upland habitat enhancement and restoration activities to benefit wild turkeys and a multitude of other wildlife species. Planning and implementing landscape-level upland habitat restoration projects make up 20% of the job. This includes working with traditional conservation organizations (e.g., NGOs, state and federal agencies) and nontraditional conservation partners (e.g., energy companies, industrial firms). Another 20% of their time is spent administering wildlife habitat grants and stewardship agreements among the NWTF, state agencies, and the US Forest Service. Administering both state NWTF budgets to assist in the acquisition of lands by government partners, and a program requesting project proposals from conservation partners seeking financial assistance for upland habitat projects are other duties that take up 10% of regional biologists' efforts. It is their responsibility to review these proposals and determine if the projects are viable, quantifiable, and have ecological benefits that impact wild turkeys and other wildlife species. Additionally, 10% of their time is spent working with state agency professionals, including upland game bird biologists, to ensure that the NWTF is emphasizing the delivery of conservation programs based on the most up-to-date scientific information available. The final 10% is focused on providing biological support, technical assistance, and information to our fundraising staff, volunteers, and members regarding effective management techniques for wild turkeys within their region. Overall, roughly 80% of their duties are devoted to working in an office or attending meetings, while the remaining 20% are spent in the field, identifying new project opportunities or verifying the completion of habitat projects. Much of the office work involves corresponding with partners through emails and phone calls. An ability to communicate effectively with people and build strong relationships is critical to a successful career as a regional biologist.

These duties have a certain amount of variability, depending on multiple aspects, such as limiting factors impacting wild turkey populations, the biologists' particular region or areas of responsibility, funding availability, and partnership opportunities. For example, assisting state or federal agencies with land acquisition opportunities may have a greater priority within one region, whereas others regions may emphasize the need for assistance with the management or maintenance of that area's public lands for the benefit of wild turkeys and other wildlife species. The differences among programs and the ability of NWTF regional biologists to demonstrate flexibility, in order to adapt and assist partners in reaching shared conservation goals, makes the NWTF very effective in conserving upland habitat throughout the country.

Background Needed

Because NWTF regional biologists are responsible for the administration and oversight of upland restoration and enhancement projects, it is important that they are familiar with the techniques needed to complete a restoration or enhancement project. Ideally, you should have experience with many of the necessary skills for these projects, even though, in most cases, you are not the one responsible for physically carrying out the work. This insight allows you to better administer the projects and communicate with partners and contractors. Most regional biologists have completed at least a few years of technician-level work, which allows hands-on experience in multiple areas, using various techniques to conduct wildlife habitat management. Internships and summer wildlife technician positions provide great opportunities to gain experience and should be taken advantage of, either after your first year as an undergraduate or as soon as is otherwise possible. The professional wildlife field is relatively small, but it is a well-interconnected group throughout the United States and the world. Keep that in mind as you move forward in your career. Begin to develop experience and thereby further your education! As an intern or technician, you should strive not just to get a good reference, but to be the employee the organization wants to hire back next season. This ultimately provides you with a strong recommendation from someone who is already well established. It is important to try and diversify your experiences, so do not be afraid to travel to positions in different places.

Experience working with wildlife is invaluable, and any opportunity to handle live animals directly is advantageous. Assisting MS or PhD students with a component of their research projects is a great way to garner this invaluable experience and gain insight into life as a graduate student and the responsibilities of this role. Another benefit of working on such projects is that it allows you to become familiar with current research needs and a university's graduate-level faculty members, should you consider pursuing an MS or PhD degree. Lastly, by way of an example, do not relinquish an opportunity to band waterfowl, even if your primary goal is to work with large mammals or raptors. Most biologists have very diverse backgrounds, as well as experience working on many different projects involving species that range from insects to large mammals.

A growing number of people are entering the wildlife field who have never hunted, fished, or trapped. It would be best for you to try each of these activities at least once (if you are not already actively involved in them), so you can relate to their recreational user groups and understand the importance of hunting, fishing, and trapping in funding conservation efforts, as well as to the North American Model of Wildlife Conservation. You might decide not to continue doing so, but that experience is still invaluable to you in understanding the importance of these wildlife conservation activities.

Education Required

The minimum requirement for regional biologists with the NWTF is a bachelor's degree in wildlife management, forestry, or another closely related fields from an accredited college or university. An MS degree, however, is highly preferred and is becoming more common as a prerequisite for consideration both in the NWTF and other NGOs. A master's degree with a research component that includes the preparation and defense of a thesis allows candidates to develop critical thinking skills and gain experience in writing within a scientific format, which are important attributes for communicating among professionals in the field. No specific level of education is needed to be eligible for a promotion.

There is an art and a science to wildlife management. Any wildlife professional can testify that work in any agency or organization often has a far greater human management aspect than wildlife management one. A few courses in areas such as business, communications, sociology, the humanities, and leadership are beneficial within the wildlife field. It is essential to become familiar with software programs such as Excel, Word, PowerPoint, and Access, as you are likely to use most of them on a daily basis. Experience with GIS is becoming an increasingly needed skill set for up-and-coming wildlife professionals. Literature and professional writing courses are important in developing your writing skills. Lastly, you should take at least one public policy–related course, in order to understand how laws are passed and how the legislative process works at the state and federal levels. Policy has become increasingly important with regard to the ways in which we manage wildlife populations and have lasting impacts in our field.

Pay Scale

The salary levels for all NWTF positions are commensurate with experience, but a good starting point for regional biologists is $40,000–$45,000 per year. The NWTF is striving to actively compete with governmental and private markets in terms of the salaries for regional biologists, and we continue to improve our internal structure to more accurately reflect this compensation range. The NWTF regional biologists have annual work plans to guide them in the correct direction to accomplish their conservation work, and this is the mechanism used to score their annual performance reviews. This provides opportunities for them to grow professionally and potentially receive additional compensation for their work efforts. Promotions due to quality performance can occur at any time during the year or over the course of a career, as staff positions become available. Salary increases take both new job requirements and the NWTF's annual budget into account, but a typical promotional increase is configured as a percentage of that person's current salary. Increases could begin at 2% and may be more, depending on overall responsibilities. In addition, the NWTF offers its full-time employees a great benefits package, including health, dental, and vision insurance, as well as retirement options.

People do not go into this field because of the money. Wildlife professionals are a dedicated and passionate group of individuals who truly love their jobs. Additionally, what you aren't paid in terms of salary is often made up for by experiences and travel opportunities. Wildlife professionals encounter amazing things in remarkable places throughout the country and the world, and that is often far more rewarding.

Regional Director
Job Description

Regional directors are entry-level to veteran field staff that oversee fundraising and volunteer relations on a day-to-day basis within a given territory. At the NWTF, these are full-time positions, under its field operations category. They are a great start for anyone who has a strong passion for an outdoor lifestyle and the preservation of our hunting heritage. The NWTF is one of the most credible and successful nongovernment organizations in the nation. It has reached this status by bringing wild turkey numbers back to an all-time high. Recently, the NWTF began a Save the Habitat, Save the Hunt initiative. Its goals are to conserve and enhance 1.6 million ha of critical upland wildlife habitat, create 1.5 million new hunters, and establish more than 202,000 ha of new hunting access. The NWTF has over 250,000 of the most passionate volunteers in the country, who want to be a part of this initiative and have it succeed. The regional directors see to it that these volunteers have the necessary tools to successfully raise funds for this initiative. The main way the NWTF does this is by volunteers holding hunting heritage banquets and other outreach events in their respective areas. These events usually consist of a dinner and exclusive NWTF merchandise items that are raffled or auctioned.

This position is very hands on and travel oriented, with trips totaling 40,000–50,000 km per year. Regional directors are assigned territories, based on the number of NWTF chapters in a given area. Some states can have as many as four or five regional directors, or as little as one per state. Travel can range from daytime excursions to overnight stays. Regional directors work with volunteers, so normally meetings are in the afternoon, when these individuals are finished with their day jobs. January through May is usually the busiest time for fundraising banquets, and often events are held every weekend. Regional directors are responsible for making sure that everything is in place, so each event can raise as much money as possible to support the Save the Habitat, Save the Hunt initiative. On average, regional directors are involved with 20–30 fundraising activities per year. During the summer and early fall months, regional directors expand their territory by starting new chapters and holding new events, such as gun and gear nights, to maximize the fundraising potential for their area. This position can be very sales oriented, as the NWTF is marketing a lifestyle that is quickly diminishing within society. Volunteer relations are also a crucial aspect of the job. Regional directors can only be as successful as the volunteers they work with, so it is extremely important to maintain a good relationship with all volunteers. All events held by NWTF chapters are budgeted to raise a certain amount of dollars. Regional directors must know how much money each event is expected to bring in, as well as try to promote yearly fundraising growth. Membership development is also essential. The NWTF, as a membership-based organization, is only as strong as its member and volunteer base. It is important to keep membership numbers up, so fundraising events can have as many people come through the door as possible. Regional directors also need to seek out and obtain large donations and enroll upper-level members.

There are four levels of seniority for regional directors, and they are all based on how quickly directors expand their territory by building new chapters and on the net amount of new dollars they raise. All regional directors start as a regional director I, and advancement then depends on that person's drive to rise to a higher level. Once hired, all new regional directors train at NWTF headquarters in Edgefield, South Carolina. Here they gain the necessary knowledge and tools to be successful within the NWTF, such as how to start new chapters, budget events, and increase membership. Regional director positions are not temporary, by any means. Many people make a career out of this job, and it can be very rewarding to know that you are promoting the longevity of our hunting heritage and the preservation of wild turkeys.

Background Needed

There is no specific degree that sets applicants apart, but having strong wildlife training is very beneficial when dealing with conservation groups and regional biologists. A background in sales is also highly desired. It is important to have a strong passion for the outdoors and for hunting. Excellent people and relationship skills are a must to succeed in this profession. The work environment is very self-driven, and regional directors must hold themselves accountable for all aspects of the job. Expertise with Microsoft Office programs and any experience as a volunteer with a NGO are a plus. A willingness to work long hours and keep a positive attitude is mandatory.

Education Required

A minimum of a bachelor's degree or equivalent experience in sales is required. Previous coursework in wildlife management and conservation is extremely beneficial.

Pay Scale

The salary for an entry-level regional director is around $40,000–$50,000 per year, depending on experience. Full health, dental, vision, and life insurance benefits are included.

Biologist
Job Description

Biologist is a general title at the NWTF that covers entry-level, term-limited positions, which are generally linked with a specific grant or agreement. NWTF biologists typically hold jointly funded posts, with a substantial degree of emphasis on partner collaboration. Their primary duties are tied to specific deliverables (a tangible product or products produced as a result of a project) with at least one partner, including state and federal agencies and other NGOs.

These positions offer an array of on-the-job training opportunities for young professionals. Supervisors are often located several states away, and biologists are based in a home office. Responsible time management and organizational skills are learned early. Though this a salaried position, with no overtime pay, biologists often find themselves working late into the evening or on weekends, sometimes due to deadlines, but usually because of a love for the job. While the majority of their time is spent in the field, a significant amount also is spent in front of a computer screen. Typical tasks for biologists include coordinating and arranging field visits to various properties; providing technical assistance and developing advanced management plans; facilitating private landowner enrollments in Natural Resources Conservation Service programs and contracts; identifying additional landowners and future sites for forest management on private and potentially public lands; communicating with and providing technical assistance to private landowners and local, state, and federal agencies on the implementation of best-management practices for specific projects; cooperating with US Forest Service foresters and resource specialists to promote, design, and implement management activities that enhance wildlife habitats and promote healthy forest ecosystems; and working to increase the effectiveness of forest management through technical assistance with project planning, design, development, implementation, and outreach efforts, under the direction of cooperating partners.

NWTF biologists have a chance to work with a wide variety of individuals in the private and public sectors. Networking opportunities for future employment within and outside the NWTF abound, but only if biologists capitalized on them. As biologists become more comfortable in their roles, there are opportunities to broaden the scope of their work, as funding allows, and focus on topics that are of particular interest to them. Additionally, numerous occasions exist to step up and help out in other departments. Assisting with chapter banquets, NWTF hunts, or youth events helps biologists gain a more complete understanding of how the organization as a whole works, and shows them that they are ready and able to take on more-complex tasks in the future.

Background Needed

Relevant field experience is regarded favorably. NWTF biologists are typically highly energetic and willing to work long days, while maintaining a positive attitude in all types of weather. Staying in excellent physical condition makes working long days in remote areas much safer, as well as more enjoyable. The ability to drive long distances on highways and operate standard four-wheel-drive trucks and all-terrain vehicles over remote back roads is a must. Practical skills that are needed to be successful NWTF biologist include but are not limited to previous experience using GPS units and ArcMap, orienteering skills, and proficiency with all Microsoft Office programs.

Although not required, the NWTF highly encourages its employees to hunt and fish. A thorough understanding of the North American Model of Wildlife Conservation, and of how that model funds conservation in America today, is critical to understanding the mission of the NWTF. Excellent interpersonal and public speaking skills cannot be overemphasized. Because of the independent nature of this job, biologists are often the first, and sometimes the only, NWTF staff representatives that landowners and some agency personnel ever meet.

Education Required

Minimum academic training for biologists is a bachelor's degree in wildlife ecology, forestry, environmental sciences, or a related field from an accredited college or university. An MS degree or a PhD, however, often is preferred. An understanding of forestry and silviculture, conservation implementation, habitat management, ornithology, or wildlife biology is desired.

Pay Scale

The starting salary for an NWTF biologist is $35,000–$40,000 per year, plus benefits and a set budget to cover specified expenses while working. Additionally, a limited life insurance policy is available. A 401(k) retirement plan is offered, and in most years there is a significant employer match for employee contributions. The NWTF's bonus system is structured to award all employees a percentage of their annual salaries, based on goals set forth for each department. Thus staff members are encouraged to work across departmental lines to help achieve the overall mission of the organization.

Promotion from one level to another for biologists within the NWTF generally happens through a competitive application process. Academic degree(s), demonstrated ability to exceed expectations, and proven leadership skills are all strongly considered during the application and interview process.

Geographic Information Systems Analyst
Job Description

GIS analyst is the standard job title for an entry-level GIS professional in the conservation profession. Analysts typically report to a GIS manager, GIS director, or a conservation director. In some cases, GIS specialist is used as the job title, but typically the duties are the same as those for a GIS analyst. Analysts with the NWTF typically hold permanent, full-time jobs that provide organizational support and assist in the delivery of specific projects or programs. Occasionally, these positions can be term-limited, depending on the funding structure for them. For example, a GIS analyst could be paid to conduct specific mapping services or particular analyses over a one-year period.

Recent graduates with a bachelor's degree or an advanced degree in the GIS field, with a specialization in conservation, are typically the individuals who look for GIS analyst jobs with a conservation organization. Since this is generally an entry-level position, analysts use their time on the job to learn how GIS is implemented within a conservation organization, while expanding their GIS skills. After five years of experience or more, entry-level GIS analysts could potentially seek advancement as senior GIS analysts, or even as GIS managers. The NWTF typically has a GIS manager, GIS analysts, and GIS interns on its active staff.

The tasks for GIS analysts can vary, depending on the needs of the organization and diverse requests that are applicable to this position, but duties typically are allocated by the GIS manager or GIS director and follow the protocols and long-term vision of the GIS program. Assignments can include but are not limited to creating maps for landowners, conducting various spatial habitat analyses, building databases that link spatial data with tabular data, editing shapefile geometry and entering tables, using GPS for data collection, and offering GIS training to colleagues within the NWTF. GIS analysts typically are responsible for inventories; the installation of and support for GIS software, GPS units, and GPS software; and other various software needs that may not be handled through the IT department but are crucial to the conservation requirements of the NWTF. While the majority of these duties are related to conducting habitat analyses, there are tedious tasks that sometimes come up, such as heavy shapefile editing or manual data entry that helps build quality databases.

GIS analysts sit for long periods of time in front of a computer, although there are opportunities to use a GPS unit for data collection in the field. They necessarily have to be computer savvy and adaptable in the use of various types of software tools. GIS analysts must be self-starters and excellent time managers, since they are working on various tasks or assignments at the same time and solving complex problems through the use of technology. GIS analysts might be limited in the technological tools available within the organization, but it is their responsibility to provide the best possible solutions and analytical resources using those tools, while also raising the organization's awareness of additional software needs.

GIS analysts must be effective communicators, since they often are required to provide mapping needs to various users and have to be able to explain how GIS technology works to anyone within the NWTF. GIS analysts may have to attend various conferences or meetings to discuss the needs of the NWTF and give presentations. They also need to exercise patience when providing GIS or GPS training or tutorials for NWTF staff, since these users have various levels of experience, ranging from novice to expert.

GIS analysts typically have a normal work schedule, from 0800 to 1700 daily, for 40 hours per week. Flexible hours are available to conduct software upgrades or run heavy analyses after normal business hours, and arrangements can be made to occasionally work from home.

Background Needed

Because GIS analysts are heavily involved with computer technology, a strong background in computers and various technologies is critical. Previous experience as a GIS intern or GIS technician in some capacity is highly desired. Time management, project management, and effective writing and public speaking skills are required. Experience with computer technology, including server technology, software installations, hardware support, and a general knowledge of various software packages outside of GIS are necessary.

Education Required

A bachelor's degree in wildlife, GIS, natural resources, or forestry, coupled with certification in GIS or computer science is required. A combination of degrees in both the conservation and GIS fields would make you a very strong candidate for a GIS analyst's position with any conservation organization. A PhD degree could be useful if the position is heavily research based.

Pay Scale

Depending on education and experience, the base salary for a GIS analyst can range from $35,000 to $40,000 per year, with a benefits package for full-time employees. This includes health, dental, vision, and life insurance, which starts after the first month of employment. Full-time employees are also eligible to contribute to a 401(k) retirement plan after the first six months on the job. This contribution is matched by the NWTF once three years of service has been reached. Promotions and pay increases depend on an employee's performance and cannot be guaranteed from year to year.

ROCKY MOUNTAIN ELK FOUNDATION
M. David Allen and Becky Bennett

Agency: Nonprofit Organization, Rocky Mountain Elk Foundation

The Rocky Mountain Elk Foundation (RMEF), formed in 1984, is the largest nonprofit organization focused on big game and these animals' habitat in the United States. Our mission is to permanently protect crucial elk winter and summer ranges, migration corridors, calving grounds, and other vital areas, while focusing on securing and improving hunter access throughout elk country. The RMEF has five primary mission priorities: (1) permanent land protection, (2) habitat enhancement, (3) elk restoration or reintroduction, (4) hunting heritage, and (5) public access to land. Each of these priorities requires specialized staff and skill sets. Our land conservation tools include land acquisitions, access agreements and easements, conservation easements, land and real estate donations, and land exchanges. The RMEF believes that healthy habitat is essential for healthy elk populations. The organization helps fund and conduct a variety of projects to improve essential forage, water, cover, and space components of wildlife habitat, and it supports research and management efforts to help maintain productive elk herds and habitat.

The RMEF has more than 205,000 members across the United States, with just over 500 local grassroots chapters that are driven by roughly 10,000 volunteer members. It operates on a $60 million annual budget, which includes land transactions. We currently have endowment funds that exceed $50 million, with strategic plans to double that by 2025. The RMEF is known for its high-quality staff capabilities and mis-

sion results. We have conserved or enhanced more than 2.6 million ha of land in our 30-year history.

Lands Program Manager
Job Description

The primary responsibility of lands program managers (LPMs) is to cultivate and complete projects under the permanent land protection program. LPMs seek opportunities for fee-title acquisitions, increased public access, purchased and donated conservation easements, assembled land exchanges, and land donations. They assist in developing and furthering focus areas and plans in which land projects are pursued. LPMs occasionally help with related habitat conservation activities, which can include habitat enhancement, conservation planning, conservation education, and public relations. Responsibilities also include seeking planned gifts, individual donations, and bargain sales (the donation of goods or services to a charitable organization) to raise funds for the RMEF's mission efforts.

The essential functions of LPMs include lands selections, lands transactions, agency and organizational interactions, budgets, public relations, and conservation easement monitoring. For lands selections, LPMs assist in developing primary focus areas for the lands program activities in the RMEF's state, regional, and national conservation planning efforts. They also may help with more-comprehensive planning. Within lands transactions, LPMs are responsible for assessing project priorities, determining the type of transaction, negotiating the terms and conditions, preparing and presenting appropriate documents, and coordinating with all parties relevant to the transaction. Contracting for appraisals and ensuring due-diligence and timely legal reviews is required. In addition, LPMs maintain complete files on active projects and work with the lands specialist and conservation easement manager to ensure that the proper files are transferred to the permanent record at headquarters. They recommend land projects to the director of lands and the vice president of lands and conservation, prepare board committee summaries, and work closely with the lands and legal departments for project reviews and advice.

LPMs are expected to effectively network with a variety of agencies, corporations, land trusts, conservation organizations, and individuals to seek projects and partnerships. Regular and frequent coordination and communication with the legal department, other related staff, regional directors, development officers, and volunteer state chairs is essential. Another responsibility is fundraising for the RMEF through negotiated opportunities, such as bargain sales and planned gifts related to the lands transactions, as well as developing relationships and seeking donations through habitat partnerships and stewardship grants.

LPMs must devise an annual work plan and budget and periodically monitor the progress of these components throughout the year. They also prepare land project summaries, assist in land project dedications, and speak at selected events to enhance the RMEF's visibility. Lastly, LPMs are required to monitor RMEF conservation easements annually and maintain a productive relationship with the owners of these lands. LPMs are responsible for ensuring that the terms and conditions of RMEF conservation easements are complied with, and they should be able to recommend stewardship projects that are mutually acceptable to the landowner and the RMEF.

Background Needed

A thorough knowledge of the RMEFs mission, goals, organizational structure, and activities, which include the biological, social, and political elements that shape them, as well as an ability to promote them are required. Having a sound grasp of lands transaction procedures; their legal requirements, documentation, and tax implications; and the relationships among the parties involved is essential. Knowledge and understanding of biology, habitat relationships, and the basic principles of wildlife ecology, range ecology, and forest management are preferred. An ability to identify funding sources and seek financial support is needed. Previous work-related experience dealing with at least one level of government is required, as is an ability to develop and manage a budget. Excellent written and verbal communication skills and proficiency with computer software, including Microsoft Office, digital photography, GIS, and mapping programs are beneficial. Because LPMs interact closely with a wide variety of people, strong interpersonal and negotiation skills and an ability to work constructively and communicate with a diversity of private landowners and other individuals are critical. A demonstrated capacity to function independently, and as a team member when needed, is essential. Previous fundraising experience, such as in planned giving and estate planning, and an ability to solicit gifts and nurture donors are mandatory. Frequent travel is required, by automobile, airplane, horseback, on foot, or in off-road vehicles. LPMs occasionally are required to travel through rough, dangerous terrain. Physical exertion in getting to and inspecting properties demands that individuals be in very good to excellent physical shape and health. Walking, hiking, climbing, and horseback riding may all be required, so your past experiences in such activities should be highlighted in a job application.

Education Required

A minimum of a bachelor's degree in natural resources or a closely related field from an accredited college or university is required. A minimum of five years of full-time professional experience in the real estate, wildlife conservation, land management, or stewardship field or in a combination of these areas is desired. Background and experience in title commitment review, appraisal review, fee-simple real estate transactions, conservation easements, and project negotiation are very important. An understanding of wildlife and wildlife habitat and knowledge of and experience in relationships with state and federal agencies and land trusts are desired. General comprehension of planned giving, tax implications, and IRS regulations as they apply to land conservation is also helpful.

Pay Scale

The annual salary for lands program managers ranges from $50,000 to $80,000, plus a benefits package that is considered to be worth an additional 34% of the salary. The RMEF provides health, life, and disability insurance. In addition, the RMEF offers a 403(b) retirement plan, a health savings account, and paid leave.

Director of Lands

The director of lands is responsible for leadership of the RMEF lands program, including specific aspects of purchased and donated easements, assembled land exchanges, gifts of real estate, the disposal of lands, management, program direction, and budget and strategic planning for all permanent land protection strategies. These efforts and specific projects flow up primarily through the LPMs. Additional responsibilities include helping with program fundraising and technical and administrative oversight in relation to the primary job functions. The director of lands reports to the vice president of lands and conservation.

Job Description

The day-to-day duties vary by priority; however, the position has eleven primary components. One, to assist in strategic planning and cost efficiency by providing analyses and recommendations for the lands program. Two, to establish short-term and long-range prioritized goals for the lands program and develop objectives and strategies to accomplish these goals. Three, to supervise LPMs (who are mostly located elsewhere within the United States), develop annual work plans and performance standards incorporating lands program and relative organizational goals, and oversee independent contractors who provide services related to the lands program. Four, to develop, recommend, and implement policies for the board and operating procedures and policies for the lands program. Five, to serve as board liaison for lands activities, as requested by the vice president of lands and conservation. Six, to prepare and manage the budget for the lands program in close coordination with other land and conservation programs and with other departments, such as marketing, development, and field operations. Seven, to develop and implement a land conservation training program for LPMs that promotes the sound use of real estate conservation practices, relationship-building abilities, and basic fundraising skills. Eight, to develop and coordinate lands development efforts to secure funds from land transactions, and to seek funding from private landowners, foundations, corporations, and state and federal grants. Nine, to work as a positive team member with lands and conservation program directors, legal staff, field operations personnel, and others to enhance coordination and increase efficiency with the RMEF. Ten, to ensure that conservation easement monitoring is conducted on all projects, as appropriate. Eleven, to oversee all related due diligence with regard to land transactions and contract provisions, to coordinate and approve the financial details of land transactions, and to work closely with the RMEF'S legal department. In addition, the director of lands ensures that LPMs coordinate regularly and frequently with regional directors and state leadership teams; represents the vice president of lands and conservation in that person's absence; oversees the management and maintenance of all lands department files, conservation program records, processes, and data storage; and represents the RMEF at local, regional, and national meetings, as well as at seminars and symposiums that deal with natural resource management, land trust activities, and related fields of resource management.

Background Needed

A strong background in natural resource management, land conservation, or conservation leadership, along with demonstrated experience in real estate practices, including sales, acquisitions, easements, and the planning and development of related transactions are essential. A minimum of five years of professional work in real estate, wildlife conservation, land management, or the equivalent is required. Demonstrated expertise in supervision, leadership, budgeting, and program and staff management is required. Excellent oral and written communication skills are essential. The director of lands must be able to work within a team, deal with multiple levels of government, think critically, take appropriate risks, and visualize the big picture. Office work and field conditions are both encountered in this position. Office work requires sitting or standing for long periods of time; the frequent use of office equipment, including telephones, faxes, and computers; and light lifting, bending, and other forms of physical exertion. Travel by automobile, airplane, or on foot is required. Occasional travel on horseback or in an all-terrain vehicle may be necessary. Evening or weekend work may be required.

Education Required

A bachelor's degree in natural resource management or a related field from an accredited college or university is required. Additionally, a law degree (JD) is preferred.

Pay Scale

The annual salary for the director of lands ranges from $78,000 to $110,000, plus a benefits package that is considered to be worth an additional 34% of the salary. The RMEF provides health, life, and disability insurance. In addition, the RMEF offers a 403(b) retirement plan, a health savings account, and paid leave.

Conservation Program Specialist
Job Description

Conservation program specialists (CPSs) manage the RMEFs project advisory committee (PAC) and state grants programs, including the tracking and processing of grant requests from multiple states. CPSs regularly communicate and coordinate with RMEF field staff to ensure program implementation, and they work closely with the LPMs on various aspects of the programs. CPSs perform and manage functions related to the overall implementation of the RMEF's PAC and state grant programs, including fielding inquiries from prospective grantees; disseminating forms; preparing pre-PAC meeting memos; processing project approval information; notifying successful applicants; reviewing contracts and collection agreements, and routing them for appropriate signatures; entering data in the mission database; maintaining manual and electronic filing systems; monitoring the PAC budget; keeping contact information for PAC members current; monitoring the state grant program budget; soliciting and processing project completion reports and photos; and maintaining the state grant website. CPSs coordinate closely with field staff to see that deadlines are met, policies are followed, and questions are answered, to ensure program implementation. In addition, CPSs coordinate and track PAC meetings established by field staff; attend PAC meetings when necessary and, on special request, assist regional directors in the preparation of PAC grants prior to PAC meetings; communicate organizational policy related to the PAC; request the director of science and planning to review PAC projects, in order to ensure that these projects meet the RMEF's mission; act as staffers to the national project review committee; prepare PAC projects and correspondence for the committee's review; schedule national project review committee meetings; and process departmental and project expenditures, including check requisition preparation and authorization, transaction record keeping, correspondence, vendor inquiry responses, and communications with the RMEF's accounting department for PAC and state grant programs and contractors. CPSs work closely with the LPMs to ensure accurate and effective communications regarding mission accomplishments associated with the RMEF's lands and conservation programs (lands and PAC and state grants), and to see that document and project naming, tracking, and reporting systems are cohesive across RMEF programs.

Background Needed

CPSs are required to have a thorough knowledge of the RMEF's mission, goals, organizational structure, and activities, which include the biological, social, and political elements that shape them, as well as an ability to promote these aspects of the RMEF. Knowledge of wildlife management, habitat management, and project and conservation planning is essential. CPSs must be extremely detail oriented and accurate, as well as able to focus on goals and priorities in the midst of constant interruptions. A demonstrated ability to communicate professionally and effectively in writing and verbally, especially to groups of individuals, and to give professional presentations is essential. CPSs must be able to organize and prioritize numerous tasks with potentially conflicting deadlines, demonstrate flexibility as priorities and deadlines change, and successfully track timelines and meet deadlines.

A working knowledge of accounting, Microsoft Office, and relational databases is required. CPSs must consistently exhibit good judgment and have a well-developed customer service ethic. A demonstrated ability to work independently, without direct supervision, is essential. Some travel is required, including on weekends and overnight.

Education Required

A bachelor's degree in wildlife management, wildlife biology, or a similar field from an accredited college or university is required. In addition, CPSs must have two to five years of progressive experience in a natural resource field.

Pay Scale

The annual salary for conservation program specialists ranges from $30,500 to $51,000, plus a benefits package that is considered to be worth an additional 34% of the salary. The RMEF provides health, life, and disability insurance. In addition, the RMEF offers a 403(b) retirement plan, a health savings account, and paid leave.

Regional Director
Job Description

Field operations are the primary fundraising structure for the RMEF. Regional directors organize local volunteer chapters, which hold fundraising events (principally banquets) and engage in major gift development activities to support the continued operation of the RMEF and finance elk and wildlife conservation projects. In addition to event-related fundraising, regional directors solicit major gifts, either directly from donors or through RMEF members, volunteers, and other contacts. Regional directors in designated elk states generally serve as the chair of their respective state PAC, from which conservation project recommendations for RMEF funding are made. Specifically, regional directors form RMEF fundraising chapter committees, which are comprised of local volunteers; organize, guide, and assist chapter committees with the planning and execution of successful fundraising events, including big game banquets, to meet annual budget projections; actively rely on key RMEF volunteers, including regional, state, and district chairs, to assist in the identification and formation of new chapters and the planning and execution of fundraising events; engage regional and state chairs to ensure volunteer compliance with the RMEF's operating and financial policies and reporting procedures; personally solicit and use RMEF volunteers to support the solicitation of major gifts, in addition to banquet revenues; prepare annual fundraising income projections and operating expense budget proposals; maintain liaisons and professional visibility with state wildlife agencies, federal wildlife and land management agencies, university wildlife departments, and nongovernmental conservation organizations, sportsmen's groups, and outfitter/guide associations within their assigned region; serve as chair of the RMEF's state PAC within their assigned region,

to coordinate the recommendation of specific conservation projects for RMEF funding; promote the goals, objectives, and accomplishments of the RMEF through media outlets and public speaking engagements; represent the RMEF at various meetings, functions, and public relations opportunities; answer correspondence about RMEF activities and programs within their assigned region; and maintain records and submit reports, as required by the vice president of field operations. Regional directors are responsible for working with key volunteers to establish a state leadership team and, together with these individuals, to formulate and follow a state action plan annually. Regional directors serve on the state leadership team and oversee the state grant program's process of solicitation, review, and recommendations for approval.

Background Needed

Regional directors are required to have a thorough knowledge of the RMEF's mission, goals, organizational structure, and activities, including the biological, social, and political elements that shape them. They need to possess excellent people skills and be able to develop effective relationships with a wide variety of individuals. Regional directors have to (1) identify future chapter opportunities and recruit volunteers to capitalize on these opportunities, in order to expand the revenue and volunteer base; (2) work cooperatively with wildlife and land management professionals in governmental agencies and nongovernmental entities; (3) develop, cultivate, and solicit major gift prospects and follow the region's development process, while working in unison with the region's major gift officer, the vice president of operations, and other staff members; (4) resolve conflicts, be flexible with changing priorities, and be able to manage multiple priorities simultaneously; and (5) work in and foster a team-oriented environment. Regional directors need to be self-starters, as well as have excellent organizational skills and be able to direct and motivate volunteers. They must be able to work independently, without frequent direct supervision, and to remain focused on goals and objectives while also supporting the regional team when needed. Excellent verbal and writing skills are necessary. Proficiency in the use of office computer programs, including word processors and spreadsheets, and an ability to understand and determine annual fundraising income projections and operating expense budgets are required.

Extensive travel is required on weekdays and weekends. Regional directors must also be able to work on any day of the week, including weekends, and be willing to work late at night (often until after midnight).

Education Required

A bachelor's degree in business, marketing, fundraising, communications, or a related field required, and a minor in wildlife ecology is preferred. A minimum of five years of experience is required. Prior work with volunteers and in fundraising is preferable.

Pay Scale

The annual salary for regional directors ranges from $50,000 to $80,000, plus a benefits package that is considered to be worth an additional 34% of the salary. The RMEF provides health, life, and disability insurance. In addition, the RMEF offers a 403(b) retirement plan, a health savings account, and paid leave.

Bugle Hunting Editor
Job Description

The *Bugle* hunting editor (HE) conceives, assigns, selects, edits, and writes hunting and human interest articles for publication in the *Bugle* and other media to enhance and advance RMEF's mission. In every issue, the HE is responsible for the generation, editing, and circulation of editorial content, including 3–7 feature stories and 10 columns and departments.

Specifically, the HE conceives story ideas for features, departments, and special sections that focus on elk hunting. The stories should illuminate and celebrate the hunting experience, while also promoting ethical field behavior and critical thinking about hunting. Above all, the stories should be a pleasure to read. The HE works closely with the editorial team to determine the most effective ways to explore critical issues facing all aspects of hunting and to define well-balanced slates of hunting features and departments for the most imminent three (and preferably six) issues of the *Bugle*. The HE compiles, prioritizes, and edits columns within the *Bugle*, such as "Bugling Back," "Base Camp Elk," "Team Elk," "Gear 101," "Hunting Is Conservation," "Scouting Report," "Situation Ethics," "Rifles & Cartridges," "Bows & Arrows," and "Carnivore's Kitchen." The HE ensures that stories are factually correct. In addition, the HE cultivates contacts with freelance writers, RMEF staff, and sources in the field to determine the best-suited writer or writers and sources for each hunting department or feature; then assigns hunting stories to freelancers or staff writers; and, lastly, ensures that the stories progress smoothly, arriving on or before the writers' deadline for a given issue. The HE collaborates with creative services to ensure that the design, photography, and illustrations in the hunting stories provide the best possible complement to the magazine's editorial content and form a coherent whole. The HE also works with the sales team to balance editorial content with maximum revenue, all the while keeping the integrity of the *Bugle* and the needs of its readers at the forefront. The HE establishes relationships with companies committed to the RMEF's mission and works to integrate them organically within the *Bugle* and on the web.

The HE receives and responds to all query letters regarding potential hunting articles, selects unsolicited manuscripts suitable for publication, and passes them on to the editor-in-chief for review, prior to purchasing them. The HE also responds to requests for articles by sending PDFs or physical copies to inquiring individuals and agencies. The HE critiques the content, style, and production process of each issue of the *Bugle* once it is published and offers suggestions for how all three of these aspects can be improved. The HE also edits copy for books, the RMEF's website, advertisements, calendars, TV shows, and other projects as needed.

Lastly, the HE represents the RMEF and the *Bugle* at events, such as the Outdoor Writers Association of America conference, the SHOT Show, and Elk Camp, in order to communicate with current and potential writers and establish relationships with RMEF licensees and potential licensees to meet their needs in the *Bugle* and on the Internet. Some travel is required, which may include by plane, car, horseback, and on foot.

Background Needed

The HE needs to have previous hunting experience and a deep appreciation of hunting as a wildlife management tool to relate to the *Bugle*'s readership. A knowledge of biological, social, and political issues influencing elk hunting, and hunting in general, is required, so reading and understanding a broad spectrum of technical and popular literature, attending symposia, and using the Internet are essential. Excellent written and verbal communication skills are mandatory. Mastery of style and grammar, along with a deep knowledge of wildlife and wildland conservation, are imperative. A portfolio containing written examples of authored articles is required, to demonstrate your writing abilities. Past experience with computers and competency with software such as Word, Outlook, Excel, Oracle, and the Internet is crucial.

Education Required

A bachelor's degree in journalism, creative writing, or wildlife ecology from an accredited college or university is essential. A demonstrated ability in journalism and creative writing is mandatory. A minimum of three years of experience editing and writing stories that focus specifically on hunting and issues affecting hunting for magazines or newspapers is required. Coursework in English composition, creative writing, journalism, publications, copyediting, advertising, and photojournalism is beneficial. If your degree is in journalism, then classes in wildlife ecology, wildlife law, and wildlife ethics are beneficial.

Pay Scale

The *Bugle* hunting editor's salary is commensurate with experience. The HE also receives a benefits package that is considered to be worth an additional 34% of the salary. The RMEF provides health, life, and disability insurance. In addition, the RMEF offers a 403(b) retirement plan, a health savings account, and paid leave.

President/CEO

I [MDA] am confident that my job description is substantially different from anyone else's, because I am very aware that I am not the typical executive of a wildlife conservation organization. It is my hope that my example provides a perspective that is outside the box, encourages creative thinking, and presents a novel perspective to those who are about to embark on a journey to develop a career in the outdoor industry.

This is a second career for me, following nearly 35 years of sports and event marketing, including managing corporate sponsorships for NASCAR, Wrangler Jeans, and the Pro Bull Riders tour since its inception, nearly 25 years ago. It is my desire for young adults to see that a combination of sound business experience and a passion for the outdoors and wildlife are as valuable as any specialized background. It is important that you recognize that there are multiple ways to engage and succeed in a wildlife career.

Job Description

The role of chief executive officer (CEO) at the RMEF requires me to wear many hats, but it makes for a stimulating and rewarding job. On both a weekly and a monthly basis, the CEO spends about 20% of the time engaging with chapter members and board directors, 15% in marketing and communications, 15% in planning and formulating a strategic vision for the organization, 15% in dealing with relevant current issues, 10% in fundraising, 5% in handling administrative issues, 5% in dealing with other NGOs, 5% in working with finances and budgets, 5% in mentoring staff, and the remainder in dealing with miscellaneous activities that always seem to develop.

The lifeblood of the RMEF comes from its grassroots chapter system of fundraising. More than 80% of the organization's operating revenue is generated through this system. The RMEF does not operate on government funding. As a result, it requires the maintenance of an efficient and supportive chapter-based system of members and volunteer members. A significant amount of the CEO's time is invested in staying engaged with the chapters and their local members. In addition, the CEO reports directly and answers to the RMEF's Board of Directors. Specifically, the CEO reports to the board's chairman, a position that can change every year, but more times than not does so every two years. As a result, the CEO has a rotating, always fluid set of board members, or bosses. The board as a whole expects the CEO to keep them informed of the status of the RMEF and holds the CEO accountable for the overall performance and results of the organization. It is best for the CEO to remain at arm's length from all board members, meaning that the CEO should not develop close friendships with board members, to avoid any areas of conflict. Board members come from diverse backgrounds, and they bring various perspectives and ideologies with them. This can be a good thing, but it also can present challenges.

All successful organizations, whether private or public, start with a focused vision. It is the role of senior management to keep this focus in front of the entire organization and require all facets of the organization to deliver results geared to this strategic framework. The CEO's role in the RMEF's strategic vision and planning process is twofold. One is to work with senior staff and the board of directors to establish a strategic plan every five years. Second is to make sure that staff members are successfully working toward these strategic goals on an annual basis. A strategic plan does no good if you do not integrate it into the entire performance of the organization.

The CEO is required to become the spokesperson for the RMEF as a whole. This requires you to get involved in many issues arising from national and state legislation that affect wildlife and land trusts. Additionally, the CEO becomes the voice of the RMEF for the media in matters such as wildlife policies, hunting regulations, collaborative partnerships with other NGOs and state and federal agencies, and controversial issues, such as predator management policies.

Fundraising is how groups like the RMEF exist and survive. Nearly 100% of the revenue the RMEF receives is generated from funds raised in one manner or another. Chapter banquets hold annual fundraising dinners. The RMEF sets financial goals for all field staff, and it has a small cadre of fundraising officers, who are responsible for major gifts. The CEO's role in fundraising varies, depending on the circumstances. The CEO attends many of the RMEF's grassroots banquets across the country, primarily to show appreciation for the chapter volunteers who put on these banquets. These volunteers are humbling when I witness the amount of passion and dedication they bring forth in supporting and promoting the RMEF. Additionally, the CEO's role is to be available as a spokesperson for the RMEF when any staff member has a major gift prospect who desires to hear from the management of the organization relative to the status of the RMEF's mission.

The CEO is accountable to the board of directors for the annual fiscal performance of the organization. The RMEF has a chief financial officer who manages the day-to-day financial operations and regularly updates the CEO on finances, cash flow, and annual budget projections.

The RMEF is structured so that only one person has to report to the CEO, and that is the chief operating officer (COO). The COO manages the senior staff. The CEO is very dependent on the skills of the COO, and, therefore, needs to hire and retain an individual who is very capable and well versed about the organization.

Maintaining a succession plan for the senior management of the RMEF is critical. No one lasts forever, and no one should be in any one job or role for too long. How long is "too long" depends on the specific job and its related results. Nonetheless, it is the role of the CEO and the COO to keep an active chart of key individuals and crucial senior management jobs, asking, "If we had to fill this specific role tomorrow, who would it be, or do we have anyone prepared for this role?" This requires keeping an inventory of RMEF staff members, their career interests, and their job skills. It also requires spending some time with each of them, one on one, to develop a rapport and create a healthy management team. The CEO needs to be cognizant of getting too close to any one individual on the staff, as there is a line that, if crossed, can result in the CEO losing objectivity regarding that staff member's performance and potential.

There is always that out-of-the-ordinary request or need when you lead a nonprofit group. For example, I [MDA] often get asked to talk to many types of groups on a variety of subjects related to conservation and wildlife. Sometimes this pertains to an area of our business and sometimes not, but I am pretty committed to responding to or engaging with most requests or expectations.

Background Needed

There is no magic formula when it comes to having a successful career in the wildlife and conservation world. It takes devotion to the resources and a passion for the people who are most engaged with those resources. In my particular case, I [MDA] grew up with a heavy sportsmen's background, as my family hunted and fished year round. I was taught that there is a certain code and an ethic that is expected of those of us who used the outdoors. The ethic is that wildlife is a sustainable resource, but only if we are good stewards of that resource. I am proud of my hunting and fishing heritage; it has taught me many life lessons. The most significant requirement for the CEO of the RMEF is to demonstrate great respect for the outdoors and wildlife, and to be mindful of mankind's obligation to and accountability for managing them well. Humans do not live in a zoo. We must keep our own sustainability in perspective, or we run the risk of marginalizing both the outdoors and the wildlife within it.

In terms of more concrete characteristics, prior managerial experience and demonstrated fiscal responsibility are beneficial. Being able and willing to communicate with a diversity of people from a wide range of backgrounds is critical, because the majority of the CEO's job deals with people.

Education Required

A broad background in wildlife science or ecology is not mandatory for the CEO of the RMEF, because the organization has plenty of solid scientific resources and is supported by a broad spectrum of scientists. An educational background in business, finance, or communications, however, can be useful, because the CEO is accountable for the RMEF's endowment fund (which today is in excess of $50 million) and is the organization's spokesperson, representing its more than 205,000 members.

Pay Scale

The CEO's salary is commensurate with experience. The CEO also receives a benefits package that is considered to be worth an additional 34% of the salary. The RMEF provides health, life, and disability insurance. In addition, the RMEF provides a 403(b) retirement plan, a health savings account, and paid leave.

THE NATURE CONSERVANCY
Joe Fargione

Agency: Nonprofit Organization, The Nature Conservancy

The Nature Conservancy (TNC, or Conservancy), founded in 1951, is a nonprofit conservation organization whose mission is "to conserve the lands and waters on which all life depends." The Conservancy is solution and partnership oriented, and it relies on science in deciding on the direction, focus, and priorities of the organization. The Conservancy's staff members work in hundreds of communities across the United States and around the world. They are supported by nearly 1 million TNC members and by state boards of trustees, made up of local leaders in conservation, business, agriculture, ranching, academia, and philanthropy. TNC has helped conserve more than 40 million ha in the United States and around the world.

Biologist
Job Description

Entry-level science positions at TNC are listed under such job categories as applied scientist I, applied scientist II, and program director I. Actual titles could be designated as prairie ecologists, forest ecologists, terrestrial ecologists, fisheries biologists, marine scientists, and program directors for a specific chapter or region.

Scientists at TNC work with colleagues and external partners to identify priority conservation areas and opportunities, develop conservation strategies and management plans, monitor the effectiveness of management practices or other conservation strategies, and assess the status of and threats to species, habitats, and ecosystem services. TNC commonly works with state and federal agencies, academics, other nonprofits, and corporations to develop and promote conservation plans and actions. Particular positions are typically focused on a subset of the above activities. For example, the jobs of some biologists may be centered on studying and monitoring the effectiveness of particular management actions, such as oyster reef restoration, longleaf pine restoration, forest thinning, patch-burn grazing (on land that underwent a prescribed burn), and the monitoring of aquatic bioindicators of reduced nutrient and sediment pollution from agriculture. Other biologists may focus on modeling species, habitats, and ecosystem services, to identify which areas are likely to be more resilient to climate changes, which ones are most important for habitat connectivity, and which ones provide important biodiversity habitat and ecosystem services. Job duties for entry-level positions often include conducting fieldwork, managing seasonal staff, writing proposals, developing collaborations and partnerships with agencies and universities, generating reports and articles in peer-reviewed publications, serving as science translators for diverse internal and external audiences (e.g., governmental relations staff, marketing staff), and engaging with stakeholders.

The precise mixture of these activities varies greatly, but all Conservancy science positions help create and communicate information from this discipline to undergird its conservation strategies, including the protection and management of lands and waters. TNC owns more than 800,000 ha of land and holds about 1.2 million ha of easements. Conservation easements restrict development on and other uses of private lands, and they are bought from or donated by landowners. Easements typically limit the subdivision of lands and the conversion of natural habitat. TNC promotes land management to improve biodiversity conservation. The organization also recognizes that the successful conservation of biodiversity requires a much broader strategy than that which can be achieved simply by buying land and easements, so it seeks to improve the conservation and management of lands through all available means, such as improved federal and state policies and corporate practices. For example, TNC often comments on federal land management policies, and it works with corporations on the location of and mitigation for new developments. Scientists often work with other TNC staff on teams dedicated to corporate practices, philanthropy, and governmental relations. Conservancy scientists provide subject matter expertise and scientific support for these efforts.

TNC does not maintain research laboratories, and it cannot fund all the research necessary to discover solutions to the challenges facing conservation today. Instead, the organization must join with academic and agency partners to conduct research that is essential to its mission. Conservancy scientists routinely initiate and participate in collaborative research to address applied questions. These studies are designed to assist with conservation actions and investments, such as the acquisition of new public lands, the purchase of conservation easements, the restoration of terrestrial and marine habitats, and the adoption of sustainable agricultural practices. For example, TNC collaborates with academics to investigate new conservation practices to reduce nutrient losses from agricultural fields, study the erosion and storm surge reduction potential of reefs and other natural infrastructure, and assess the effectiveness of novel grazing techniques for maintaining prairie diversity. Consistent with developing these collaborations and their research capacity, Conservancy scientists also are encouraged to serve as adjunct faculty members in local universities, to facilitate networking and research partnerships.

Background Needed

The required background for TNC biologists varies greatly, depending on the position. Important attributes for all such positions, however, are good verbal communication skills, prior field research, experimental design skills, a track record of scientific publications, and experience in applied conservation. Previous use of GIS is highly beneficial. An understanding of local species and natural communities; knowledge of current trends in the ecology and management of a particular natural system of interest; an ability to synthesize, interpret, and communicate scientific information to influence conser-

vation practice; and the capacity to collect, manipulate, analyze, and interpret scientific data and prepare reports on these findings are essential. A demonstrated ability to use statistical software packages, such as SAS or R, is needed. Successful written, spoken, and visual communication experiences with diverse audiences are desirable. A track record of peer-reviewed publications is required. Project management experience, including with budgets, is beneficial. This can include supervising a field crew, or organizing research workshops.

Education Required

Most full-time science positions with the Conservancy require an MS degree. PhD candidates, however, generally have an advantage for permanent science jobs. In addition to coursework related to your primary area of expertise, knowledge of statistics, experimental design, modeling, GIS, and ecosystem services may be required, depending on the position.

Pay Scale

The salaries for entry-level positions for PhD-level scientists (or those with an MS degree and equivalent work experience) can range from $45,000 to $70,000 per year. Pay increases are based on merit. Promotion to a new position generally requires applying for that position within the Conservancy.

THE PEREGRINE FUND
Richard T. Watson

Agency: Nonprofit Organization, The Peregrine Fund

The Peregrine Fund (TPF) was founded in 1970 at Cornell University by Professor Tom Cade, as a project to prevent the extinction of peregrine falcons in the United States through their captive breeding and release. It was one of the most successful species restoration efforts ever conducted in North America, with more than 4,000 peregrines bred and then released. Through the efforts of many, the species was removed from the US endangered species list in 1999. Early in the project's history, it became evident that similar needs existed worldwide, including for the critically endangered Mauritius kestrel (a species that later became another success story), so TPF was formed as an organization, and its scope and reach of mission were expanded to encompass the globe. TPF built new facilities when it moved to Boise, Idaho, in 1984. Its headquarters is known as the World Center for Birds of Prey.

Today, TPF's mission is to work throughout the globe to conserve wild populations of birds of prey. Conserving raptors provides an umbrella of protection for entire ecosystems and their biodiversity. The Peregrine Fund is a nonpolitical, solution-oriented, hands-on, science-based organization. Goals are achieved by restoring and maintaining viable populations of species in jeopardy, studying little-known species, conserving habitat, educating students, fostering local capacity for science and conservation in developing countries, and providing factual information to the public. Since beginning its work in 1970, TPF has assisted with raptor conservation projects in more than 40 countries on six continents. Positions with TPF include project biologists, hacksite attendants, field biologists, project managers, and project directors. Positions can range from short-term, temporary jobs, such as hacksite attendants or field biologists, to long-term career positions in charge of at least one conservation and research project anywhere in the world.

Hacksite Attendant
Job Description

Hacking is a training method that teaches young birds of prey how to hunt, while also providing the young birds with exercise and experience. Hacksite attendants usually work for a few weeks or months with one principle task: to supply food to young raptors in the field (at the hacksite location) while they transition from being dependent on humans for food to hunting independently. Undergraduate students or recent graduates in the biological sciences often use these opportunities to gain valuable field experience both to build resumes and to enjoy total immersion in fieldwork, often in remote and beautiful landscapes. For many, it can be the most rewarding experience of their lives. The Peregrine Fund has employed thousands of hacksite attendants since first beginning releases of captive-bred peregrines in many of the lower 48 states in the 1970s (Burnham et al. 2003). Since then, hacksite attendants have supervised the release of species such as aplomado falcons and California condors in the United States, and Mauritius kestrels, orange-breasted falcons, harpy eagles, and Ridgway's hawks in Mauritius, Belize, Panama, and the Dominican Republic. TPF hacksite attendants have gone on to pursue careers in medicine, obtain graduate degrees, and write books. Some became directors of TPF, and one was the director of the US Fish and Wildlife Service. After completing postgraduate training or additional work experiences, former hacksite attendants who worked hard and demonstrated the important qualities of problem solving and perseverance are often preferred candidates when openings for more-senior positions occur within TPF.

Background Needed

Hacksite attendants must be at least 18 years of age. They are usually undergraduate students or recent graduates with a biology-related degree, though other kinds of experience are also considered, such as being a licensed falconer. Some ornithological or other wildlife experience in the field is desirable but not mandatory. Most hacksites require attendants to know how to work outdoors in rugged terrain and in all kinds of weather, from extremely hot to very cold. Feeding raptors usually occurs in the early morning and late afternoon hours, with observation taking place in between. Due to both the short time period in which the hack can be completed and to the remote locations where hacking occurs, hacksite attendants usually work seven days per week, unless arrangements are made in advance for time off.

Hacksite attendants provide food to young raptors in the field until the birds learn to hunt independently (photo: Scott Henke)

Education Required

Hacksite attendants can be current undergraduate students or recent graduates with a bachelor's degree in wildlife or the biological sciences. They must have hands-on experience with raptors.

Pay Scale

Hacksite attendants receive a subsistence stipend, with the amount depending on the site's location, as well as other benefits, such as housing and transport.

Field Biologist
Job Description

Field biologists working for TPF conduct field studies, as directed by their immediate supervisor. Their work can include tracking the movements of free-flying raptors using radiotelemetry, taking detailed behavioral notes, observing breeding activity and prey brought to the nest, counting raptors or their prey populations, and taking detailed habitat measurements. The job requires long, sometimes strenuous days in the field. Applicants must be able to carry heavy equipment and supplies (which include food for raptors and water) over rough terrain, sometimes in the dark. Interacting and sharing information on these charismatic birds with the general public can be an important part of the job. Field biologists are responsible for the management of project equipment, including vehicles, radios, and optical and computer equipment. Employees can expect to have unusual work schedules, such as four 10-hour days on, followed by three days off per week, although schedules are adjusted to meet project needs. Most of the biologists' time is spent in the field.

Background Needed

Field biologists usually need a biology-related undergraduate or graduate degree, in addition to relevant fieldwork experience. This is not a desk job. Applicants must be highly energetic and willing to work long days, while maintaining a positive attitude. Successful field biologists must be able to tolerate climate extremes, be in excellent physical condition, and be prepared to work independently in remote areas. They should be confident in driving standard four-wheel-drive trucks over remote back roads. Previous experience in using telemetry and GPS units and making behavioral observations and notes is desired.

Education Required

A bachelor's degree in wildlife ecology or related biological sciences from an accredited college or university is required, as is demonstrated field experience, such as in summer jobs or internships while in college.

Pay Scale

The salary for field biologists starts at around $20,000–$30,000 per year (depending on other benefits, such as provided housing), plus a benefits package. Benefits for full-time employees include health, dental, vision, and life insurance. Beginning after the first year of employment, TPF makes a contribution of up to 10% of an individual's salary into a 403(b) retirement plan. Promotion from one level to another within TPF depends on academic degrees and a demonstrated ability to meet the required skills of project and staff supervision and leadership. People who do well in this organization put others' needs before their own and produce tangible results at a plausible rate.

Project Manager
Job Description

Project managers generally run one of TPFs raptor research and conservation projects by supervising a field team, usually under the guidance of a project director. They must be able to design and adapt field studies to understand the distribution and abundance of raptors and their limiting factors; conduct ecological field studies on raptors, in habitats ranging from the Artic to the tropics; and train and lead a research team by personal example. They recruit, train, and supervise students, as well as contribute to publishing and presenting project results in peer-reviewed journals and at professional meetings. They work with the project director, administrators, and other staff to develop and direct meaningful raptor research and conservation projects in locations throughout the world. Roughly half of their time is spent in the field.

Background Needed

Project managers are usually required to have an MS or PhD degree in wildlife science, animal ecology, or a closely related discipline; experience with wildlife research field techniques; a proven ability to perform quantitative data analyses and publish in peer-reviewed scientific journals; and more than two years of postdoctoral experience or the equivalent. An understanding of and experience with raptor biology are important in this position. Proven abilities in project management and

coordination are essential, along with excellent communication skills, such as in writing and in public speaking.

Experiences that are often considered valuable for project managers include proposal writing and success with obtaining competitive extramural grants; college teaching; a track record of scholarly publications in wildlife science; collaborative research experience with wildlife management agencies or NGOs; and an ability to contribute to research, teaching, or engagement with the public. Another helpful attribute is experience studying patterns of spatial and temporal variation in avian distribution, abundance, and other ecological and population parameters (e.g., breeding success and survival) at continental or regional scales. Familiarity with spatial analysis software, such as ArcGIS, and statistical software in R is useful. Language skills that are appropriate for the location are essential (e.g., French in Madagascar, Spanish in Panama, Portuguese in Brazil).

Education Required

An MS degree in wildlife, ecology, or related biological sciences from an accredited college or university is required, as well as demonstrated field experience, such as that which may be gained during an MS degree and part-time or temporary seasonal jobs.

Pay Scale

The salary for project managers starts at $30,000–$40,000 per year, plus benefits.

Project Director
Job Description

Project directors lead one or more of TPF's raptor research and conservation programs, sometimes at a continental scale. Projects typically involve students, researchers, and other partners in the study of raptor ecology. They have multiple levels of engagement, including online and field experiences. Project directors shape their research agendas to answer significant conservation questions and publish the results in the scientific literature. They coordinate all aspects of project delivery and supervise staff, including those project managers who recruit participants, provide volunteer support for field observations, and offer guidance in entering data. Project managers help design, develop content for, update, and oversee the maintenance of the TPF's interactive website and data-entry systems; collaborate with web and software programmers, designers, and communications staff to ensure data accessibility; and promote TPF and its projects online and in other media, including popular literature. Project directors assist development and marketing staff to disseminate information about the projects and raise funds for a project's continuation, and keep the president and vice president of TPF apprised of new developments with and changes to the projects. They annually submit and manage a project's budget. They also collaborate with all other TPF projects and perform other tasks as requested and needed. As senior staff, project managers spend more time with administrative responsibilities and at the computer, analyzing data and writing papers and reports.

Background Needed

Project directors are usually required to have an MS degree or a PhD in wildlife science, animal ecology, or a closely related discipline; experience with wildlife research field techniques; a proven ability to perform quantitative data analyses and publish in peer-reviewed scientific journals; and more than two years of postdoctoral experience or the equivalent. An understanding of and experience with raptor biology is important. Proven skills in project management and coordination are essential, along with excellent communication skills, such as in writing and in public speaking.

Experiences that are often considered valuable include proposal writing and success with obtaining competitive extramural grants; college teaching; a record of scholarly publications in wildlife science; collaborative research experience with wildlife management agencies or NGOs; and an ability to contribute to research, teaching, or engagement with the public. Another helpful attribute is experience studying patterns of spatial and temporal variation in avian distribution, abundance, and other ecological and population parameters (e.g., breeding success and survival) at continental or regional scales. Familiarity with spatial analysis software, including ArcGIS, and statistical software in R is useful. Language skills that are appropriate for the location are essential (e.g., French in Madagascar, Spanish in Panama, Portuguese in Brazil).

Education Required

A PhD in wildlife, ecology, or related biological sciences is required, as is demonstrated field experience and a record of publications. Promotion from one level to another within TPF depends both on academic degrees and on a demonstrated ability to meet the required skills of the projects and staff supervision and leadership.

Pay Scale

The salary for project directors starts at $40,000–$50,000 per year, plus benefits. Promotions and pay increases are dependent on performance and do not follow a predictable scheme. Usually more than 12 months of employment is required before consideration is given to either.

THE WILDLIFE SOCIETY
Ken Williams

Agency: Nonprofit Organization, The Wildlife Society

The Wildlife Society (TWS) works to inspire, empower, and enable wildlife professionals to sustain wildlife populations and habitats through science-based management and conservation. Established in 1937, TWS represents and works on behalf of professionals dedicated to excellence in wildlife stewardship through science and education. TWS is based in Bethesda, Maryland, and has volunteer-led chapters, sections, and working groups throughout North America. The organization publishes the *Wildlife Professional* magazine and three peer-reviewed scientific journals (*Journal of Wildlife Management*, *Wildlife Society Bulletin*, and *Wildlife Monographs*), hosts an annual conference, and has a robust governmental affairs program. TWS employs individuals with diverse expertise, including those who have wildlife science and management backgrounds. Careers with TWS that benefit from a wildlife science education and experience include positions on our science writing, governmental affairs, and operations teams. TWS is led by an executive director and is organized into three departments (publications and communications, governmental affairs and partnerships, and operations), each with its own director and supporting staff.

Governmental Affairs
Job Description

TWS's governmental affairs and partnership program aims to have wildlife professionals and the knowledge they provide play an active role in the formation of wildlife management and conservation policies, laws, and regulations, thereby ensuring that these are both scientifically sound and practical. The team works to empower, enable, and assist wildlife professionals with science-based management and conservation. Governmental affairs staff provide timely, usable, scientifically grounded input to policy makers; see to it that wildlife professionals are aware of the impact policy has on their work and that they are up to date on current issues in wildlife policy; and serve as unbiased reviewers of wildlife management and conservation policy and strategies.

Specific tasks within the governmental affairs and partnership team include tracking the status of legislation, writing letters to members of Congress, analyzing the impact of policies on wildlife populations and wildlife professionals, representing wildlife professionals on various coalitions, and writing articles.

Background Needed

There are a multitude of qualities that equip individuals to be effective members of TWS's governmental affairs and partnership team. Staff members need to have a working knowledge of wildlife science and policy, and an understanding of how policies can impact wildlife management activities. An ability to multitask and work on several different items simultaneously is essential. Interpersonal and leadership skills also are beneficial. Staff members must be able to effectively interact with wildlife professionals, governmental officials, policy decision makers, and other stakeholder groups. Excellent written and verbal communication abilities and organizational skills are required. Dedication to a scientific approach to the conservation of natural resources is needed. Experience conducting scientific research is preferred, but not necessary.

Education Required

A bachelor's degree is required. An MS or other advanced degrees are preferred. One of these degrees should be in natural resources, wildlife policy and management, or a related field. A demonstrated understanding of legislative and regulatory processes is required. Preference is given to those who have TWS certification as a Certified Wildlife Biologist or Associate Wildlife Biologist, or can become certified.

Pay Scale

The pay for governmental affairs staff is commensurate with background and experience. Typical starting salaries range from $35,000 to $65,000.

Science Writer
Job Description

Science writers works with a team of editors to create feature articles and short news items for the *Wildlife Professional*, a quarterly publication that goes to all of TWS's members. In addition, science writers are responsible for composing short pieces and creating multimedia presentations and other content for TWS's website (see appendix B). These responsibilities require writers to know how to scan news outlets and scientific journals for appropriate topics in the wildlife and conservation science communities. To obtain more information on these findings, science writers interview researchers and managers, who may be academic, federal, or state wildlife professionals. Asking the right questions and taking accurate notes or recording interviews helps writers establish the integrity and accuracy of an article or news story. Further duties include proofreading, sourcing photos, and working with the editorial team to ensure that the pieces follow the standards of good journalism.

Background Needed

Excellent written and verbal communication skills are a must. An ability to work on a tight deadline and produce clean, accurate prose that meets style guidelines is required. Science writers also need to be able to translate complex science into readable and engaging articles. Excellent interpersonal skills and a willingness to work cooperatively with a small editorial team are mandatory. An ability to juggle multiple tasks, meet deadlines, and work both independently and collaboratively is essential.

Education Required

A bachelor's degree or an MS in wildlife, natural resources, journalism, mass communications, or English, or one to three years of experience reporting and writing on scientific topics is mandatory. A combination of a wildlife science background and strong writing and communications experience is preferred.

Pay Scale

The pay for science writers is commensurate with background and experience. Typical starting salaries range from $35,000 to $55,000.

Wildlife Programs Coordinator
Job Description

Wildlife programs coordinators are part of TWS's operations team. They serve as the primary liaisons to TWS's chapters, sections, student chapters, and working groups (*organization units*). Wildlife programs coordinators work with the organization units on a daily basis to ensure that all annual reporting requirements are met on time; provide advice on membership recruitment, meetings, workshops, and internal operations; and ensure that each unit is as successful as possible. Occasional travel to chapter and section meetings and student conclaves is required. Coordinators interact with TWS partners to ensure that a partnership is successful and that all deliverables (a tangible product or products produced as a result of a project) are being met on time and have the partners' approval. They assist with other aspects of the operations team as needed, including development activities, social media oversight, and website management. Coordinators also work with various publications to promote the organization units through web articles.

Background Needed

An ability to work independently as well as part of a team is essential. Excellent verbal and written communication skills are required. Interpersonal, organizational, and leadership skills are necessary.

Education Required

A bachelor's degree in wildlife management or a related field is required.

Pay Scale

The pay for wildlife programs coordinators is commensurate with background and experience. Typical starting salaries range from $35,000 to $55,000.

WILDLIFE MANAGEMENT INSTITUTE
Steven Williams

Agency: Nonprofit Organization, Wildlife Management Institute

Founded in 1911, the Wildlife Management Institute (WMI) is a private, nonprofit, scientific and educational organization dedicated to the conservation, enhancement, and professional management of North America's wildlife and other natural resources. The WMI provides services to the conservation profession through science and management expertise, program review and policy development, information, education, and conservation project administration. The WMI works primarily with federal, state, and provincial fish and wildlife agencies, other private conservation organizations, and professional associations. This entails the evaluation of the science used to establish population estimates and harvest strategies, an assessment of organizational structure and function, and program performance audits that ascertain coherence among strategic plans, budget plans, and operational plans. The WMI also advises and provides educational services on timely, wildlife-related issues, such as the Land and Water Conservation Fund, transfers of federal land, chronic wasting disease, the recruitment, retention, and reengagement of hunters, and the Young Forest Initiative.

Most WMI staff members have served for more than 20 years in federal or state governments. Our staff has found that working for a small, nonprofit organization is liberating and professionally satisfying. The bureaucracy found in larger governmental agencies does not hamper our organization. The WMI handles a relatively small portfolio of exciting and rewarding topics and projects and operates with a focused and flexible approach to wildlife conservation issues that we believe are significant for states, regions, and the nation.

Field Representative
Job Description

Field representatives are the face of the WMI in the states affiliated with the regional association of the Association of Fish and Wildlife Agencies, as well as at a national level. Field representatives work with federal and state agency staff, conservation organizations, and individuals to promote the scientific management of wildlife populations and their habitats. The WMI gives its field representatives the flexibility to prioritize their efforts, based on conservation projects relevant to their region of the country. For instance, midwestern priorities may include agricultural policy and its implementation, western priorities may include the impacts of energy development on wildlife and their habitat, and northwestern priorities may include forest management. Field representatives are expected to be self-motivated and highly productive. This productivity is measured by progress made toward conservation objectives, collaborative relationships that are established with conservation partners, and revenue generated to help sustain the WMI's operations. Field representatives must be able to work with only limited supervision and produce conservation successes at the state, regional, or national level. The WMI is engaged with some of the most important conservation issues of our times (e.g., climate changes, habitat management, hunter recruitment and retention, wildlife policy, recovery plans, energy development). It employs people who can hit the ground running and provide immediate service to conservation and to the WMI's many partners.

Specifically, WMI field representatives develop population and habitat management plans; conduct reviews of agency plans, programs, projects, policies, and organization; create informational and educational materials to advance conservation; develop and coordinate regional conservation plans and programs; serve on committees, boards, and working groups; prepare letters, reports, and competitive grant proposals; provide administrative support as required; and submit monthly travel and activity reports to WMI officers. In addition, field representatives, both independently and as part of a WMI team, are expected to develop or identify actual and potential conservation projects that would generate revenue to support the WMI's administrative and project expenses. These projects may take the form of securing conservation grants, funding from foundations, or contracts for services provided to agencies or organizations. WMI officers approve those projects that conform to WMI's mission and provide productive collaboration with our partner organizations and agencies. Field representatives are required to travel to regional and national conferences (semiannually) and to meet with conservation partners, as needed.

Background Needed

A job as a field representative with the WMI is not an entry-level position. Field representatives typically join the WMI with 15–20 years of mid- to upper-level career experience in a state or federal fish and wildlife agency, or commensurate experience with a national conservation organization. To be considered for one of these positions, you must be able to demonstrate knowledge of state, regional, and national wildlife resource policies and issues; possess communication, technical, and analytical skills; and be able to work productively both independently and in a team environment. Examples of your writing and speaking skills may be requested, in addition to references from current or past employers. Field representatives work out of home offices and must be self-motivated and entrepreneurial.

Education Required

A minimum of an MS degree in wildlife biology, wildlife management, or a closely related natural resource management–related field from an accredited college or university is required.

Pay Scale

Salaries for field representatives are commensurate with experience and competitive with mid- to upper-level management positions in state and federal fish and wildlife agencies. An average salary is approximately $85,000 per year. Also, certain office expenses and approved travel expenses are paid for by the WMI. A benefits package includes medical and dental plans, life and disability insurance, and a 401(k) defined retirement contribution plan.

WILDLIFE REHABILITATION
Kai Williams

Agency: Nonprofit Organization, Wildlife Rehabilitation Center

Wildlife rehabilitation centers are nonprofit or governmental agencies that provide care to injured, ill, and orphaned wild animals and assist area residents with human/wildlife conflicts. Organizational goals and missions focus on the conservation of species, conflict resolution, public education, the relief of animals' pain and suffering, and the monitoring of anthropogenic issues (influences of humans on nature), including lead ammunition, rodenticides, and climate changes.

Wildlife Rehabilitator
Job Description

Wildlife rehabilitators are quick thinkers who work well with people and animals. They have a passion for wildlife, but the job is more far-reaching than feeding and caring for individual animals. Many centers have limited staffs, which require their employees to be jacks-of-all-trades, ranging from construction and maintenance to veterinary nursing and habitat design. On an annual basis, rehabilitators can expect to spend 35% of their time caring for animals, 35% working with the public, 15% handling administrative tasks, and 15% managing the facility. The duties in each of these areas vary seasonally, as do the expected hours worked per week. Spring and summer months see baby animals brought to the centers, with at least 12-hour days of feedings and public education to prevent the kidnapping of young wildlife that do not need assistance. Intakes in summer and, especially, fall involve many immature species venturing out on their own and having accidents with cars, windows, diseases, and poisonings. Winter is traditionally a quieter season, with time to concentrate on records and continuing education, while also caring for a smaller number of juvenile and adult animals that are more critically injured.

One of the most important aspects of this work is interacting with the public. Rehabilitators are ambassadors between wildlife and the public. A conversation with one person is shared with friends and family and will reflect the way they handle wildlife situations in the future. Rehabilitators humanely resolve human/animal conflicts, from squirrels nesting in the attic to woodpeckers that are busy removing termites from the siding of a house and, in the process, damaging that siding. A busy center may get over 100 phone calls on a spring day, which need support from skilled animal caregivers to assess whether an animal is exhibiting natural behavior or if it may need to be admitted. Every animal that stays in the wild and does not need to come into a wildlife rehabilitation center is a success story.

Animal intakes require human interactions and wildlife knowledge. Intake rehabilitators are the public face of the wildlife center. These rehabilitators obtain the necessary history on the animal, gathering information that assists in its diagnosis and care. Often this happens at the center, but in some circumstances this occurs out in the field, where the rehabilitators deal with on-site conflict resolution or rescue and capture operations. Members of the public are usually in an emotional state during their initial interactions with a wildlife rehabilitator. They may be scared of the animal, as well as scared for the animal's welfare. Part of the rehabilitators' regular job is to counsel these individuals and help them make the best choice for the animal.

The second part of an animal intake is an initial exam and triage. Rehabilitators follow wildlife center protocol, which often includes a quick exam for immediately life-threatening problems, followed by triage care for blood loss, dehydration, and hypothermia. Once the animal has been stabilized, a more thorough examination is completed by a lead wildlife rehabilitator or veterinarian.

Additional wildlife care duties include follow-up treatments, daily rounds and observations, the feeding of young nursing mammals or the hand feeding of altricial birds (young hatchlings), and assisting with veterinary examinations and surgeries. Some interactions have a strong emotional component (e.g., euthanasia, cadaver management). Rehabilitators perform necropsies and ensure the appropriate disposal of deceased animal remains. Rehabilitators also release healthy wildlife into suitable environments.

Many of the tasks rehabilitators do on a daily basis for animals that are in a center's care are indirect. Entry-level wildlife rehabilitators can expect to spend most of their time preparing food for the animals and cleaning laundry, dishes, and cages. This unglamorous group of tasks is critical for both the animals' and human health. Rehabilitators also perform cage management, to ensure that these areas are appropriate to an animal's age and health and provide proper substrates, enrichment, and exercise options for that animal. A surprising amount of time is spent in food acquisition. This can include foraging for wild insects and plants, raising and caring for farmed insects and rodents, and soliciting grocery stores and other companies for donated produce and seeds.

Rehabilitators do extensive research on and planning for each species that enters the center. For example, when faced with a new species, I have spent countless hours reviewing natural history texts, especially volumes that contain accounts of direct observations, and being on the phone with biologists and other wildlife rehabilitators who have prior experience with that species. Such research supplies information about the diet, caging, and release criteria for each animal brought to the center, and this is an essential aspect of the job for wildlife rehabilitators.

Each individual patient has a treatment plan, created in conjunction with the center's veterinarian. The treatment plan is the culmination of subjective and objective observations, examinations, and laboratory results. Often rehabilitators' duties include blood and fecal analyses for parasite identification, packed cell volume, white blood cell counts, and differential blood cell counts, while more in-depth work in this area generally is sent out to a lab by the attending veterinarian.

Wildlife rehabilitators often participate in research, either within the center or in conjunction with a university. Topics may include patient case histories, disease identification, parasites loads and identification, release rates, post-release monitoring, and the success rates of new and novel treatments. For examples of such research, see the *Journal of Wildlife Rehabilitation* website (appendix B).

Administrative aspects of wildlife rehabilitation include keeping records, maintaining organizational health (account balancing, public relations, board and staff relationships, and the revision and care of organizational documents, such as bylaws and strategic plans), and managing human resources. Most wildlife centers do not have large staffs. Therefore, administrative tasks often are performed by the people caring for the wildlife. Record keeping is done both for the center's information and for governmental reporting requirements. (In the United States, wildlife centers are regulated by state departments of natural resources and the US Fish and Wildlife Service.) Extensive records are kept on each intake, from data on the citizen who found the animal to the final disposition of the patient. Records also must be kept for controlled drugs licensed to the wildlife center veterinarian, donations received, and staff members. Accounting, budgeting, and fundraising might feel like intrusions, shifting time away from the care of animals, but they are a necessary component in keeping an organization solvent and functioning. Rehabilitators have a responsibility to continue their professional development, in order to maintain an excellent standard of organizational and animal management. Upper-level staff members also are expected to interface with the media and the wildlife center's board of directors.

Facilities management also is a duty for most wildlife rehabilitators. Expect to do some of the same maintenance work you do at home (e.g., landscaping, maintaining electrical equipment, replacing light bulbs, troubleshooting plumbing, painting). Additionally, you become proficient at basic woodworking while building and repairing cages.

Rehabilitators are also people managers, working with that most challenging and valuable of human resources: volunteers. Volunteer management and training are tasks for mid- to upper-level rehabilitators. Some large centers have a specific volunteer coordinator to manage the hours worked by volunteers and to teach them various skills.

Wildlife rehabilitation is not a 0900–1700 job. The work varies from 4 to 5 hours during the winter to 14-hour days during the summer. Wildlife rehabilitation is an exhilarating and exhausting career choice, requiring total commitment but providing many tangible and intangible rewards. The best ones are to witness the bird you've spent the last five months caring for fly free, or to oversee the release of a beaver that took two years of care before it was independent and ready for the wild.

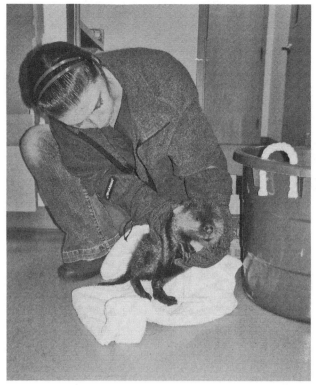

A wildlife rehabilitator examines a beaver to assess its health when the animal is admitted to a rehabilitation center (photo: Kai Williams)

Background Needed

Successful wildlife rehabilitators have knowledge of and experience in ecology, business, medicine, public policy, and construction. Wildlife rehabilitation is still an emerging field and much can be learned on the job, but the greater the preparation and the number of skills you have beforehand, the more likely you are to obtain a paid position. Useful hands-on skills include animal handling; knowledge of wild animal behavior; basic wound management; animal rescue techniques; an ability to identify and use basic medical supplies, including common bandage materials, syringes, and needles; experience with basic construction and maintenance tools; expertise in microscopy; an excellent telephone presence; and conflict resolution skills.

As a prospective wildlife rehabilitator, you should not be surprised that the list of required knowledge includes wildlife conservation and medical ethics, natural history, basic pathology, parasitology (especially zoonoses, which are diseases transmitted from animals to humans), anatomy, nutrition, and animal behavior. Often rehabilitation centers are quite small entities, and staff and volunteers must perform multiple tasks. Be prepared to assist with the general management needs of a small nonprofit business, including bookkeeping, fundraising (winter hours may be spent submitting numerous grants and planning events to gather support from the local community), human resources, facility maintenance, and all the policies that go with these critical functions. You also will be responsible for understanding and following governmental mandates related to wildlife rehabilitation, at levels ranging from local municipalities to the federal government. For example, the transportation of white-tailed deer between counties might be illegal in one state, to prevent the transmission of chronic wasting disease, or special dispensation might be needed for transport between countries for a Swainson's hawk that missed migration, due to a car accident.

Education Required

At this time, a formal education is not necessary in the wildlife rehabilitation field, but you should expect to need a bachelor's degree or a associate's degree as a veterinary technician for paid positions. States and provinces may also require a specific level of education, certification, or the passing of certain exams before issuing a license to rehabilitate wildlife.

Pay Scale

Most wildlife rehabilitators are volunteers. Paid positions do exist, however. The general annual pay range is between $20,000 and $40,000, with senior positions at large facilities having salaries of up to $75,000 per year. The pay scales in wildlife rehabilitation depend on the resources and fundraising ability of each organization.

SUMMARY

Incoming freshmen wildlife majors historically enter college with a limited knowledge of employment opportunities within their fields of interest. We have provided descriptions of nearly 100 wildlife jobs from 35 agencies and organizations where a person interested in a wildlife-related career can be employed. The majority of these positions require a minimum of a bachelor's degree in wildlife science or a related discipline. To get beyond an entry-level position, however, typically an advanced degree (e.g., MS, PhD) is either recommended or required. Common links among potential jobs emphasize the need for competence within a particular discipline, a positive and hard-working attitude, and excellent verbal and written skills. The majority of these descriptions highlight the fact that about 50% of the time on a job is spent filling out or managing paperwork, rather than being outdoors. Virtually every wildlife-related position requires the preparation of some type of written document (e.g., reports, memos, briefings, papers directed toward either a general audience or the scientific community). Therefore, have a good grasp of English grammar and develop concise writing skills. In addition, much of wildlife management is people management, so hone your

interpersonal skills. Wildlife-related jobs deal with a variety of people from a diversity of backgrounds. Learn to be confident and comfortable speaking with individuals or groups that range from children to adults, and from the general public to dignitaries. The typical salary for an entry-level wildlife position ranges from $30,000 to $40,000 per year.

LITERATURE CITED

Apple, D. D. 1997. Changing social and legal forces affecting the management of National Forests. US Forest Service Policy Analysis Publication, www.fs.fed.us/research/publications/wo/wo_1996_apple_d001.pdf.

Burnham, W., J. P. Jenny, and E. Levine. 2003. The role of field personnel in recovery and research of the peregrine. Pp. 229–259 in T. Cade and W. Burnham, eds. Return of the peregrine: a North American saga of tenacity and teamwork. The Peregrine Fund, Boise, ID, USA.

Cutler, M. R. 1982. What kind of wildlifers will be needed in the 1980s? Wildlife Society Bulletin 10:75–79.

Georgia Department of Natural Resources, Law Enforcement Division. 2016. Current Georgia nuisance wildlife trappers list, http://gadnrle.org/sites/uploads/le/pdf/Special-Permits/Nuisance_Wildlife_Trappers_List.pdf.

Gerber, M. E. 2001. The E myth revisited: why most small businesses don't work and what to do about it. HarperCollins, New York, NY, USA.

North American Bird Conservation Initiative, US Committee, 2011. The state of the birds: 2011 report on public lands and waters. US Department of the Interior, Washington, DC, USA.

Office of Personnel Management. 2014. Federal employee viewpoint survey results. 2014. Governmentwide Management Report, www.fedview.opm.gov/2014FILES/2014_Governmentwide_Management_Report.pdf.

———. 2016. 2016 general schedule (GS) locality pay tables, https://www.opm.gov/policy-data-oversight/pay-leave/salaries-wages/2016/general-schedule/.

Organ, J. F., and G. R. Batcheller. 2009. Reviving the public trust doctrine as a foundation for wildlife management in North America. Pp. 161–171 in M. J. Manfredo, J. J. Vaske, P. J. Brown, D. J. Decker, and E. A. Duke, eds. Wildlife and society: the science of human dimensions. Island Press, Washington, DC, USA.

Skillen, J. R. 2009. The nation's largest landlord: the Bureau of Land Management in the American West. Kansas University Press, Lawrence, KS, USA.

Thomas, J. W. 1985. Toward the managed forest: going places that we've never been. Wildlife Society Bulletin 13:197–201.

———. 1986. Effectiveness—the hallmark of the natural resource professional. Transactions of the 51st North American Wildlife and Natural Resources Conference 51:27–38.

Professional Societies

The Inside Track to Career Success

WINIFRED B. KESSLER

What a great career I've had in the wildlife field! This is largely the result of having held interesting jobs with American and Canadian universities and the US Forest Service. But reflecting on the past 40 years, I see that some of the most valuable and rewarding experiences came from my involvement in professional societies, especially The Wildlife Society (TWS). That's how, as a young student, I met accomplished wildlifers whose advice and support gave me a solid footing on my early career path. It's how I tapped into an extensive network of colleagues who inspired and fueled my career growth. It's how I discovered, applied, and honed my leadership skills. My participation in these societies gave me a global view of wildlife conservation, greatly increased my knowledge, and made me part of a larger voice for science-based policies and the future of wildlife. I may have held the same interesting jobs without my involvement in professional societies. But I would not have enjoyed the same amazing career.

Over the years I have hired many people, including recent graduates in wildlife and other natural resource disciplines. Time after time, I saw real differences in applicants who had been active in the student chapters of professional societies, compared with those who had not. The former had superior resumes, a better understanding of career options, more hands-on experience, and more-impressive letters of reference. Many had worked alongside state wildlife biologists, assisted in graduate student research, given wildlife talks to elementary school children, competed in regional wildlife conclaves, worked on habitat restoration projects, presented posters or talks at professional conferences, served as student chapter officers, and much more.

Professional society involvement can give you a real edge in competing for your first job. But that is just the start. Involvement in these societies is an engine for growth. It makes you part of something much larger: a community of like-minded individuals committed to science-based management

and a future for the natural resources you care about. The friendships and professional contacts you form will support and enrich you throughout your career.

The purpose of this chapter is to show why professional society involvement is so important for both getting a solid start and continuing to thrive in your career as a wildlifer.

WHAT IS A PROFESSIONAL SOCIETY?

Most people who pursue wildlife careers have a keen interest in this area. Thus it is common for wildlifers to belong to a variety of related organizations, such as the Audubon Society, Ducks Unlimited, the Rocky Mountain Elk Foundation, and The Wildlife Society, which are described in chapter 5. All of these nonprofit groups, and others, work hard to ensure a future for wildlife. But among them, only one, TWS, is a professional society. What is the difference, and why is it important for those planning a career in the wildlife field?

A professional society stands apart with respect to who its members are and what its fundamental purposes are. Most participants in such organizations practice in that area of expertise or have professional ties to it. Some of these societies are exclusive, admitting only those who meet specified requirements in their levels of education and experience. Others are more inclusive, but their aim is still to represent and serve those who work in that field and to advance the profession overall. As a parallel example, consider the American Medical Association (AMA), the professional society for medical doctors. Its mission is "to promote the art and science of medicine and the betterment of public health" (American Medical Association 2016). As patients, it gives us confidence to know that our doctors belong to an organization that sets standards and ethics for practice, enforces them, and works hard to advance the quality of medical practice for all. Similarly, professional societies for wildlife and other natural resources exist to

Numerous professional societies exist to guide their membership in educational and ethical standards, such as the ones with their logos pictured here: the American Society of Mammalogists (a), the Society for Range Management (b), the Society for the Study of Amphibians and Reptiles (c), and The Wildlife Society (d)

regulate aspects of their practices and ensure that the public's interests are well served.

One way that professional societies do this is by setting and enforcing standards. Most of these organizations have an important role in guiding and evaluating educational programs, to ensure that students are well prepared to enter the profession. In addition, some have formal accreditation programs, to verify that what universities offer meets specified standards. Professional societies may also have certification programs, to verify whether individuals have the requisite qualifications to practice as trained individuals in a discipline or specialty. In some cases, the society may be responsible for issuing a license as a requirement for practice within a jurisdiction. For example, in Canada, it's necessary to be registered by the professional forestry association in a province in order to practice in a forestry career there. In other cases, such as TWS's Certified Wildlife Biologist (CWB) program, this certification is not required by most employers, but it may be an important factor in evaluating and selecting job candidates. A professional society may set continuing education requirements for its certified members, offer opportunities for meeting those requirements, and monitor compliance.

Another important role of professional societies is to function as learned bodies, helping those in the profession share information and disseminate research findings. Most of these societies produce journals (reporting peer-reviewed science) and magazines or newsletters (to keep members informed of developments of interest to the profession). Many hold regular conferences, where research is shared through oral presentations, posters, workshops, and field trips. These conferences are also important for addressing issues of concern and expanding professional networks.

WHAT PROFESSIONAL SOCIETIES DO WILDLIFERS BELONG TO?

The preeminent one for the wildlife profession is TWS. But, given the variety of jobs in the wildlife field, it's not surprising that many wildlifers are members of multiple professional societies. Some key ones are described in the following sections, but the range of societies that wildlifers belong to is very broad. That's because it is beneficial to join organizations that align with your particular interests and specialties. For example, members of TWS may also belong to the American Society of Mammalogy, Society for the Study of Amphibians and Reptiles, American Ornithologists' Union, Society for Ecological Restoration, Native American Fish and Wildlife Society, Ecological Society of America, or at least one of many other professional societies.

The Wildlife Society

The Wildlife Society is nearly as old as the profession itself. In 1930, Aldo Leopold, considered the father of wildlife management, chaired a group appointed by the American Game Institute (now the Wildlife Management Institute) to draft a comprehensive policy for reversing the widespread declines in North American wildlife. The 1930 American Game Policy called for a broad program of restoration, to be carried out by trained professionals—the first reference to wildlife management as a distinct profession. Three years later, Leopold published the discipline's first textbook, *Game Management* (Leopold 1933), and joined the University of Wisconsin–Madison as North America's first wildlife professor.

It's one thing for a specialized field to self-declare as a distinct profession, and quite another to gain recognition as such. The early wildlifers quickly saw a need to create professional standards for this new discipline, as well as develop a means through which practitioners could interact and share knowledge. Some of them met at the first-ever North American Wildlife Conference, held in Washington, DC, in 1936, forming the Society of Wildlife Specialists, with Ralph (Terry) King as president. The next year, at the Second North American Wildlife Conference in St. Louis, they renamed their organization The Wildlife Society, installed Rudolf Bennitt as president, and launched the *Journal of Wildlife Management* as the profession's outlet for peer-reviewed science.

Initially, TWS set out to do four things: (1) establish professional standards, (2) develop wildlife management along sound biological lines, (3) publish papers that advance those goals, and (4) protect the interests of its members. Those functions still remain important today. In addition, TWS has a strong role in providing good science and advocating its use to inform policies and decisions that affect wildlife.

Today, the mission of TWS is "to represent and serve the professional community of scientists, managers, educators, technicians, planners, and others who work actively to study, manage, and conserve wildlife and habitats worldwide" (The Wildlife Society 2016). TWS has nearly 10,000 members, spanning the full panoply of career pathways covered in this book. TWS is organized into eight geographic regions (called *sections*) spanning the United States and Canada; 56 chapters, usually at the state or provincial level; and more than 130 student chapters, based at universities. Many members join at least 1 out of 26 working groups that focus on subject areas. While heavy in North American participation, TWS membership includes individuals from nearly 60 countries.

A primary goal of TWS is to support the education, training, and ethical practice of wildlife professionals throughout their careers. It collaborates with universities to ensure that the various wildlife curricula are effective in preparing graduates for wildlife careers. It has a code of ethics and procedures for investigating alleged violations and for censuring or even suspending members who breach this code. It has a certification program to evaluate the credentials of TWS members and verify that high standards are met. This includes Associate Wildlife Biologist (AWB) certification, granted to applicants who have completed the TWS's rigorous academic standards but whose professional experience is limited. TWS members completing the organization's requirements in this area, including AWBs, are eligible to apply for Certified Wildlife Biologist (CWB) status. AWB and CWB are registered trademark designations and may be earned and used only through the rigorous certification process of TWS.

The Wildlife Society disseminates knowledge in three peer-reviewed journals: the *Journal of Wildlife Management*, the *Wildlife Society Bulletin*, and *Wildlife Monographs*. All TWS members receive frequent newsletters and the society's magazine, the *Wildlife Professional*, to stay current on issues affecting wildlife and the profession. TWS uses social media to keep members in the loop. In addition to TWS's annual conference, many sections and chapters bring their members together regularly to share information and facilitate networking.

The Wildlife Society has a strong emphasis on students and early career development. These programs and activities, so important to those entering the field, are examined in a later section.

Society for Conservation Biology

Many people in the wildlife field, especially those working on at-risk species and ecosystems, are members of the Society for Conservation Biology (SCB). Established in 1985, the SCB has more than 5,000 members and is "dedicated to advancing the science and practice of conserving the Earth's biological diversity" (Society for Conservation Biology 2016). The SCB is organized into seven geographic regions (Africa; Asia; Australia, New Zealand, and the Pacific Islands; Europe; North America; South and Central America and the Caribbean; and the marine realm). The SCB also has chapters, with coverage

that ranges from individual university campuses to entire countries, as well as topical working groups.

Conservation Biology is SCB's peer-reviewed journal. All members receive its online publication, *Conservation Letters*, and have access to the SCB news blog. SCB holds an International Congress for Conservation Biology every other year, with section meetings or symposia in the intervening years.

The SCB Education and Student Affairs Committee promotes conservation education at all levels, encourages student participation in meetings, and creates opportunities for students to interact with experts in their fields. Other resources for students and early-career professionals include a jobs board, a free online textbook, student awards, and postdoctoral research support through the David H. Smith Conservation Research Fellowship program.

Society of American Foresters

Many jobs in wildlife involve the management of forest habitats and conservation of forest-dwelling wildlife. Wildlife biologists hold positions with the US Forest Service, state and provincial forest management agencies, forestry companies, consulting firms, and many other forest-sector employers. These wildlifers are often members of the Society of American Foresters (SAF). Established in 1900, the SAF has a mission "to advance the science, education, technology, and practice of forestry; to enhance the competency of its members; to establish professional excellence; and to use the knowledge, skills, and conservation ethic of the profession to ensure the continued health and use of forest ecosystems and the present and future availability of forest resources to benefit society" (Society of American Foresters 2014). The SAF has a code of ethics and a certification program for forestry professionals, and accredits undergraduate and master's degree programs in forestry.

The SAF is organized into chapters at the state or multistate level, and student chapters at universities that offer forestry degrees. Its peer-reviewed journals, *Forest Science* and the *Journal of Forestry*, as well as regional versions of the *Journal of Applied Forestry*, often contain research relevant to the management of forest wildlife and habitats. Members obtain news and information through SAF's newsletters, *Forest Source* and the *e-Forester*. The SAF's Wildlife and Fish Ecology Working Group consist of members with a particular interest in the relationship of forests and forest management to fish and wildlife resources. Other working groups on fire ecology, remote sensing and geospatial applications, and other subject areas may be of special interest to wildlifers. The SAF's annual conference, periodically held jointly with the Canadian Institute of Forestry, includes many sessions and topics relevant to the management and restoration of forest wildlife habitats.

Society for Range Management

The Society for Range Management (SRM) is "dedicated to supporting persons who work with rangelands and have a commitment to their sustainable use" (Society for Range

Management 2016). Founded in 1948, the SRM has more than 4,000 members in 48 countries. Many wildlife experts who work with species in grassland, desert, shrubland, and other rangeland habitats find benefits in participating as members of the SRM. The society has a code of ethics, standards of conduct, and programs to certify professionals in rangeland management and range management consultants. It also accredits university programs in Range Management Education. The SRM's publication for peer-reviewed science is *Rangeland Ecology and Management*, and its members' magazine, *Rangelands*, includes science and opinion articles. An SRM conference is held every year.

The SRM supports members at all stages in their careers. It promotes student recruitment and participation through its International Student Conclave, Mason–Range Science Scholarships, High School Youth Forum, and graduate student competitions at the annual conference. Its Young Professionals Conclave seeks to integrate young or less experienced members into SRM and the profession. Since 1993, the SRM has had a Wildlife Habitat Committee to address wildlife issues and provide a forum for interested members.

American Fisheries Society

Some wildlifers have job responsibilities that span the wildlife and fisheries fields, so it is not uncommon for them to have membership in both TWS and the American Fisheries Society (AFS). Started in 1870, and with more than 9,000 members today, the AFS is the world's oldest and largest fisheries professional society. The mission of the AFS is to "improve the conservation and sustainability of fishery resources and aquatic ecosystems by advancing fisheries and aquatic science and promoting the development of fisheries professionals" (American Fisheries Society 2014). Members must abide by the AFS's Code of Professional Conduct.

The AFS is organized into four major geographic divisions, with 48 chapters in North America, although members live all around the world. The 21 sections of the AFS are defined by subject-area interest. Its scientific journals include *Transactions of the American Fisheries Society*, the *North American Journal of Fisheries Management*, the *Journal of Aquatic Animal Health*, the *North American Journal of Aquaculture*, and *Marine and Coastal Fisheries*. All members receive the magazine *Fisheries*.

The AFS has a professional certification program, as well as 58 subunits based at universities. It encourages student participation through travel grants to attend the AFS annual conference; student awards; social media; a jobs board; and the Hutton Junior Fisheries Biology Program, a paid summer intern program for high school juniors and seniors.

The Specialty or Critter Societies

Many people in the wildlife field join professional societies that specialize in particular kinds of wildlife. Such organizations exist throughout the world. Examples based in North America include the American Society of Mammalogists, Canadian Herpetological Society, American Ornithologists' Union,

American Society of Primatologists, Raptor Research Foundation, Society for Marine Mammalogy, American Society of Ichthyologists and Herpetologists, and many more. The missions of these organizations are similar to the societies profiled above. Key elements include the advancement of science, education, conservation, and professional practices. Their more specific focus fosters a tighter community of specialists who share an interest and expertise in a taxonomic grouping of wildlife. Typically, their individual missions also strive to both increase the profile of that particular grouping through public education and improve its status by advocating for science-based conservation. Members may represent a range of professions—from researchers to environmental education specialists to naturalists and science writers—united by their desire to advance an understanding of and conservation for the biological category of interest.

By participating in the critter societies, wildlifers benefit from access to species experts, professional networks, and specialized knowledge that more directly aligns with their own areas of interest and endeavor. These specialty societies can be an important catalyst for career development, particularly for those aiming to go deep rather than broad in their work with wildlife.

FAST-TRACKING YOUR CAREER

As a student, how can you put your best foot forward toward a career in wildlife? Because TWS regards this as a very important question, they periodically conduct a comprehensive evaluation of wildlife education offerings at universities and colleges; the most recent was completed in 2009 (Baydack 2009). From time to time, TWS also surveys employers to evaluate whether graduates are being adequately prepared for entry-level jobs in wildlife (Coalition of Natural Resource Societies 2012, The Wildlife Society 2012).

The findings in these evaluations are mixed. On the positive side, universities continue to do a responsible job in delivering core subject areas, such as ecology, principles of wildlife management, mammalogy, and ornithology. According to employers, most graduates are well grounded in ecological theory and the basic sciences that provide a solid foundation of knowledge in the wildlife field.

On the negative side, employers reported that most new graduates are weak in their ability to communicate effectively, work in teams, and interact with the public. Often those very competencies are emphasized in job descriptions as necessary for today's positions in wildlife. Employers do not expect new recruits to be highly proficient in these areas, but neither can they accept new hires who are clueless about the demands of the job. Well-prepared graduates will have a clear picture of what their particular profession entails and understand that knowing principles and facts is not enough. Once on the job, they will have to apply their acquired knowledge and skills to real-world problems, which often require working in teams and interacting with diverse segments of society.

Professional societies host quiz bowl competitions at their professional conferences to promote student involvement (photo: Wini Kessler)

The Wildlife Society Ad Hoc Committee on Collegiate Wildlife Programs (Baydack 2009) also reported that many university curricula are weak in providing practical, hands-on experience. This finding is consistent with employers' concerns that many new hires arrive on the job without field savvy or practical experience in using basic field equipment.

This is information you can use! If there are skills or abilities that employers say are important but are not emphasized in the wildlife curriculum at your university, then you'd better seek other opportunities to fill those gaps. A prime way is through early involvement with one or more professional societies. The student chapters of TWS exist exactly for that purpose: to help students develop a clear picture of what the profession is about and provide opportunities for them to interact with people in diverse wildlife careers, gain hands-on experience, build teamwork abilities and a sense of community, and develop public outreach skills.

As an example, let's highlight some of the activities of the 2014 Student Chapter of the Year, an annual award given by TWS to recognize outstanding accomplishments in the previous year. Most members of the student chapter at this small college (Abraham Baldwin Agricultural College, ca. 3,000 students) attended the annual meeting of their state's TWS chapter and the regional annual conference of the Association of Fish and Wildlife Agencies. Twenty-two members of this student chapter competed in the Regional Wildlife Conclave and finished fourth overall. In cooperation with their state's wildlife agency, the student chapter worked at a wildlife management area to remove invasive species and help restore habitats. They constructed 75 wood duck boxes, predator guards, and mounting poles for landowners enrolled in the state's Forest Stewardship Program, and 100 more boxes for state wildlife management areas. They performed monitoring and stewardship activities for natural areas around their campus, and cleaned roadside areas on behalf of the Adopt-a-Mile Program. They excelled in public outreach, assisting their state's Forestry Commission with its annual Forestry Field Day. Working with their state wildlife agency and the National Wild Turkey Federation, they held a Jake's Day event that engaged 780 school children in outdoor and conservation activi-

ties. They staffed check stations for 17 hunts on state wildlife management areas. They held public talks and workshops on habitat restoration and wildlife research. The student chapter was active in the policy arena, linking up with the Conservation Affairs Committee of their state's TWS chapter and submitting comments on wildlife issues under consideration by the state legislature.

These many activities show that a student chapter, even at a small college, can play a substantial role in the professional and local communities of which it is a part. In return, students gain knowledge, skills, and experience in the very areas that employers want strengthened.

Learning also extends beyond a local student chapter. The Wildlife Society student chapters located within one of TWS's five geographic regions come together each year for a Regional Student Conclave. Participants receive valuable hands-on training in wildlife management and conservation, network with one another and with wildlife professionals, participate in field trips and workshops, and compete in many events, including the Regional Wildlife Quiz Bowl. In addition, many student chapters send delegations of members to TWS chapter and section meetings and to TWS's annual conference.

Not every university or college has a student chapter of TWS. If that is the case at your school, why don't you approach faculty members and fellow students about the idea of starting one? You won't know what leadership qualities you may possess until you seek an opportunity to discover them!

Resources to help jump-start your career can be found at all levels of a professional society. Travel grants to conferences, scholarships, mentoring, and jobs boards are among the student-focused services provided by these organizations. Many students assume that the main benefit of an annual conference is the opportunity to attend technical sessions in which research is presented. While the science content of a conference is important, do not underestimate the importance of meeting face to face with other wildlifers and forming networks vital to your own career development. The Wildlife Society works hard to increase the number of students attending its annual conference and the quality of their ex-

TIPS FOR ATTENDING A PROFESSIONAL MEETING
Jim Heffelfinger

PLANNING

- Look through the conference schedule and highlight the talks and functions you don't want to miss (or use the TWS planner). The gaps in your highlighting are times free for networking.
- Don't be afraid to miss talks if they overlap with networking. You can always catch up with the presenter later or by email.
- Carry a notepad, portfolio, or small, soft-sided document case for papers and notes.
- Before or during the meeting, maintain a list of people you want to connect with there or talk to before leaving.
- Prepare an elevator speech (chapter 8). You typically only have about 30 seconds to make an impression when first encountering someone important, and you want to have a brief set of comments prepared to deliver naturally and confidently when the opportunity arises.
- Before the meeting, leverage social media to get on the radar of individuals you would like to interact with at the meeting. Many agencies, as well as professors and their labs, are on social media or online networks. "Like" them on Facebook, retweet their tweets, comment, share posts, or "like" their open-access journal articles or interviews. This will get you noticed and give you an easy icebreaker to start a conversation at the meeting.

NETWORKING

- A classic study reported that 56% of employees found their jobs through a personal connection.
- Even brief connections you make at a meeting can have profound effects on your career.
- A big network of weak personal ties is more useful than a small network of close friends. Meetings are for expanding your network.

PERSONAL CONTACTS

- Be not afraid. The most famous people at meetings are just people, and they like to talk to students.
- Google a person's nametag to get some background information if you want to meet someone. That's not creepy.
- Demonstrate warmth and competence. Research shows that people evaluate everyone they meet by these two things.
- Listen. Be excited. Ask for details. Smile.
- Offer a conversation catalyst. Ask people about themselves or their work. People love to talk about themselves, and it's a subject they know a lot about.
- Mingle. Don't sit with a herd of friends during the whole meeting—disperse and connect. Eat separately or in pairs. Meals are the best network hubs.
- Introduce yourself to people. Don't walk up and butt in on an ongoing conversation. Linger nearby and seem to be preoccupied, so it doesn't look as though you are waiting to interrupt. Be close enough so you are strategically placed to take a couple of steps and introduce yourself when there is a lull in the conversation or when it comes to an end. Looking at your phone or studying the conference literature is a good way to appear busy and not predatory. If the conversation you want to join stalls, you can step up and say, "Excuse me, I don't mean to interrupt, but I just wanted to introduce myself." They may be happy for the diversion. Don't wait until the end of the session breaks to speak with someone. Be mindful of when talks are starting, because the person you'd like to talk to may want to get back to them.
- If someone you want to speak with gives a talk, meet up with them at the end of the presentation or shortly thereafter to compliment that person. Say that you enjoyed the talk, and then ask a question or share information. If a person you'd like to speak with has a student or collaborator giving a talk at the meeting, go to it. That might offer a good opportunity to meet both of them afterward and introduce yourself. Likewise, chat with a person presenting a poster.
- Have a faculty member or other professional introduce you to people. That's the whole purpose of building a network.
- Choose an effective ambush spot: a coffee/refreshment stand, a breakfast buffet table, sit-down tables at meals, a hallway, a waiting line. Not bathrooms!

- Establish a solid connection by following five important steps.
  1. *Introduce yourself.* Also, give the people you meet a business card before you separate.
  2. *Make a tie-in with what that person has done.* Say, I'm familiar with your work on _____, [even if you just googled it when standing off in the corner], and we have a similar situation in my state, where [insert something relevant]." Or, "I know about your work on _____, and this summer I was part of a related project, to determine the effects of caffeinated soft drinks on spider monkey activity levels, where we saw [insert something relevant]" Or, "I am somewhat familiar with your research on _____. Have you seen the work my state's game and fish agency is doing on [insert something related]?" This way, you have something to offer them, not the other way around.
  3. *Touch base a second time.* Do this before leaving the meeting.
  4. *Send something relevant when you return.* Either send an item that would interest the person (a link to a news story, a resume, a manuscript) or ask a follow-up question. Questions are a powerful way to engage people.
  5. *Touch base again a month later.* Ask another question or send something of interest you just ran across.

perience. The conference program includes a student/professional mixer, a student work-in-progress poster session, group and one-on-one mentoring, workshops for student chapter success, Student Development Working Group meetings, and the very popular and fun Wildlife Quiz Bowl. The sidebar describes a number of things students can do to get the most out of their attendance at conferences.

Professional societies help with the transition from student to early-career wildlife employees by offering opportunities for career advancement and leadership development. For example, TWS has an Early Career Professional Working Group dedicated to addressing the challenges and opportunities that wildlifers encounter at the beginning of their careers. The society's Leadership Institute provides selected early-career members with basic training in this area and intensive mentoring to prepare them to move into such positions, both in their workplace and in TWS.

A highly valuable activity for people at early stages of their careers is volunteering for the many committees, officer positions, working groups, special projects, and other openings that professional societies have at all organizational levels. There is no substitute for jumping into the arena where wildlifers work on real issues and projects that affect the future of wildlife and the profession. Think of it as high-quality training for growth in your career—and it's free!

MAKING A DIFFERENCE

I've saved one of the more important functions of professional societies for last: their role as advocates for the principles, values, and resources at the core of the profession they represent. Let's again consider the mission of the American Medical Association: "To promote the art and science of medicine and the betterment of public health." The AMA does not exist just to represent doctors and their interests. Its members share a commitment to advancing the field of medicine for society's benefit.

Similarly, the mission statements of professional societies in natural resource fields show that they seek to do more than represent and support their members. They talk about "conserving the earth's biological diversity" (SCB), as well as supporting efforts to "improve the conservation and sustainability of fishery resources and aquatic ecosystems" (AFS), "conserve wildlife and habitats worldwide" (TWS), and so on. Conserving natural resources is a core value of these societies, as is their guiding principle that policies and decisions should be informed by the best available science. Most professional societies recognize an important role for disseminating scientific information and advocating its use in decision making.

This desire to make a difference is certainly widespread in the wildlife profession. If truth be told, few people enter this line of work merely out of scientific interest. Most are driven by a deep personal interest, often a passion, for wild animals and the environment.

The Wildlife Society was not always clear on what role, if any, it should play in influencing policy choices. More than 30 years after its formation, opinion remained divided on whether TWS should exist primarily to benefit its members, or whether it should serve a larger purpose in promoting wildlife stewardship. Perhaps inspired by the environmental movement of the 1970s, TWS made a bylaw change in 1974 to include "an active role in preventing man-induced environmental degradation." Periodic member surveys since that time reveal strong support for TWS's role in informing and influencing policy choices. Today the Government Affairs & Partnerships Division of TWS works with the society's chapters, sections, student chapters, working groups, and many partner organizations to bring the best science into the policies and decisions that affect wildlife.

TWS and similar societies offer a collective, trusted, and influential voice for science-based conservation. An individual wildlifer, especially one just beginning a career in the profession, may hesitate to speak out on issues he or she thinks are important. Perhaps such novices doubt that the opinion of one young wildlife biologist will matter. They may worry about a possible conflict of interest with their employer. Or they may be concerned that speaking out on an issue may call their scientific objectivity into question. Professional societies provide a better path for weighing in on issues. They enable their members to be a collective and more effective voice for wildlife. These individuals contribute their specialized knowledge to technical reviews and position statements, comment on policy proposals, and assist with many other policy-related activities. Together with their colleagues, they are making a positive difference for wildlife, while furthering the development of their own careers.

SUMMARY

The earlier your involvement in professional societies, the more advantages you will have in preparing for your first wildlife job. You will have access to practitioners in your field who can advise and mentor you, and opportunities for hands-on experience in the real world of wildlife conservation. The tar-

geted networks you begin to build will assist your transition into the profession and support and enrich you throughout your career. You may distinguish yourself from others in your field by becoming certified by a professional society. You will have many opportunities for continuing education, career development, and leadership training. Volunteering for committees, leadership positions, and special projects will enhance your own skills and knowledge, while also helping professional societies advance their conservation mission. You will be part of a trusted and influential voice that promotes the use of best-available science in policies and decisions that affect wildlife. Participation in these societies can be a powerful engine for growth and advancement, taking your career to heights that otherwise would remain out of reach.

LITERATURE CITED

American Fisheries Society. 2016. AFS home page, www.fisheries.org.

American Medical Association. 2016. AMA home page, www.ama-assn.org.

Baydack, R. K. 2009. The Wildlife Society ad hoc committee report on collegiate wildlife programs: summary report to The Wildlife Society Council, http://wildlife.org/wp-content/uploads/2015/03/Ad_Hoc_Collegiate_Wildlife.pdf.

Coalition of Natural Resource Societies. 2012. Natural resource education and employment conference report and recommendations, www.tandfonline.com/doi/abs/10.1080/03632415.2012.698159?journalCode=ufsh20/.

Leopold, A. 1933. Game management. Charles Scribner's Sons, New York, NY, USA.

Society for Conservation Biology. 2016. SCB home page, www.conbio.org.

Society for Range Management. 2016. SRM home page, www.rangelands.org.

Society of American Foresters. 2016. SAF home page, www.eforester.org.

The Wildlife Society. 2012. The Wildlife Society blue ribbon panel final report: the future of the wildlife profession and its implications for training the next generation of wildlife professionals. The Wildlife Society, Bethesda, MD, USA.

———. 2016. TWS home page, www.wildlife.org.

The Resume

An Important Tool in Your Career Kit

JOHN P. O'LOUGHLIN, PAUL R. KRAUSMAN, AND KERRY L. NICHOLSON

There is no single format for a resume or curriculum vitae (CV) that is appropriate for all jobs. When employers are looking for employees, they will come across an array of resumes that include the good, bad, and the ugly. In our experience, we have looked at literally hundreds of them. JPO has been involved with human resources for decades. Part of his job has been to hire employees for large and small companies, and his initial evaluation of the candidates is based solely on their resumes. PRK has worked with wildlife undergraduates and graduate students at major universities around the world. And KLN is a wildlifer who has successfully applied for jobs using the information contained in this chapter. The three of us, with our wide-ranging experience, want to offer you proven tools to develop a resume or CV that will assist you in getting that first job—and subsequent jobs—as you develop professionally.

HISTORY AND DEFINITION

Webster's Dictionary defines resume as "a record or summary of one's personal history, educational background, and employment experience." While the terms *resume* and *CV* are often used interchangeably in most workplaces, by strict definition and use, a CV is longer (generally several pages, compared with one to two for an average resume). CVs are prevalent in fields that require long years of study, such as academic, medical, and scientific professions, and they are much more comprehensive than resumes. In this chapter, we will be focusing on resumes, because they are more relevant to beginning wildlifers; a CV will be more appropriate as you progress professionally. After you have secured your first job, you will learn additional information about formatting your CV, and some organizations even have specific guidelines that you are asked to follow.

WHAT DO POTENTIAL EMPLOYERS LOOK FOR?

Modern technology is expanding, and the role of resumes is growing at an accelerated pace. A resume is a critical component in a larger tool kit that is now part of integrated social media, Internet marketing, and branding campaigns. Therefore, your resume needs to be well planned, concise, focused, aligned, and executed. While no two employers have the exact same criteria or hiring processes, common areas that resonate among them include subject-matter expertise in a field related to the position; relevant work experience, such as summer jobs, volunteer work, and internships; significant contributions you have made, such as offices held; and major accomplishments, such as awards, scholarships, and grants. In addition, employers like to see progress, usually in the form of enlarged roles or responsibilities, key special project assignments, and significant advancement. They are also looking for examples of qualities that will be needed in the job, such as someone who can be an effective team player, and someone who, when required, can work independently and productively. This is especially important when professors evaluate prospective graduate students. Ideally, professors want brilliant students, but all students do not have to have this trait. What is important is demonstrating that you have the desire, drive, and determination to succeed, as well as the ability to work both independently and in teams. If you also take the time to get acquainted with the mission of the organization and the details of the position, this will impress prospective employers. If you have limited paid experience, volunteer work will resonate positively with potential employers if it pertains to the position you are seeking, demonstrates a strong work ethic, or shows potential for leadership. This is extremely important in wildlife professions, because wildlife in North America belongs to the public, and active managers and biologists need to have positive interactions with all stakeholders in order to be effective. Craft your resume so these criteria match up with what agencies, organizations, and employers desire.

Resumes for individuals who work their way to the top of the candidate list are ones that are capsulized and concise, rich with data and quantification, and feature the prioritized ranking of key experiences and accomplishments. It is rare that you would be hired based solely on your resume. Personal interactions, job interviews, phone follow-ups, letters of recommendation, and other references are used by employers to make sure they hire the right person. But it is a first impression—the resume, in many cases—that gets your name into the hat and lets you continue on to the other steps in being considered for the job. Some positions may not require a resume, but instead ask candidates to fill out forms. In those cases, use elements borrowed from your resume to get the best information into the forms. Take the preparation of your resume seriously. The simple fact is, when someone is responsible for hiring new employees, ranging from employers in big government offices to those in small consulting firms, the first impression they receive is important. In everything from your cover letter to the last page of your resume, hiring managers will look for something to eliminate you from the stack of applications they are reviewing. If anything is questionable, it becomes an easy way for an employer to reject your application. This can apply to something as trivial as spelling errors, forgetting to include a requested item with the application (e.g., transcripts, a cover letter, names of references), or writing the wrong name on the cover letter that accompanies your resume. A spelling error might be overlooked, but anything more than that will be an indicator of laziness or carelessness, which would make an employer question your dedication to the job and the potential meticulousness of your data collection. If you cannot be bothered to check your spelling and grammar on a job application, then how trustworthy are you to collect and record data? Further, a cover letter speaks to an applicant's ability as a writer. Consider any written material as an example of your writing skills, a highly desirable quality in employees.

Now here's a punch line that you are not going to like (at least initially): you need to edit and personalize your resume for each and every opening for which you apply. A resume is not a one-size-fits-all entity. The hardest part in creating a resume is when you put all of your pertinent information down on paper for the first time. Once you have done that, it gets easier to customize your resume for each desired position. Every job and each organization is unique, which begs for some careful rewording, so that your resume features what is going to resonate with different employers. Many job seekers—even beginning ones—have experiences and accomplishments in multiple fields, and all of this information can be contained in a single master resume. You will, however, want to modify this master document when you send out a resume for a particular job. For example, if you have been both a field biologist and a project manager, you would want one resume that begins with and emphasizes your field experience, and another that fully develops your project management experience. You would send the first to employers with openings for fieldwork, and the other for positions in project management. We strongly suggest that you take this one step further and recognize that you may want to begin *each* resume going out your virtual door with a customized version that is the best representation of you and your experience related to that particular opportunity. For example, KLN started by compiling one master list that included all the jobs she ever held; every grant and scholarship she earned; each volunteer opportunity and training session she participated in; every computer program she was familiar with; and possible references for various types of jobs. Then, when she applied for a field technician position on boreal owls or some other species, she went to her master list and pulled out the most pertinent information. Likewise, if she applied for a supervisory position leading a small mammal sampling crew, she extracted and emphasized different information. Thus she could cut and paste various elements from her master document and assemble a carefully targeted resume (or CV, once she had more experience to share with employers) for the specific job that was advertised.

One mistake beginning wildlifers make when looking for their first professional position is to include every single job they've held, as though each was a lifetime occupation. As a general rule, only list jobs where you have been employed for at least three months. Many of you will have familiarity with several types of work without really mastering them. That does not mean that these experiences are unimportant; they just need to be placed in the proper light. For example, working with telemetry for a month or so is not the same as using this tool for an entire project (e.g., one to two years) and determining how the data from it will be used. Instead of presenting minor jobs as though they were full-length ones, simply list your brief experiences and the organizations you have worked with for short periods, along with mentioning the tools you used. Elsewhere in the resume, you can list the equipment you have mastered and the computer programs in which you are competent. Also include a segment on training opportunities you have had.

We recommend that you use Word or a similar word-processing program to create a master version of your resume and then save it. This will allow you to easily revise and edit your resume for each different job opportunity. Make it a practice to keep your master resume up to date as changes occur, especially as you advance professionally. Always email your resume as a PDF document. Some experts recommend also creating a text version of your resume. Save your document using the title of the position (or at least part of it). It could be embarrassing if you don't, as evidenced by a story from one of our students:

One time I applied for a job with a nongovernmental organization, and six months later they called to schedule an interview. I forgot I had applied, forgot the job description, and was too embarrassed to ask for it. I went searching, couldn't find any posting (as the job had closed, and they had removed the announcement). So I couldn't prepare for the interview. It was very evident that I didn't prepare about three minutes into the phone interview. The job was for a field tech position on fire and salamanders in the Southeast, that part I knew, but what I was going to be doing with the salamanders, I had no idea. At the time of application, I was applying for anything and everything I could, regardless of my experience or lack of experience. This was one of those jobs that would have stretched my training, as I had none in the reptile and amphibian world. So, six months later, when they finally got around to interviews, I had moved on mentally. As it turned out, the job was to help with prescribed burns and study the response of the salamanders . . . At the time of applying I had just finished a course on fire ecology at the university I was attending. I'd been on fire field crews and everything. However, I looked very foolish when I couldn't answer some basic ecological questions about fire and salamanders, the study area, or about how and why fire might be an important tool for salamanders. From that point on, I've always kept a copy of the job description, along with my tailored copy of the CV and the cover letter specifically addressed to that hiring person in a notebook. I also kept the notebook near the phone, so I would never be caught off guard again by a potential employer.

An additional component that will allow you to create the ultimate resume for your most desired job opportunities is to look for key phrases being used by a prospective employer and insert them into your application. Key words and phrases are important in several contexts. First, using them demonstrates knowledge and experience in the targeted field. Incorporate the most critical and current phrases and words both in the general discipline for the job and the specialized subfield. Second, such words signal action and leadership. Third, resumes are often scanned by computers in an initial review. The computer may kick out your resume if it lacks the key words for the job. Of course, if you are applying for your first job and only have limited experience, you will not be using many key words—nor would they be expected, at least early in your career.

If an employer describes some job that requires a specific tool and you have experience with it, mention that ability using the same terms appearing in the job description. For instance, if the ad or announcement indicates a need to use ArcMap, don't simply write that you have geographic information system (GIS) experience, say that you have experience in ArcMap and list all the versions you've mastered. If you haven't used ArcMap specifically, then list the other programs you are familiar with, such as ArcView, GRASS, OpenGeo, or QGIS. This points out that although you do not have direct ArcMap experience, you still do have some related and very applicable computer skills.

Use verbs that shout out at readers, such as "executed," "improved," "influenced," and "implemented." McCarthy and Southam (2014) discuss winning words for resumes. Examples for wildlifers include words such as "accomplished," "allocated," "analyzed," "collaborated," "conducted," "delegated," "drafted," "edited," "enhanced," "equipped," "founded," "governed," "guided," "mastered," "prioritized," and "surveyed."

BUILDING THE ULTIMATE RESUME

You are probably getting a little impatient at this point and are eager to begin developing your resume or enhancing your existing one. But we still need to discuss more planning steps before we get to the writing phase. Proper planning and preparation will yield substantial dividends for the finished product. Yate (2014:12) suggests that you begin your resume with the end result in mind, and he goes on to illustrate his point with a story:

> Once upon a time, long ago, there lived a man who spent his days watching life go by. He lived in a town ravaged by bears, and one of his dreams was that if he could shoot bears, he could travel the world as a bear slayer. Every day he sat on his porch and waited for a bear to go by. After weeks of watching and waiting he thought he might go looking for bears.
>
> He didn't know much about them, except that they were out there somewhere. Full of hope, he loaded his single-shot musket and set out. Arriving at the edge of the forest, he raised his rusty old flintlock and fired blindly into the trees. Then, disappointed at hitting nothing, he slouched back to sit on the porch.
>
> Because our hero couldn't tell dreams from reality, he went hunting unprepared and earned exactly what he deserved: nothing. The moral is: When you go for bear or job-hunting, get a grip on reality, and don't go off half cocked.

Countless opportunities exist out there in the forest of the professional world. Even in times of severe economic downturn—and remember, such periods are cyclical and are likely to come by every seven to ten years throughout your career—there are always jobs out there. Yes, they are going to be harder to find, and yes, the competition is tougher, but someone is going to find those openings and land those job offers. It can be you.

Resume Format

There are different types of resume formats, but most wildlifers use a chronological one. This presents your work history, beginning with your most recent job and going backward in time. It should include the organizations you've worked for, the respective positions you've held, and the dates when you

CHARLIE D. WINN

Current Address
1234 Wildlife Way
Spotlight, MT 59808
Phone: 406.238.4987
Email: CDwinn@wild.com

Permanent Address
2564 Flick Street
Utopia, IA 50841

Dedicated college graduate with excellent experience beginning a career in wildlife management and conservation. Willing to work with birds or mammals anywhere in the United States.

Education

University of Arizona, Tucson, AZ, 2014, 3.69 GPA
Bachelor of Science, Wildlife Biology

Core Competencies

Telemetry, GPS, R, SPSS, speak and write French and Arabic, excellent communicator orally and in writing

Professional Organizations

The University of Arizona Student Chapter of The Wildlife Society
The Wildlife Society
The Society for Conservation Biology

Employment History

United States Air Force, Holloman Air Force Base, NM, 2000-2004.

Field technician, Nevada Department of Fish and Wildlife, Reno, NV, 2004-2006.
I assisted field biologists with plant and animal sampling in remote areas working independently and in teams.
Supervisor: V. C. Blister

Undergraduate research assistant, University of Arizona, 2006-2014.
I worked with different graduate students studying the habitat relations between collared peccaries and their predators, measured vegetation, animal locations, survival, and entered data into models and assisted with data analysis.
Supervisor: Dr. O. S. Krispman

Volunteer Positions

Annual deer drive at the Three Bar Enclosure, 2000–2004
Assisted BLM with removal of fences to assist with pronghorn movement, summer 2001
Assisted USFWS with stream sampling in the Lower Colorado River, Arizona, summer 2002
Coordinated the annual western TWS Quiz Bowl, Albuquerque, NM, 2013
Organized, coordinated, and conducted a wildlife job fair on the University of Arizona campus, summer 2013

Publications, Awards, and Positions

Winn, C. D., B. B. Bob, and O. S. Krispman. 2015. Relationships between collared peccaries and desert predators. The Nature Journal 14:205–216.
Best poster award, The Arizona-New Mexico Section of TWS, 2014
Associate Certified Wildlife Biologist, TWS
Who's Who Among Students in American Universities and Colleges, 2014
President of the University of Arizona Student Chapter of TWS, 2013–2014

Interests

Hiking, hunting and fishing, birdwatching, reading, public speaking

References

Dr. O. S. Krispman, University of Arizona, Tucson, AZ. Email: OSKrispman@uarizona.edu
Dr. B. B. Bob, Nevada Department of Fish and Game, Reno, NV. Email: B3@NV.gov
Mr. V. C. Blister, Nevada Department of Fish and Game, Reno, NV. Email: VCB@NV.gov

The University of
Montana

Ms. K. L. Birdsong
Cushman's Conservation Consulting
P.O. Box 12345
Bozeman, MT 59717, USA   20 February 2016

RE: Application for field technician project (Job No. 321XZ)

Dear Ms. Birdsong:

I was pleased to see the application for the field technician position advertised in your monthly newsletter of 20 January 2016. This job calls for an individual with a B.S. in wildlife management and one year of field experience working with upland wildlife and the ranching community. I have these qualifications and others outlined in my resume, which is included with this letter. In particular, I have taken classes in upland game management, have mastered the capture techniques for upland birds and understand their habitats, and have effectively worked with ranchers on similar projects. I also know how to expertly operate ORVs, 4X4 pickups, and chainsaws, and I have leadership skills for working on and leading field crews.

This job provides an excellent opportunity to begin a career in wildlife management and conservation and I am excited at the prospect of being considered for the position. I have listed three references at the end of my resume that can verify my academic qualifications, knowledge of upland wildlife, and positive communication skills. I wish you the best with your search and look forward to hearing from you.

Respectfully submitted,

Charles D. Winn
Associate Certified Wildlife Biologist®

Examples of a professional resume *(opposite)* and a cover letter *(above)*. These items provide potential employers with a first impression of you. Make sure that impression is of a dedicated person who pays great attention to details.

were employed. Provide summaries of your roles in those positions, with an emphasis on your major responsibilities. Don't hesitate to include as many facts and what types of expertise you have gained as is reasonable. Potential employers want to see successes. A chronological resume format is the most prevalent one used and is especially impressive when it shows solid career progression and a respectable work history, such as beginning as a field technician and gradually working up to becoming a supervisor of field teams.

Kursmark (2012) presents important steps for developing an effective resume, including identifying possible jobs, listing your key requirements for the jobs, and providing evidence as to why you should be hired for them, such as enumerating skills you have mastered and those areas in which you have some experience. Each resume should begin with your name and contact information and include your educational background—especially when certain degrees are required in the job for which you are applying. You also should include awards and grants you have received, publications, volunteer positions, workshops and other special training, and anything else to describe your expertise. Job applications often require a list of three references whom your potential employee can call for further information. This list is usually placed at the end of a resume. When you select people to act as your references, always ask in advance if you can list them on your resume, and choose those who can offer a positive assessment of you and your work. Anyone can provide a recommendation, but it ought to be a good one. Also, make sure these chosen individuals know you are using them as a reference in each of the jobs for which you apply. This is particularly important at the beginning of your career, when you may not yet have established a long-term relationship with them. It is appropriate to remind these people that they have agreed to serve as a reference, particularly if there is a time lag between when you initially asked them and when you send out a resume. This is important, for two reasons: it brings who you are back into focus for them, and it gets them mentally prepared to talk with a potential employer or write a positive letter about you.

Once your resume is completed, make sure you proofread and edit it, something you should continually do throughout your career. Kursmark (2012) has over a hundred professionally written resumes in her book, which present strong examples of well-developed documents. The sidebar provides a sample resume format that beginning wildlifers may want to follow when summarizing their professional accomplishments.

Other than questions regarding the content or format of resumes, a common concern is the ideal number of pages it should have. We contend that a one- or two-page resume is best. Either is acceptable, but most beginning wildlifers will only need a single page. The majority of other resumes are two pages long, but three may be required for candidates with several years of experience or expertise in several fields. A CV is longer and typically does not have a page limit.

Another frequently asked question about resumes is in regard to fonts and colors, which are intended to highlight items or otherwise have them stand out. We suggest using no more than two to three colors in a resume. For example, you could have a color highlight or boldface font for your personal contact information, and then use another color to highlight your employment experience.

In addition to customizing your resume for each different job opportunity, you also should thoroughly review the various versions of it regularly, depending on your career progress and other adjustments you may want to add. Editing your resume to reflect your very latest experiences, using the most recent best-practice formats and phrases, will ensure that it represents you in the best possible light. You also should seek to improve your resume by asking for input from professionals. These can be university professors, supervisors, or other successful individuals in your network. They can provide great feedback, as well as tips on details you may have overlooked. You also should ask those who have been successful job applicants to assist you in developing a winning resume. Ask to see copies of their resumes, to get some other helpful examples (most people have their resumes posted on their web pages). In addition, many state and national wildlife conferences have workshops for young professionals on resume writing. Take these training opportunities to help you craft the best possible resume.

You should have multiple hard copies of your resume, printed on high-quality stock, to have on hand at all times. You never know when someone will ask for one! Many candidates email an electronic version of the resume and also send a hard copy (if requested) for each opportunity they pursue. A mailed hard copy allows you to attach a personal note, reinforcing your interest in the position and presenting a rationale for why the job is a strong fit for your background and the organization's needs. A strong cover letter can have the same effect, and one should accompany your resume when applying for a job (or to graduate schools). A cover letter allows you to expand on why you are well suited for the position, as well as to provide any additional information asked for in the job announcement. (See the sidebar for an example of a cover letter.) You should always bring hard copies of your resume with you to every meeting and interview and offer one to each person you meet that has a job in which you are interested. The only exception would be organizations that have made a commitment to go paperless. For those groups, perhaps bring a flash drive to hand out, or at least get the appropriate contact information for the job, so you can apply electronically. (This type of application approach has recently gained traction.) Attach flash drives to your business cards, resumes, and cover letters, and hand them out in person when you meet with prospective employers, especially to those with whom you have an interview. Just like you would with a resume, have trusted colleagues look at and edit your first proofs and then tweak them accordingly.

Items to Consider

Yate (2014) offers suggestions to follow when preparing a resume:

1. Always list a target job title or professional objective that the reader can use to quickly understand your goals. This should be at the top of the resume, immediately following your contact information.
2. Always include a career summary or employment history. This should represent and capture your ability to successfully perform the job.
3. Always have a core competencies section—a concise list of your primary skills. This section of a resume is a good place in which to insert key words and key phrases used in the job description.
4. Do not mention salary. This is normally not an item to be included on a resume. Salary negotiations can take place during the interview stage.
5. Keep your resume focused. The short page limit demands that only the most important information be presented.
6. Emphasize your achievements, skills, and any unique qualifications you may have, so the resume reader is favorably impressed and believes that you belong on the list of candidates to be advanced to the interview stage of the hiring process.

The Final Product

Incorporating all of this advice and expertise into your resume should result in a strong and favorable response to your job application. If these are not meeting your expectations, you should reevaluate your resume. Send it out again to trusted colleagues, soliciting their edits and constructive criticism. Embrace and incorporate their feedback accordingly. Revisit all of the tools you are working with in order to secure a position that will serve as a launching pad for your career. Those include the components discussed in the other chapters of this book, such as interviews (chapter 8).

One final bit of advice: two actions that always yield positive results are perseverance and face-to-face meetings. Perseverance can demonstrate your dedication and drive with respect to the position you seek. Live meetings are far more productive than emails or phone calls. Even if a meeting does not go any further, request advice regarding your job search from the people you've spoken with. Even more importantly, ask them for referrals and introductions. (Chapter 6 contains additional information about interactions at professional meetings.) We contend that if you heed the advice in this chapter, you will successfully complete your job odyssey with drive, determination, and dedication. Do not expect to obtain a position without allocating a significant amount of time to expanding your circle of contacts and connections. Solicit their support, guidance, and counsel for what is one of the most important undertakings in your professional career.

SUMMARY

Your resume is what makes a first impression on potential employers. To be a successful applicant, first and foremost you must have the qualities a hiring manager is looking for. Develop a concise, accurate, engaging, and appropriate record of your accomplishments. There are many types of resumes, and no one size fits all! Resumes should be customized for each separate job application. Most employers will want to see a one- to two-page resume that begins with your name and contact information, states your professional goals and objectives, presents your qualifications for the job to which you are applying, and lists your awards, grants and scholarships, core competencies, and other pertinent information. Your resume should end with the names and contact information for at least three references.

LITERATURE CITED

Kursmark, L. M. 2012. Best resumes for college students and new grads, 3rd edition. JIST Works, St. Paul, MN, USA.

McCarthy, A., and K. Southam. 2014. Writing resumes and cover letters for dummies, 2nd edition. Wiley, Milton, Queensland, Australia.

Yate, M. 2014. Knock 'em dead resumes. Adams Media, Avon, MA, USA.

The Professional Interview

Preparing for Success

JOHN L. KOPROWSKI AND KAREN E. MUNROE

Congratulations, you now have an interview for the position to which you applied! After the elation subsides, the reality of the situation and the question "What do I do now?" become imperative, and time is of the essence. There is considerable work to be done, and how you approach everything from an initial phone call inviting you to interview to post-interview communications will influence your success (Bolles 2012, Molidor and Parus 2012). Interviews can take place for various position levels, including entry-level field assistant jobs, agency or nongovernmental organization (NGO) employment, graduate school admission, or academic appointments. We hope this chapter will be of use to a broad range of wildlife biologists who will interview for a diversity of positions. In it, we review the types and purpose of interviews; go over the entire process, from pre-interview preparation to post-interview decisions; and provide helpful advice to navigate this hiring stage and yield success.

For an employer, the purpose of an interview is to find the person who best meets the qualifications for the position, works well in a team with others, and brings some unique or valuable attributes to the work environment (Hein and Bates 1983). Your objective, as the candidate, is to demonstrate that you are the best person for the job by showcasing your knowledge and skill set, your strengths in communication, and your enthusiasm, and by conveying your ability to work well with others. Additionally, one of your goals should be to find a place where you can be productive and content in your professional aspirations.

THE GENERAL INTERVIEW PROCESS

Interviews consist of four or five stages for you, as the job candidate: (1) an invitation to interview, (2) pre-interview research and preparation, (3) a screening interview (not always

included), (4) a final interview, and (5) a post-interview response. Understanding the process from the perspective of your potential employer also is important. This starts long before a job announcement is posted. Typically, authorizations for new positions are not easy for an agency or department to obtain, due to budget limitations. A supervisor generally is required to make the best case for why a new one is necessary. As a result, people who will be involved in the decision-making process have a personal stake in the success of the search. After an opening has been approved, a search committee is often organized to draft a job description (in some state and federal agencies, general descriptions already exist, but often details specific to a particular position will be added). Each word is usually carefully discussed, including the specific skills necessary for the job. A group will complete an initial review of the applications received and whittle down the list to a manageable subset. More details may now be requested from the remaining candidates, or an examination of skills or knowledge may be required. This produces a short list of applicants, which is then reviewed in considerable detail and may include an initial screening interview. Finalists for the position are then selected, and they are interviewed, generally at length. The search committee often ranks the finalists after conducting these interviews, but the ultimate decision on who to hire typically lies with the head administrator, who considers the search committee's feedback. Your task as an applicant is to be ready for each of these stages so you can capitalize on your opportunities. Breaking your preparation into separate steps can be helpful in organizing your approach.

TYPES OF INTERVIEWS

Interview procedures can vary (Kennedy 2011), and three kinds are typically conducted in the wildlife field.

Parallel Processes

Candidate		Hiring organization
	Begin	
		Identify need
		Position description
		Convince leadership
		Advertise
Identify positions		
Submit application		First cut
		Short list
Invitation to interview		
Pre-interview preparation		Screening interview
Screening interview		
		Interview list
Final interview		Final interview
Post-interview response		
		Rank & recommendation
		Leadership extends offer
Offer & negotiation		
	End	

A diagram of the parallel processes undertaken by a job candidate and a hiring organization. The shaded box encompasses the stages emphasized in this chapter.

When interviewing for a job via Skype, make sure the room is quiet and the background appropriate (photo: Scott Henke)

The Phone or Video Conference-Call Interview

These two interview formats permit the hiring entity to assess the qualifications of many applicants through conversations lasting for one hour or less. Usually the questions are scripted and shared among the members of the search committee, each of whom asks one or two of the prepared items. Frequently this type of interview is used to screen candidates and derive a list of who will be invited for an onsite interview. In some circumstances, given budget restrictions, this may be the only interview that occurs. Success depends on conveying your enthusiasm and qualifications in a professional manner, using a communication medium that is often informal for most people.

More and more frequently, video conference-call software, such as Skype or GoToMyMeeting, is used in place of the telephone. Interviews using this technology offer many benefits. For example, you can see the interviewer's body language and facial expressions and take advantage of that to clarify and expand your answers. Video calls also pose added concerns, however, in that you will need to look as well as sound professional. Don't forget to glance behind you and assess the background beforehand, because that's what the interviewers will see. As with a phone interview, be sure to pick a place that is quiet and undisturbed. It is important to practice first with a friend or colleague, to ensure that you know how to use the software and can be clearly seen and heard.

The Airport or Conference Interview

These interviews are the least commonly used ones in our profession, but they are more usual in other disciplines. At large conferences, one or more members of a search committee will interview potential candidates, who generally have been identified by applications submitted prior to the conference. For some positions, most frequently for higher-level administrators, winnowed candidates will be flown to an airport, with

interviews conducted in an adjacent hotel. These interviews are generally short, often lasting only 15–60 minutes (though they can take up to two hours for administrative positions), and enable the search committee to see a large number of candidates over a short period of time. Success is based on a brief interaction with the search committee, which requires the candidate to stay focused. This type of interview does not give you an opportunity to cultivate much of a relationship with committee members. Projecting your personality, leadership style, acumen, and enthusiasm are particularly important facets to convey in such a short time period.

The Onsite Interview

During what is the classic type of interview, conducted on site, a candidate interacts in person with the entire search committee. For many entry-level jobs, these interviews can be short, 30–60 minute conversations, whereas for more-advanced agency, NGO, or academic positions, the interviews may last from one to three days and involve many meetings with various stakeholders. A variation on this kind of interview is to have candidates come together in a group to answer questions and even, occasionally, deal with hypothetical situations. Onsite interviews allow the search committee to focus on a smaller group of finalists and get to know individual candidates in considerable detail.

THE INITIAL CALL TO INTERVIEW

This part of the hiring process begins with an initial call to inform you that you have been selected for an interview. Be prepared for this call, and make it more than a cursory experience. Keep a file or a list of the jobs for which you have applied readily available, so you know exactly what position is being discussed when you receive an interview call. In such situations details can blur, so having an easily accessed job summary can be very helpful. Carry a notebook or create a file on your portable electronic device, with details that can serve as a reference point for you. For example, as the interviewer on several occasions, I [JK] made the initial calls to set up future interviews. Very quickly into the conversation on one such call, it became obvious to me that the applicant had confused the position I was phoning about with another job for which he had applied. Having to clarify this mistake did not leave me with the best first impression of the candidate. This is not how you want your initial contact with a future employer to transpire. Also, always start out in a professional manner. Even if you are expecting a call from a friend or your significant other, don't answer the phone in a cute manner. Unless you have caller ID, you never know who may be calling. Similarly, check to be sure that your voice mail message is appropriate.

The call to set up an interview also gives you an occasion to influence initial perceptions of your candidacy, as well as allows you to learn a bit more about that particular position. Besides demonstrating your enthusiasm and professionalism, you have an opportunity to ask a few questions about the nature of the upcoming interview (e.g., its duration, expected dress, will you get a schedule in advance, will there be any exams). This initial conversation also enables you to indicate what is important to you. If the interview is one that will involve more than a single meeting, you can ask to meet certain professors, researchers, or unit heads, which both demonstrates your knowledge of the organization and conveys your interests. For example, you might ask if you will have a chance to meet with undergraduate students, if interviewing at a university, or meet with classified staff (administrative assistants) and field biologists, if visiting a state agency. Asking for suggestions of realtors in the area also can demonstrate your genuine interest in the position.

PRE-INTERVIEW PREPARATION
The Elevator Speech

An *elevator speech* is a 30–120 second introduction about yourself, your interests, and your qualifications (Johnsen 2010). The name derives from the short period of time a person has to chat in an elevator and make a positive impression (Kwok 2013, Settele 2013). Write and practice a concise elevator speech in front of friends so you will have a polished introduction regarding who you are, your background and skills, your professional interests, and your goals. Use terminology from the job announcement, as this will be the jargon that will help demonstrate you are an appropriate fit for the position. You probably will be able to use this brief speech on several different occasions during your interview. By rehearsing and honing your introduction, you remove the pressure to create something instantaneously during your interview, avoid reinventing a response multiple times, present focused and vetted initial comments, and buy time for yourself to prepare additional responses.

Homework

Learn as much as you can about the position, key people in the organization, its history, and the city and community where you might be working. Try to know the educational background of the people whom you are likely to meet during the interview. If they have published papers, read some of their work. All of these preparatory steps will assist you in conversations during your interview. You might want to ask a few leading questions. What are the major land management agencies in the vicinity with which you may be working or which might have staff members participating in the interview process? What are important wildlife, land use, and natural resource issues in the region? Often these aspects will be incorporated into questions asked by the search committee. Table 8.1 presents some commonly asked questions, but be prepared to answer queries on topics that are not listed, because you will undoubtedly be asked some.

For certain agency positions, you may be taking an exam to demonstrate your general or specific knowledge or assess your skills, and now is the time to get ready for it. Pre-interview preparation reduces the surprises you might otherwise encounter. It also lets your interviewers know that you have done your homework and are serious about obtaining the position. One suggestion is to print out and bind together all of your pre-interview research into a booklet, so you can study it on the airplane, in the lobby, or at the hotel. Also download this information onto your personal electronic device, to refresh your memory at any time.

Draft a list of questions that you would like to have answered (table 8.2). Several of these will probably be covered during the interview process, but you might want to request clarification or assess the level of agreement among the search committee members. Use your questions as a way to highlight your own strengths in a follow-up response to the committee's answer.

Although you may not be able to control very much of the interview process, you have complete control over your preparation for it. This is your easiest opportunity to gain an advantage over your competition.

Practice, Practice, Practice

Practice your elevator speech and your responses to questions in the mirror, on camera, and with friends. Often universities have a career center that conducts and critiques practice interviews to make sure you use appropriate body language and have prepared as much as you can. Don't forget to review the job announcement. As you practice, emphasize the specific terms that are used in it and address all of the preferred qualifications. Remember that a committee argued over every word in the job description, so use this insight to help you prepare. Furthermore, do not assume that everyone present at the interview is intimately familiar with your resume, so emphasize your strengths repeatedly. Practice doing this now, so it comes naturally then.

Dress for Success

This old adage carries some truth. No hard and fast rule exists for how to dress for a particular interview. In your pre-interview research, you can inquire about the appropriate

Table 8.1. Common questions asked by wildlife-related organizations during professional interviews. Some of these are more likely to be asked in an interview for entry-level positions, whereas others generally are targeted for supervisory or academic positions. Consider the purpose of each question and why it is being asked. Try to imagine how it could be asked in your upcoming interview.

Why are you interested in this position?

What are your career goals?

Where do you see yourself five and ten years down the road?

What special skills do you have that make you qualified for this position?

Why are you the right person for this position?

What are your strengths?

What are your weaknesses?

How do you handle working alone?

What are the challenges of working in a team?

What is your experience with living and working outdoors?

How do you address problems that you might have with a team member?

How do you address difficulties that you might have with a supervisor?

How do you handle an upset client/student/landowner?

What are the greatest challenges for wildlife in our region?

What are the greatest challenges facing our field nationally or internationally?

In what professional societies do you maintain membership?

Are you a Certified Wildlife Biologist®?

What would be your first action on the job?

What do you view as the most challenging aspect of the position?

How would you handle the following situation?

What part of the position is most exciting to you?

When will you be available to start the position?

What will your first few months at this position be like?

How do you approach a new task that has been assigned to you?

How do you prioritize and organize your time?

What are your skills in organizing and planning?

How do you work to affect change as a new hire?

What intrigues you about living in the location for this position?

How have you overcome adversity in the past?

Tell us about a difficult time that you had in a previous job and how you handled the situation.

Tell us about a success that you previously had and describe what contributed the most to this success.

Discuss an instance when you had to give feedback to a colleague.

Where and in what situations are you most comfortable/uncomfortable and why?

What do you appreciate in a supervisor?

What do you seek in an employee?

What makes a good team member and coworker?

What do you think that your references said about you?

How do you like to spend your free time?

What is a recent book that you have read?

Who is a historical figure that you admire?

Compare yourself to a historical figure / species of animal / type of tree.

What additional questions do you have?

type of attire for the interview, which can be quite helpful and lessen your anxiety. The most important generality is that you should be the best-dressed person at the interview, and so err on the side of being a bit overdressed. Wear an appropriate yet comfortable outfit, so you can be as relaxed as possible. Try not to wear a new shirt or shoes that will literally and figuratively cause friction. Think not only about your appearance, but also about the weather conditions you are likely to encounter. A wool jacket on a hot day with high humidity may not serve you well. Heels that provide little traction in ice and snow might be a poor choice. Will you need an umbrella that day? Check the forecast.

Dress appropriately for the person who will be interviewing you. For example, wearing a suit and tie or a dress may be too much for a ranching position but appropriate for an interview with a wildlife agency (photos: Scott Henke).

THE INTERVIEW

Remember that, first and foremost, the interviewing agency is interested in finding a professional whose talent is capable of fulfilling the required duties and enhancing the position (Campion 1978, Harolds 2014, Hein and Bates 1983). Professionalism is key. Be certain to address the interviewers in a respectful manner and recognize their titles. Refer to them as "Doctor," "Professor," "Director," "Ms.," or "Mr.," unless they specifically introduce themselves by their first name (although it still does not hurt to use their title) or ask you to refer to them by their first name. Shake hands firmly and make eye contact during introductions. Be sure to acknowledge all members of

Table 8.2. Questions that a job candidate might ask when prompted. Some of these are more appropriate in an interview for entry-level positions, whereas others are more likely helpful for supervisory or academic positions. Consider the purpose of each potential question and imagine how you could modify it for your upcoming interview.

What are the most challenging aspects of the position?
What are the most satisfying aspects of working here?
How would you describe the atmosphere and camaraderie in the workplace?
What do you enjoy about living here?
What is the potential for advancement?
Are employees encouraged to participate in public outreach?
Can you review the fringe benefits, such as medical and dental care, that come with the position?
Please tell me about the annual review process for employees.
Are there support personnel and/or training available to hone employees' skills in supervision and proposal writing?
Are there opportunities for continuing education or meeting attendance?
What are you looking for in the ideal candidate?
Why is this position open for hiring at this point in time?
How dependable are annual budgets?

the search committee, where possible. Do not react only to those who are the most vocal, but demonstrate that you have command of the situation by being inclusive of all members of the interview group (Kennedy 2011). It is appropriate for you to introduce yourself, rather than wait for someone else to do so. Speak clearly and use appropriate terminology to indicate your grasp of the discipline. Remember, this is your chance to demonstrate your technical knowledge and assure the search committee that you meet the specific qualifications listed in the job announcement. Stay on point and remain focused. Two critical errors in responding to questions are being too short with your answer or being vague and meandering in your response. Unless the question calls for a simple answer, take every opportunity to emphasize your strengths, but do not ramble. Take a moment to organize your thoughts before responding, and try to think in terms of an outline as you craft your reply. Although that moment may seem lengthy to you, it will only last a few seconds and will seem perfectly appropriate to the interviewer. That brief pause will be critical in synthesizing a thoughtful and organized response.

A secondary function of an interview is to see if you will be a valuable teammate and collegial member within the workplace. While you wish to demonstrate your professional knowledge, communication skills, and political acumen in what you say, also realize that you should interject your personality into the interview. At this point, the search committee would not be conducting an interview if they did not think you were a qualified candidate. Show enthusiasm in your responses; be sure to smile; make eye contact; and use strong, positive body language. Mild humor is fine, but make sure that it is politically correct, appropriate, and retains a professional tone. Demonstrate that you would be an enjoyable addition to the team, as chemistry, a positive attitude, and a sense of camaraderie can make the difference in the organization's final decision (Fox and Spector 2000). This could counterbalance the effect an extroverted candidate may have (Burger and Caldwell 2000). Key behavioral predictors of success in job interviews include conscientiousness (Caldwell and Burger 1998), extroversion (Boudreau et al. 2001), and confidence in your own abilities (Tay et al. 2006). These qualities extend beyond the strength of your CV or resume and your biographical accomplishments (Campion 1978, Harolds 2014, Hein and Bates 1983). Agencies and private-sector wildlife organizations also tend to agree that personal characteristics—such as cooperation, enthusiasm, conscientiousness, empathy, diligence, and confidence—are crucial factors (Hein and Bates 1983). In short, your state of mind and how you represent yourself will have important influences on your success in the interview process.

Toward the end of the interview, the search committee will generally ask if you have any final questions. Rather than simply saying, "No, I believe you have covered everything," be prepared to ask a few questions of your own. At this point, it is best to focus on the work environment, job duties, or substantive aspects of the position (table 8.2). Ask for clarification of any substantive contradictory information you may have received during the interview process, but be sure to do so in a manner that does not offend anyone on the search committee. Be certain to frame your questions in a positive manner and not sound negative or nitpicky about details that are not relevant. For instance, you might ask, "Are employees encouraged to participate in public outreach?" with the foreknowledge that you have some capabilities that might be of interest to the search committee. After the interviewers weigh in with a reply, you will be able to mention your previous experience in this area and convey some added enthusiasm for the position. This also gives you an important final opportunity to emphasize your skills and elaborate on any previous responses that you wish had been more effective.

When the interview is over, be certain to thank each member of the search committee and let him or her know that you remain interested in the position. If you have not yet been informed of when they will make their decision, this is an appropriate point at which to ask them about the timing of the remainder of the hiring process.

The unique setting for a phone or video conference-call interview requires some modification from the classic face-to-face interview format (Kennedy 2011). Major advantages of this kind of remote interview are that you can (1) have a set of notes on important points you want to emphasize and questions you wish to ask placed within your view, (2) jot down the questions that are asked, and (3) make notes about items you want to return to. Be certain to arrange the interview at a time period that works well for you, so you are not rushing to your computer or phone, and in a location where you feel comfortable and will not be interrupted. Test your Wi-Fi strength and local cell phone reception (or use a land line) well in advance of the call. If possible, try to stand when you talk on the phone or during the video conference call, in order to project your voice with maximum effect. There is a reason why opera singers stand rather than sit! In addition, smile and use your normal hand gestures (even during a phone call), as this will help convey your enthusiasm for the position.

Another advantage to this type of interview is that the distance between you and the interviewers often results in a little less nervousness on your part. This distance and a dearth of visual feedback, however, make it more difficult to get a feel for how your responses are perceived and increase the challenge of conveying your enthusiasm and personality. Realize that there may be some unnatural pauses, and you will feel a bit less connected with the people who are interviewing you, but do not let this reduce your efforts to share your thoughts and interject your personality. Some people find it more difficult to remain formal on the phone. Be certain to safeguard yourself against too much informality and maintain an appropriate level of professionalism.

For a video conference call, make sure that the camera on your computer gives a good shot of your head and shoulders, as well as of your hands. Remember that a good portion of understanding comes from body language and other nonverbal cues, so make sure that the upper half of your body is showing. The lighting and background are also important aspects to consider. You will need to maintain eye contact, which means focusing on the camera—not on the view from your screen. If you find it distracting to see yourself while trying to look at the interviewers, you may want to minimize or turn off that portion of the video feed (most programs allow this). You can have notes in front of you, possibly even attaching them to a spot behind your computer, to help you maintain eye contact with the interviewers, but avoid relying on your notes. You need to maintain eye contact and connect with the people you're talking to. Keep all other programs on your computer closed, especially if they might make noise during the call.

Helpful Hints during Interviews

We have eight suggestions for key points to remember during your interview. You may wish to carry them with you, as the first page in a small notebook:

1. Be certain to demonstrate your enthusiasm for the position through your demeanor, strong voice, eye contact, and positive body language, as well as present evidence of the homework you did prior to the interview.
2. Never let your guard down, as every instant is part of the interview. From the moment when you meet the first search committee member—at the airport, in a car on the way to the interview site, in the conference lobby, or

at a hotel—until you leave the room or close the vehicle or hotel door, you are going to be judged. Take this opportunity to enhance your candidacy.

3. Stay on point. Remember your mission and be sure to highlight your key skills, especially in reference to the job description.

4. Have a short phrase or battle cry that reminds you to turn on your energy, get in the proper mood, and focus. It helps to say something like "show time" to yourself, either to get back into the moment before you enter the room, or to recover after you were caught off guard by an unanticipated question.

5. Know something about everyone on the search committee, and let each of them speak. Demonstrate active listening, and do not try to be the constant center of attention. You need to make your best case, however, so do not let them do all of the talking. Nod and respond effectively, and use your polished answers to highlight your skill set.

6. Pause momentarily before answering questions and organize your thoughts. For difficult questions that have you momentarily stumped, prepare a couple of general responses that will buy you time. Saying something like "That is one of those difficult challenges" or "What a great question, which gets directly to the kind of challenge" is a nice transition and gives your mind a few seconds to focus. On occasion, taking a moment to write the question in your notebook can provide you a bit more time.

7. Don't forget the restroom. A bathroom stall is the only place where solitude is guaranteed. No one will follow you there, and you will have a moment of down time to refocus. Be certain to visit the restroom before your presentation, and accept the opportunity when offered during the course of your interview. This way, you can regroup, remember what comes next in the interview, take a deep breath, get the adrenaline pumping, and exit, ready for "show time."

8. Be certain to make a case for why the position is an excellent fit for you, in addition to showing why you are well suited for the position. This demonstrates your sincerity about the job and suggests that you are not simply looking for an immediate stepping-stone to another position.

AFTER THE INTERVIEW

A professional email or handwritten note to the head of the search committee is appropriate, particularly if you request that it be shared with the rest of the committee. Similar expressions of gratitude can be sent to individual committee members if additional materials were requested, or if a member went above and beyond the call of duty during your interview, such as taking you into the field for a prolonged period. Be professional and genuine in your thanks. In the note, include an affirmation of your continued interest, but do not go overboard in ways that might suggest ulterior motives for your thank-you note.

Should you return from an interview and realize that the position is not what you are searching for, the most appropriate way to handle this is to contact the search committee chair. Express your gratitude for their interest and efforts, but request that your application be withdrawn from further consideration. This action will save the search committee time in debating your candidacy and will be appreciated. The professional world is small, and such efforts will ensure that you are viewed positively in the future.

COMMON MISTAKES IN INTERVIEWS

There are at least eight common mistakes job candidates make during the interview process:

1. Underestimating the time required to get to location of the interview, and therefore being tardy.
2. Exuding too much confidence and acting as if the employer would be lucky to have you.
3. Missing opportunities to interact with search committee members and emphasize your strengths, especially if offered the opportunity to do so by being asked if you have any questions.
4. Preparing poorly, so each of your responses has to be crafted on the spot and lacks polish.
5. Focusing too much on "I" and taking credit for accomplishments that are clearly the "we" victories of a team.
6. Letting your guard down during less formal moments, such as while sharing a meal or in the car on the way to the hotel or airport.
7. Not knowing important issues facing the agency or university.
8. Falsely assuming that it is important to name drop or demonstrate that you know minutia and trivia about the organization and its people. Don't try too hard to display your knowledge!

A FEW FINAL WORDS

Although an interview means a great deal of work and is likely to create a fair amount of stress, do not forget that being selected for an interview is an affirmation that your efforts and hard work have started to bear fruit (Bolles 2012). It shows that a wildlife-oriented professional organization has tipped its hand, indicating that you are a very competitive candidate for their staff opening. Revel in this news, and use it as motivation when you prepare for your interview. Should the result not be favorable, remember that you are clearly very close to being hired and redouble you efforts to capitalize on the next opportunity!

SUMMARY

The interview process consists of several stages, and it is important that you are prepared for each step. Homework and practice are essential for each phase, in order to influence how you are perceived as a candidate and to let you learn a bit more about the position. Remember that above all, the interviewing agency is interested in finding a professional with talents that fit their needs, as well as someone who will be a valuable teammate and collegial member in the workplace. Do your homework and learn as much as you can about the position, the key people in the organization, and where you might be working. Good preparation for an interview is the easiest way for you to gain an advantage over your competition. Practice an elevator speech, as well as answers to potential questions, and be prepared to ask a few questions of your own. In this chapter, we have presented strategies for the three kinds of interviews (by a phone or video conference call, at an airport or conference, and on site) and offered specific advice for each interview type. Remember that despite the work involved and anxiety it may cause, an interview is an affirmation of your efforts and hard work.

LITERATURE CITED

Bolles, R. N. 2012. What color is your parachute? A manual for job-hunters and career-changers. Ten Speed Press, New York, NY, USA.

Boudreau, J. W., W. R. Boswell, T. A. Judge, and R. D. Bretz Jr. 2001. Personality and cognitive ability as predictors of job search success among employed managers. Personnel Psychology 54:25–50.

Burger, J. M., and D. F. Caldwell. 2000. Personality, social activities, job-search behavior, and interview success: distinguishing between PANAS trait positive affect and NEO extraversion. Motivation and Emotion 24:51–62.

Caldwell, D. F., and J. M. Burger. 1998. Personality characteristics of job applicants and success in screening interviews. Personnel Psychology 51:119–136.

Campion, M. A. 1978. Identification of variables most influential in determining interviewers' evaluations of applicants in a college placement center. Psychological Reports 42:947–952.

Fox, S., and P. E. Spector. 2000. Relations of emotional intelligence, practical intelligence, general intelligence, and trait affectivity with interview outcomes: it's not all just "G." Journal of Organizational Behavior 21:203–220.

Harolds, J. A. 2014. Tips for a physician in getting the right job, part XI: important things to do for an interview. Clinical Nuclear Medicine 39:531–533.

Hein, D., and S. F. Bates. 1983. Criteria for hiring wildlife employees. Wildlife Society Bulletin 11:79–83.

Johnsen, C., ed. 2010. Taking science to the people: a communication primer for scientists and engineers. University of Nebraska Press, Lincoln, NE, USA.

Kennedy, J. 2011. Job interviews for dummies. For Dummies, New York, NY, USA.

Kwok, R. 2013. Communication: two minutes to impress. Nature 494:137–138.

Molidor, J. B., and B. Parus. 2012. Crazy good interviewing. John Wiley and Sons, Hoboken, NJ, USA.

Settele, J. 2013. Job seeking: two sentences to impress. Nature 496:169.

Tay, C., S. Ang, and L. Van Dyne. 2006. Personality, biographical characteristics, and job interview success: a longitudinal study of the mediating effects of interviewing self-efficacy and the moderating effects of internal locus of causality. Journal of Applied Psychology 91:446–454.

Being a Professional and Acting Professionally

WILLIAM F. PORTER AND KELLY F. MILLENBAH

Browse the Internet or look at a newspaper any day of the week and you will be bombarded by at least one example of poor behavior in the workplace. Often these examples leave you shaking your head in disbelief over how a person could display such blatantly bad judgment. We may laugh and say, "I would never do something like that," or "How did they not know better?" For someone on the outside, such questionable behavior is easy to recognize. Yet many of us make subtle and sometimes not-so-subtle missteps during our careers. The principles of professionalism exist, in part, to help us recognize highly effective behavior and avoid those missteps. But the question is, where do you find out about these principles? Many are especially challenging to figure out, because they are dependent on context. Others are basic behaviors that are essential in nearly all situations. We present seven principles for embracing professionalism in the workplace that have served us well in our more than 60 years of combined experience. Their execution and adoption require awareness and practice. Yet we believe these principles will serve you well, both today and throughout your career. We are certain that your professional success will be dependent on them.

STUDY YOUR JOB

Stepping into a new job can be exciting and nerve-wracking, all in the same breath. You want to get started on making a positive impact, but you are nervous about doing your job well. All along the way, you hope to not make too many mistakes that everyone will see. Confidence is good, but overconfidence can have several negative ramifications. Walking into a job thinking you already know everything about the position, your supervisor, and the people with whom you will work is dangerous territory. It is imperative that you take the time to truly understand the your responsibilities. The highest priority is finding out what your supervisor expects. In today's team-

based environment, however, getting to know your coworkers is also vital. Conversations with them, and your subsequent actions, will be the foundation for building greater competence in your position. They also will give your supervisor and coworkers greater confidence in you. More importantly, the time spent getting to know the people you work with and their expectations will give you more confidence in yourself, which you will need in moments of great stress and significant challenges.

> KM: I have had the privilege of trying out many different positions in my career. Some of them I felt confident about pursuing. In others, I was less confident but had a great mentor pushing me to take a chance on myself. Without a doubt, one of the best decisions I made when entering each new position was to not assume that I knew everything. In fact, the more I learned about a job, the more I discovered how much I did not know. A recent position I accepted put me at the helm of a college with a $34 million dollar budget and over 4,800 students. It was a daunting challenge, and I was admittedly scared that I would mess up in some very public manner. But the one skill that I brought to the position at the very start was my ability to ask questions. I asked a lot! And I still do. By asking questions, I was able to build trust and foster relationships with people as we worked together to solve some of our most challenging problems. Over the years, I have learned that it is better to say "I don't know, but give me a couple of days and I will get back to you" than to make a potentially erroneous statement that would be hard to correct in the future.

Can asking questions make you appear to be less competent? Quite the contrary. People will respect you for acknowledging that you are capable of knowing only so much at a given point in time. They will respect you even more for making sure that you have a good understanding of the issue.

In addition, you might be surprised by what you learn. Experience teaches us that there are three sides to every story: person X's side, person Y's side, and what really happened. By asking questions, you may even convince stalwart opponents to reconsider their position on an issue. In most cases, the more you know about your organization, your supervisor, the people around you, and the issue at hand, the better your decisions will be, and the greater the impact you can have in the long run.

KNOW YOUR AUDIENCE

There are few jobs that do not require you to talk to at least one other nonaffiliated person at some point during your day-to-day activities. Certain jobs require you to speak with K–12 students, college students, business leaders, state and federal agency staff, administrators, legislators, or all of the above, albeit at different times. *How* we communicate is just as important as *what* we communicate to each audience. Your job does not end with sending a message. Rather, you are responsible for ensuring that the audience gets the message.

KM: Before any event or class where I make a presentation, I always ask for background information about the audience to whom I will be speaking. How many will be there? What is their exposure to the topic? What have they been told about why I am coming to talk with them? Will the group be contentious because of the topic? How is the room set up? How much time do I have to get my message out? Without knowing who is in the audience, it is difficult to craft the best way to deliver my message. In a college-level introductory wildlife management course I used to teach, I would invite a wide range of guests to present different topics to the class. I wanted experts talking about their particular subjects, because they could better communicate—with great enthusiasm—why students should care about that topic. Without fail, those who did not ask me questions about the background of the students missed the mark in their presentation. I can vividly remember watching my students disengage with a presenter when he droned on about an issue that was beyond their comprehension, based on their past experiences. Cell phones were out; students were checking social media; and, not surprisingly, no one did well on the test questions related to that speaker. While the speaker was passionate about his subject, he failed to connect with his audience, because he prepared his material for listeners at a level that was clearly not appropriate for an introductory wildlife class. It was a lost opportunity to connect with a large group of undergraduates still looking to find their compelling career interests.

A comical video exists online about how an annoyed professor tears a student's email apart, word by word. The professor reacts to the lack of a salutation, poor sentence structure, poor punctuation, and the use of texting language in the student's plea for help with coursework. While the video itself is

Biologists in nearly every wildlife job are required to speak in front of students, business leaders, state and federal agency staff, administrators, and legislators (photo: April Conkey)

intended to be entertaining, what it portrays happens all too often with students when they interact with faculty and employers. Even when you know someone well, if it is a professional situation, proper grammar and sentence structure are imperative. Failure to present yourself as anything less than professional quickly erodes your ability to be taken seriously. It damages your chances to show that you are competent and care about the work in which you are engaged. Similar statements apply to verbal communication. "Ums" and "ahs," cursing, and other casual verbal interactions should be saved for your friends and close companions. Remember that whatever the situation, people always evaluate who you are as a person. What you say and write speaks (no pun intended) volumes about your character.

WP: Among my most lasting lessons as a young professional was learning to write a memo. I, like many, thought: "How much can there be to know about writing a memo? It's just a short description of a need." My first memos to my department chairman came back to me bloody with red ink. He reorganized them, corrected the grammar and punctuation, and caught all the typos. He taught me to pay ruthless attention to style and substance. I can still remember his admonition: "Put your request in the first sentence! You want me to be sure to see it." Only when the *presentation* was absolutely correct would he talk with me about the *substance* of my memo. It's surprising how easy it is to be more successful just by paying attention to details like these.

BE PREPARED

We cannot say enough about the importance of being on time—every day! You should always show up when expected and be upbeat and ready to engage the day. Nothing looks as bad as being late, grumpy, and unprepared. Be the first to arrive. Dress appropriately, whether you are in the field or the boardroom. (Wearing flip-flops for winter field activities is a no-no, and yes, that has actually occurred). Know what is happening, and refresh your memory, if necessary. Before any event, learn the techniques to be used, the concepts to be discussed, or the issues to be debated before you arrive. If, for any reason, you think you will be late for an event, let others know about the delay and extend your regrets. Call, text, and email until you reach someone. Let's face it, even with the best-laid plans, we have all been late and unprepared. Others will forgive that once or twice. What is inexcusable is making this as a habit. Such behavior is disrespectful to others with whom you are working, because it wastes their time.

KM: When I was a master's degree student, I was exceptionally nervous for my first big international professional presentation. I had given up spending a holiday with my family so I could work on my talk and refine my slides for the umpteenth time. I wanted to impress my faculty advisor, who had been telling me horror stories of people who gave presentations at professional meetings and either forgot what they were going to say or literally passed out at the podium from stress. I knew I only had 20 minutes to speak and needed to get everything said in that time, as well as answer questions from the audience before leaving the stage. When I stood up to give my talk, my knees were knocking and my mouth tasted of cotton. The lights were dimmed, and my first slides popped up. I sailed through the first slide and began to relax, aware that I knew this material better than anyone. But I wasn't prepared for what happened next. The audiovisual system died. Panic set in. Do I keep going? How will I ever explain my statistical analysis without the figure? I kept hearing my advisor in my mind, reminding me that "whatever you do, you only have 20 minutes to get it done." So I moved forward with my talk. Luckily, I had come prepared with a typed copy of the entire presentation. I was able to deliver the necessary information—with some humor—and complete my talk successfully. Being ready to handle the unexpected was well worth the investment of my time and energy. Shortly after I finished my presentation, I was offered a job with a state agency. One of the reasons they were interested in me was my poise under pressure and ability to complete a task, even when things were not going well. This all happened because I had taken the time to be prepared.

FOLLOW-THROUGH

Success, in the professional world, is finishing a project on time and getting it done right. Working with others on team projects is among the experiences students often intensely dislike. Some people on the team invariably make promises but don't deliver. In the professional world, nearly every project requires teamwork. When others fail to get a task or an as-

signment done on time, or it's not quite right, this presents the entire organization in a bad light and reflects poorly on everyone. Consequently, follow-through is one of those qualities that is widely recognized and quickly evaluated.

WP: My younger son completed an undergraduate degree in English. He moved with a girlfriend to Boston, where he eventually landed a job at a university, helping a dean work with people interested in making large donations to that institution. His primary task was to write letters for the dean. He not only met the deadlines, but he followed up on behalf of the dean with people who expressed interest in making a donation. He learned that this subsequent contact was the key to actually getting the funds. As a result, the dean gained confidence in him, and my son was able to build a diverse resume. When my son and his wife moved to Chicago to begin her career, he asked the dean for a letter of recommendation. He got a strong, positive letter and was offered a bigger job in Chicago.

We think following through is a relatively simple professional quality and should always occur. Yet, as we grow as professionals, we find that even when we focus nearly every moment of long workdays on completing tasks, we cannot get everything done. We are forced to make decisions about what gets finished on time and what gets delayed. As professionals, it is wise—and often required—to share our prioritizing with our supervisor. Good supervisors recognize that time crunches happen, and they will help you reset priorities. To miss a deadline, without having that conversation, is always going to reflect poorly on you.

Nevertheless, prioritizing is much more than meeting multiple deadlines. It is a work style that recognizes the full range of activities occupying our time and requires us to manage the portion we allot to each. A time-management matrix (Covey 1989) describes how or, in this case, where we spend our time.

Are we prioritizing so that quadrant I is where we spend most of our time? This quadrant applies to deadline-driven activities, or actions we must take so others on our team are able to accomplish what they need to do. When should we place more emphasis on the important but not-so-urgent activities? It is better to undertake these quadrant II actions after we have done more research and thinking about them. Much of quadrant III is filled with urgent requests: telephone calls and emails. If we are not conscientious, these distractions will absorb significant portions of our day, often to the exclusion of our work in quadrants I and II. Quadrant IV should be the one of most concern to a professional. These are the computer games and long conversations with coworkers that are not related to the job. Effective professionals who will rise in an organization develop strong abilities to focus on quadrants I and II, manage III, and minimize IV.

As our careers mature and we begin to demonstrate strong professional skills that complement our technical capabilities, we advance to supervisory positions, managing what others do. We assign work objectives and evaluate underlings on

	Urgent	Not urgent
Important	I	II
Not important	III	IV

A time management matrix

their ability to complete tasks promptly and well, especially because this aspect of their work affects our own. Given that our employees may have neither the experience with large projects that we do (after all, we are in a supervisory position) nor the larger perspective necessary to assess priorities, we have a responsibility to ensure that they are following through where it counts. Thus the job of a supervisor becomes one of monitoring each member of his or her team to make sure the team is collectively doing what it should.

MAINTAIN INTEGRITY

Integrity is about trust. Whether it involves friendship or marriage, a sports team or a military combat unit, trust is at the heart of every human relationship. It runs deep, because its roots can be traced to mankind's early evolution. Human survival was dependent on being able to trust others in a social group. Consequently, society is highly sensitive to detecting reasons for mistrust and generally invokes severe penalties when trust is broken. There has been considerable exploration of the integrity of personal relationships in other studies, but let's think about it here from a professional perspective.

WP: One of my strongest memories of professional trust, which was built up and then shattered, was working with Jon, an undergraduate wildlife student. One spring, he came to see me about a summer job as a wildlife technician. We had a large white-tailed deer project underway, and Jon met the qualifications. The job required that he monitor the survival of about 60 radio-collared deer, and he was given authorization to drive a university truck. One day the director of the university's motor pool called to say that the truck we were using had been damaged. An entire side panel needed to be replaced, at a cost of about $2,500. I spoke with Jon about the damage, and he claimed that it must have been caused by another driver when the vehicle was in the parking lot. The motor pool then asked the university police to investigate. The officers pieced together the complementarity of scrapes and dents on two trucks in the parking lot and were able to demonstrate how the damage had occurred. Unfortunately, the investigation pointed to the truck Jon was driving as the cause of the damage. The university police interviewed Jon, and he admitted that he had lied earlier. Damaging the truck was embarrassing and would have resulted in a stern lecture. Had he come to see me immediately to report the accident, however, it would not have broken our trust. The real

problem was that Jon had lied, so now there was reason to distrust him. Could we be confident he would tell us the truth in the future? Could we be sure he was really finding every one of the 60 radio-collared deer each day? That was a lot of deer, and we began to wonder if he was recording some of the deer as being alive on a day when he really had not found them. He no longer carried the integrity we needed in order to send him out alone. He had broken our trust, and I was forced to fire him.

If there is a difficult lesson to learn in life, it is that trust is built slowly, over time, but can be broken quickly. If that trust has grown into a strong relationship, there is sometimes an opportunity to repair the damage without a complete fracture. When an individual has been on the job only a short while, however, reestablishing trust can be difficult or impossible. In either situation, the act may be bad, but the cover-up is worse.

ADMIT MISTAKES AND LEARN FROM THEM

Owning up to mistakes is more than a question of integrity. No one is perfect. Some mistakes are factual errors (e.g., including the wrong numbers in a report). In other instances, we have poor judgment. Or we cut a corner because we did not think anyone would know or that it would matter. As much as we try to avoid mistakes, and are often embarrassed when they occur, they are part of being human. As a professional, we are at our best when we focus on learning from our errors. To do that, first we have to admit them. We often can see our own mistakes, but perhaps equally as many times, we don't. We need a mentor or confidante to help us spot them.

WP: Early in my career, I was given a significant leadership role for a large research program. I was young and inexperienced, and in the first six months, I made one mistake after another. While I knew a lot, I didn't know all the rules and regulations that guided what I could do within university policy. I was not sufficiently acquainted with the people in the organization—those who would help me, and those who could not or would not. I met with my supervisor frequently to discuss successes and failures. One day, I expressed my frustration with constantly erring, and he offered wise counsel: "To be a professional is to learn from mistakes. It's not so much about how many mistakes you make, but how often you make the same mistake twice."

We often hear athletic coaches say that the secret to top-level performance is to put each blunder behind you. To do that, you have to be willing to admit your mistake to those who are harmed by it. Do not obsess over your error, and do not permit others to wonder if it was their slipup instead. When you have time, reflect on the situation, consider what would have been a better option, and learn from it. Of course, that is the fundamental purpose of practicing, whether it is learning music or engaging in wildlife management.

KEEP THE TEAM FOREMOST

Being a professional often means that you are part of an organization. Your success within it is dependent of the actions of the other members, and vice versa. Consequently, it could be argued that you have a selfish reason for wanting the other members of your team to succeed. There are certainly people who take advantage of this duality. A professional, however, recognizes the personal advantages but chooses to put others on the team—and the team itself—foremost. We hear about this less self-centered and more organization-centered philosophy in team sports.

What we often fail to recognize is that there are many subtle ways in which this plays out. Most people are aware of how an organization's reputation can make a big impact. In wildlife management, we are often trying to persuade stakeholder groups to partner with us in accomplishing an objective that will further wildlife conservation. Our ability to make the case depends, in part, on what others think of our agency. That's why organizations are so compulsive, not only about the substance of what is being presented, but also about its style. The image that is projected by every member is a reflection of—and reflects back on—the organization.

> WP: Brian was an exceptionally capable specialist in Geographic Information Systems. Once a year, he and several other members of his team would attend a large conference where the dress was business casual, that is jackets, a collared shirt, and slacks for men, and dress slacks and blouses or sweaters for women. Brian did not believe such attire was necessary. He argued that the content of his presentation at the conference was far more important than what he was wearing. He dressed in blue jeans and an untucked flannel shirt. What he failed to realize is that other people are always evaluating you: what you say, how you look, and how you act. We know that at least 60% of communication is nonverbal, so the question is, what was Brian attempting to communicate?

The welfare of an organization is also influenced by how you engage with others. As professionals, we often are involved with other people, either on teams or in carrying out decisions handed down from above. As humans, we tend to want our needs to be met and our ideas accepted before listening to others. Consequently, young professionals tend to arrive at meetings in transmit mode: they cannot wait before offering their views. Experienced professionals recognize that understanding others and listening to their ideas helps in two important ways. First, hearing what others have to say provides an opportunity to quickly recast your own ideas, using phrasing that will resonate with others. Second, by better understanding what others need, you are in a stronger position to do something to help meet those needs. You therefore have a greater chance of suggesting a win-win solution and dramatically increasing the likelihood of success.

SUMMARY

When boiled down to its essence, professionalism is about conveying—and commanding—respect. Everywhere we turn, we see evidence that respect is central to social interaction: for the office a person holds, for an individual as a man or a woman, for someone's time, for people's trust in us. These are the plots of the best stories, from blockbuster movies to urban legends. We encounter this nearly every day in our workplaces. For example, first-year students send us emails that lack any formality. There is no salutation, no punctuation, and the messages often request us to call them back to set up a meeting with them. By the time these student are seniors, they are aware of the principles of professional behavior. In seeking our help, they understand that showing us respect will, in all likelihood, gain a greater share of our time and attention in addressing their needs.

In the end, we learn professionalism because we want to be effective. Everyone often faces the challenge of seeking to accomplish something that requires the time and energy of others. Highly effective people understand that working with others, and working as a team, are both about respect. They realize that, at its core, respect consists of a series of traits that should become second nature to us. The seven traits described in this chapter are a cornerstone to that professionalism. We think you will be amazed if you use them to watch and evaluate the varying degrees of professional behavior displayed by those around you, as well as to pay attention to your own actions.

LITERATURE CITED

Covey, S. R. 1989. Seven habits of highly effective people. Simon and Schuster, New York, NY, USA.

Graduate School: A Professor's Perspective

FIDEL HERNÁNDEZ, NOVA J. SILVY, AND KELLEY M. STEWART

Your time as an undergraduate student can be one of great uncertainty. You may ask yourself: What will I do when I graduate? Where will I get my first job? Should I attend graduate school? Although such questions may be daunting, this phase of your life can be the most exciting, because it is open to endless possibilities. Later, you will rarely have such flexibility in making choices about the course of your life as you do when graduating with an undergraduate degree.

In this chapter, we discuss information that you should consider in the decision-making process. We present, from a professor's perspective, general advice on factors to ponder when contemplating whether graduate school is right for you, as well as discussing the skills and qualities that will assist you in becoming a successful graduate student, and information to facilitate your transition from graduate student to employee.

DECIDING TO GO TO GRADUATE SCHOOL

There are two general types of undergraduate students: those who know the exact career path they wish to pursue, and those who figure it out along the way. The former group may already be aware of whether they wish to pursue a graduate degree, the kind of career they want, and the type of organization (e.g., state, federal, nonprofit) for which they wish to work. For the latter set of students, the idea of graduate school may never have been a formalized thought. They may not know exactly what career they desire, or who their preferred employers might be. Neither approach is necessarily better. As an undergraduate, you have time to explore options and discover your interests. Regardless of which group you may fall into, there are still some general guidelines that you should keep in mind if you would like graduate school to be a potential option.

Reasons to Attend Graduate School

There are many legitimate aspects to why a student may wish to pursue a graduate degree. These may include a desire to expand your knowledge, acquire new skill sets, or enhance the likelihood of obtaining a job or a future promotion. For all of these reasons, graduate school can be helpful in deepening your knowledge of wildlife management, conservation, and general ecology; developing your critical thinking skills; and enhancing your verbal and written communication skills. Graduate school also offers many opportunities for professional development. You may have a chance to conduct research in stimulating places, present these findings at national or international conferences, and network with professionals from around the country and the world.

While graduate school can offer you excellent opportunities, there are also reasons why it can be a poor choice. Oftentimes, students will consider graduate school simply because they do not have any definitive employment after graduation. With no immediate job opportunity on the horizon, going to graduate school sometimes becomes a default decision for students. This rationale almost always ensures failure. Graduate school involves a strong commitment to your coursework and to research. Classes can be difficult, and research demanding. Maintaining enthusiasm and the energy level necessary to excel in a graduate program can be hard, even for dedicated, passionate students. Students' earnest interest in graduate school and research, however, will allow them to persevere through the tough times and ultimately be successful. If your decision to attend graduate school is not based on a sincere desire for additional education, but is simply due to a lack of immediate employment, your enthusiasm will wane once the challenges of graduate school emerge, which they surely will. Our advice is to pursue graduate education only if it you have a serious commitment to it. If this is not where your interests lie, then continue looking for a job. If you have been a diligent undergraduate student, you will find employment that is right for you.

Applying to Graduate School

Ideally, you should begin to consider graduate school during your junior year. There are many universities offering wildlife graduate programs throughout the country, and you will need time to explore the schools and the aspects of their particular programs that best align with your interests. This does not imply that you need to enter graduate school right after finishing your undergraduate degree, however. Many students wish to obtain a few years of professional experience, either through jobs or internships, before enrolling in graduate school. This is perfectly fine. The important point is to allow yourself plenty of time (at least six months) to apply, once you decide that graduate school is for you. Each university will have different entrance requirements and deadlines for applications. Gathering the necessary application materials, such as a cover letter, resume, official transcripts, graduate record exams (GREs), and letters of recommendation, is a lengthy process. The more familiar you are with the application process and the entrance requirements of the various programs that interest you, the better prepared you will be to submit a timely and strong application.

Acceptance into a graduate program often involves a two-step process. First, you must apply to work on a specific graduate research project. Second, once the professor spearheading the project has chosen you to participate in it, you must apply and be admitted to the university's graduate program. Most graduate programs in wildlife require the professor who selected you for a particular graduate research project to agree to serve as your major advisor, mentoring you through the completion of the project.

How can you submit an application that will be competitive? Different professors emphasize different components, but some generalizations can be made about the characteristics of a strong applicant: a graduating grade point average (GPA) of at least 3.0 (based on a 4.0 scale), good GRE scores (at least 150 each in verbal reasoning and quantitative reasoning), well-rounded experience, and strong letters of recommendation. Students often wonder if excellence in one area can compensate for a deficiency in another. For example, what if a student has a low GPA (e.g., 2.75) but plenty of relevant experience; will he or she be competitive? This depends on the professor selecting the student and the needs of the project. Although you might still be chosen with such credentials, you generally do not want to find yourself in this situation. Give yourself the best chance of being accepted to graduate school, which entails taking your education seriously during every phase of your life. As an undergraduate, this means excelling in your coursework, volunteering for or being employed in various wildlife-related jobs, and delivering high-quality work, all of which can result in persuasive letters of recommendation. A good reference goes hand-in-hand with a strong work ethic, which consists of a positive attitude, enthusiasm, dependability, self-initiative, motivation, and a tireless work capacity. Possessing all these qualities will go far toward earning you excellent recommendations from your professors or employers. In addition, be sure that your application contains a well-written cover letter. Think about why you wish to attend graduate school, why you are interested in the position, and what assets make you a valuable candidate. Make your cover letter concise, informative, and personal. An application consisting of a well-written cover letter, an excellent GPA, good GRE scores, extensive professional experience, and highly supportive letters of reference will make you a very competitive applicant. Naturally, you have to earn these qualifications. Keep in mind that even if a professor does not select you to participate in a graduate project, this does not necessarily indicate that you are not qualified to be a graduate student. It may just signify that the hiring professor thought another person was better qualified for that position. The selection process often has considerations that go beyond simple qualifications, ones that may be specific to a particular project. In addition, generally only one position is available, and a professor receives applications from prospective students throughout the world. Do not allow a closed door for one project to discourage you from another, or from graduate school in general.

Once you have successfully been admitted to a university's graduate program, an advanced degree will involve formal coursework and a research project. The latter includes a written thesis (for an MS degree) or dissertation (for a PhD degree). You will have greater flexibility in selecting your graduate coursework than you did as an undergraduate. The department may require all graduate students to take a few specific classes, but in general, you will have some latitude in selecting your courses. The exact ones you will need for a graduate degree will depend on your major professor, your graduate committee (three to five professors or professionals that assist with your graduate development), and the university. The following are a few considerations to keep in mind when selecting your courses. Classes should be tailored to strengthen how you conduct your research project and analyze your data. You also should select graduate courses in areas not or only tangentially covered by your undergraduate degree, such as enrolling in a wildlife nutrition course to round out your education. The same can be said when selecting a graduate committee. The purpose of this committee is to assist and guide you through your research project. Committee members should be selected based on their expertise and how that can strengthen your project.

Obtaining a graduate degree is very different from obtaining a bachelor's degree. As an undergraduate, you are expected to take a predetermined number of courses within a set program curriculum. When you successfully complete every course, you then earn your degree. Graduate school, however, consists of two parts: coursework and research. One without the other is useless, and both are required to earn a graduate degree. A few graduate programs in wildlife will offer a non-thesis option, where a student may earn a degree through the mere completion of coursework. If they do, they usually require an on-the-job training internship or the development of a professional paper.

The process of applying to graduate school is very much like applying for a job. There are many places where you can find position announcements for graduate research projects, but one of the most popular, where many wildlife professors in the United States post graduate openings, is Texas A&M University's job board for the Department of Wildlife and Fisheries Science (see appendix B). If you have not visited this site yet, we encourage you spend a few hours becoming familiar with it. The job board will assist you in finding the right graduate program and research project for you.

What to Look For in a Graduate Program

Once you decide that graduate school is in your future, there are a few general factors you need to consider in narrowing down your choices, listed here in no particular order of priority:

1. Do you feel comfortable on the university campus and with its location?
2. Are you genuinely interested in the project's research question?
3. Is the graduate assistantship and research project fully funded?
4. Does the professor's mentoring approach align with your learning style?

Interviews for a graduate research project may not take place at the university itself, so you should visit the campus and meet with the professor leading the project prior to accepting an offered position. America is a melting pot of cultures, ways of life, perspectives, and tastes. Although there are commonalities across cities and universities in the United States, each one has a character unique to its region and origin. Some are more rural, and others are more urban. Some are predominated by one culture, and others are a medley of many. Some are located in arid deserts; others are situated along coastlines, where the climate is moderate and humid. Visiting a campus and its locale at some point during the interview process, even if at your own expense, will help you decide if the setting is one in which you feel comfortable and can thrive. No campus or region will be perfect. They all have their pros and cons, and the most successful students are the ones who are easily adaptable. An onsite visit, however, can provide additional information to help you make a well-informed decision about your future.

A second consideration when evaluating a graduate research program is whether the research question being asked truly appeals to you. Is your interest in birds, mammals, insects, genetics, human dimensions, or habitat restoration? Select a project that aligns with your preferences, whether for a taxonomic group (e.g., ungulates) or a topic (e.g., predation).

If your desired specialty is ornithology, look for a research project dealing with birds. Do not accept one investigating species' relationships of rodents. We have observed many students accept a graduate research position simply because their preferred research projects were not available, but these students often are not truly fulfilled. People are at their best when interested and passionate about the task at hand. Be patient and wait for a research question that truly interests you. The research, and you, will be the better for it.

A third consideration is whether your stipend and the research project are fully funded, with a grant or other source of money in place for the duration of your degree. In general, an MS degree requires two and a half to three years to complete, and a PhD degree, four to five years beyond that. Because graduate school and your research project will require the same level of commitment and time as full-time employment, a monthly stipend to help offset your living costs for the entire time you are in graduate school is imperative. Without such funds, you will be forced to find employment outside of your research project and not have the time or energy for your education. In addition, committed financing for the entire project ensures that sufficient resources are in place to complete the research, and therefore your degree. Be cautious about accepting a graduate research position for which the stipend or research costs are only partially secured. There are no guarantees in *potential* future funding. You should be focused on being successful in graduate school, not worried about being financially unable to complete your research and your degree. Do not, however, accept a particular assistantship just because it pays more than one for another project in which you have greater interest. If you are not passionate about your graduate project, more money for an assistantship will not help you produce a better research product.

A final consideration in choosing a program is whether you can meet the faculty member who will be supervising your graduate education. People have different styles of teaching and learning. Some professors take a hands-off approach to graduate mentoring, whereas others choose to be more fully engaged. Some students thrive on near-complete autonomy and independence, whereas others yearn for one-on-one interactions and frequent discussions with their mentors. Know your learning-style preferences and make sure they align with the professor you are considering as a mentor. Ask that person's current and former graduate students about their experiences with him or her. This information will help you decide if that professor will be the correct mentor for you, one with whom you can have a productive and supportive advisor-student relationship.

By choosing a university setting, research project, and major professor that fit your interests and learning style, you are on the right path to being successful in graduate school. All four factors are important considerations, but the value placed on each is up to you. The main point is that when you are accepted into a graduate program, you should be confident and comfortable with your choice.

HOW TO SUCCEED IN GRADUATE SCHOOL
Expectations of Graduate Students

What will your graduate advisor expect from you? And what do you believe you will get out of the program? The answers may be similar in many ways, yet different in others. Students generally are excited about beginning a project, working with their species of interest, learning new techniques, and interacting with professors and professionals. Advisors look at their students' management of individuals on the project, the quality of their data collection, their analyses of those data, and their ability to produce agreed-upon products, such as reports or publications. If funds were provided by a state wildlife agency, then the expected products may be annual or semiannual reports. To your advisor, however, the products are going to be joint authorship on peer-reviewed publications, not just a thesis.

A research project will involve not only your collection of field data, but also your involvement in all other aspects of the scientific method. A student generally is not expected to design the entire project (the professor in charge normally has done the majority of the planning, in order to obtain funding for it). A doctoral student usually has more input in a study's design than a master's-level student. Some professors, however, may expect graduate students to develop some original research of their own, regardless of whether they are MS or PhD students. Some of the specifics of on-the-ground data gathering will be expected of all students, as well as the collection of quality data, analyses of those data, and attention to detail throughout the process. Students are expected to be meticulous and careful with data collection.

Your advisor's role (along with your graduate committee) is to educate you, show you ways to become a professional, and disseminate the results of your research. When funding has been acquired for a project and you have collected the necessary data, a question that sometimes arises is, "Who does the data belong to?" In general, this is the funding agency and the person who obtained the funding, usually your major professor; data rarely belong to the students, although permission may be granted to use that information in future work. We are certainly not saying that you do not have an important part in data collection or a vested interest in the outcome, but remember, your advisor obtained the funds for the project and hired you, and that money generally was provided by an agency or an organization. Your priority is to write your thesis and publish the results of your study. Your work does not end until that material appears in a peer-reviewed journal.

Although most of your time generally will be devoted to your own research and coursework, you also may have other responsibilities. Stipends provided to graduate students designate them as university employees, with their major professor as their supervisor. Professors therefore are the ones who establish their graduate students' job duties, just as any other supervisor would for an employee. These may include assisting professors with classes and helping other students with their projects.

Characteristics of Successful Students

Along with enthusiasm and a passion for their research project, some other characteristics we have observed in successful graduate students include perseverance, creativity, motivation, resilience, and independent thinking. We have heard many professionals comment that completing a graduate degree has more to do with perseverance than any other single quality. Most students reach that goal because they refuse to quit, even when things became difficult—often extremely difficult. Those who do best will figure out a way to make a task or a situation work, even when the odds are against them. And things always will go wrong in some way.

Creativity plays a large role in perseverance, and it is a huge asset for success in graduate school. Being able to figure out how to solve problems—ranging from "How do I collect the right data to answer the questions or hypothesis I wish to test?" to "How can I collect the data that are needed in light of logistical challenges?"—is a valuable skill. Motivation is another aspect of perseverance. You should learn how and when to seek information on your own, but you also should know when to ask for help if you can't find an answer. Resilience and adaptability when things go wrong are additional hallmarks of success.

We also would argue that successful students think independently, adding their own dimensions to a project, even if that project was designed by their major professor. Within the boundaries of an established project, such students are able to fashion a portion of the research into something that is unique to them. This may occur to a greater extent with PhD students, but we have witnessed very impressive master's degree candidates do so, too.

Another useful characteristic is to know *your* endgame, as well as that of your advisor. When a professor designs a project and obtains funding for it, there are goals and obligations that need to be met. An advisor who is at an early career stage will be seeking tenure, part of which involves publications and grants. Although your obvious aim is to obtain an advanced degree, your advisor's goals are broader. When you start your graduate program, have a conversation with your potential advisor and discuss expected products and timelines. How long do you and your advisor think it will take you to complete your degree? Will you be working with minimal support in the field, or will you be supervising a large field crew? What is the order of authorship in future publications? All of these questions are important conversations to have early in your graduate program, perhaps even when you are initially interviewing for graduate school.

Communication with your advisor is very important. Some professors have an open-door policy, while others prefer you to make an appointment with them. Faculty members are busy individuals, and their free time generally is limited. Students often wonder why so many college teachers appear to be absentminded. We did, too, until we became professors. Most advisors are juggling several projects, multiple graduate students, various courses, and numerous administrative duties. It therefore is really not surprising that you may some-

Writing often is the least favorite aspect of being a graduate student, but it is a critically important attribute in becoming a wildlife professional (photo: Scott Henke)

times need to remind your advisor about specifics, such as the details of your project prior to an in-depth discussion on improving field methods. Or you may need to send a polite email reminder about reviewing your paper in a timely manner. Being aware of your major professor's stress level—and your own—can make both of your lives easier. Our students have become good at knowing that if they walk into our offices (we have an open-door policy) and observe us busily engaged with the task at hand, it would be better to choose a later time to discuss whatever question brought them there. When you are able to visit with your advisor, we encourage you to provide a brief email summary afterward, to make sure everyone is on the same page regarding what was discussed.

Data collection and fieldwork can be difficult and logistically challenging, but being outdoors is often engaging and entertaining. Data analysis usually is neither quick nor easy. This part almost always takes longer than most students expect. You are not going to get your data thoroughly scrutinized in a week, or a month, or even possibly a semester, depending on the complexity of your analyses. Prepare ahead, and give yourself plenty of time to complete this stage. Writing up your results is perhaps the hardest of all, and students often abhor this phase. Recognize beforehand that data analysis and writing may be difficult aspects of your degree.

Part of being a graduate student is learning how to communicate effectively, both verbally and in writing. The latter is where criticism becomes a lot more common. Critiques of your analyses, compositions, and oral presentations are not easy to take, but this is an important part of the educational process. Your first paper will probably require numerous iterations. This is not intended to degrade you as a student, but rather is a natural part of the process of becoming a better writer. Everyone, including us, has undergone similar experiences.

How to Handle Life if Graduate School Is Not for You

In an article entitled "The importance of stupidity in scientific research," Schwartz (2008) addresses one of the issues that we think makes many students give up on graduate school,

even though they are obviously intelligent enough to finish. Schwartz notes that professors do not do a very good job of imparting to students that research is hard work. He also suggests that faculty do not teach them to be what he calls "productively stupid." Schwartz is not discussing what he refers to as "relative stupidity," where students may fail because they did not read the material before class. Productive stupidity means investigating something that no one else has done, and where no one knows the answers. Schwartz notes that as students, we often became fascinated with science in high school and in our undergraduate programs, but "science," in this sense, meant taking courses and getting the correct answers on tests. Consequently, we felt smart. Research, however, rarely has that result. We generally do not know if we are asking the right questions or doing the right experiment. Often, it is not until we compile the results that we realize we may have missed certain crucial elements. Designing a good experiment is difficult, and a field setting (which is where most wildlife studies occur) adds to the challenge. Collecting detailed data on everything is just not possible. Krebs (1998) described an ecologist trying to collect information on everything, who ended up not having really good and effective data on anything. Schwartz (2008) suggests that the more comfortable we are with being productively stupid, the deeper we will wade into the unknown and thus be likely to make important discoveries.

Some students may decide that graduate school is not meant for them, because of frustration with either their professors or their projects. This disillusionment often occurs when they are collecting data, analyzing what they found, or writing up the results. On the other hand, there certainly are times when students and advisors simply do not connect. We believe this can be avoided by doing your homework prior to entering graduate school, in order to select the correct project, major professor, and university, and by maintaining good communications with your advisor once you are in graduate school. These steps are essential when deciding on a graduate program that is right for you. By accepting a position that aligns with all these considerations, you do yourself, your advisor, and the research project a favor, as well as having a higher likelihood of finishing your degree.

When things simply are not working out, students may feel that the best solution is to quit the graduate program. If you really must do so, try to remain professional. Never burn bridges if you can possibly avoid it. Leave a program amicably, with no hard feelings on anyone's part. In the field of wildlife biology, everyone knows each other, especially in the subdisciplines. For example, professors and biologists who work with big game, waterfowl, or upland game birds nearly all know each other well within their respective areas of interest. You do not want a conflict that could have been avoided to affect you later in your career.

Another consideration when leaving a graduate program is to remember that your advisor obtained funding for the project, with timelines, obligations, and specific goals that still need to be met. If you sincerely feel that you cannot remain there as a graduate student, then plan your departure to affect the project as little as possible. For example, it is inappropriate to leave in the middle of a field season, or when people or technicians are depending on you. Try to find the best solution that both works for you and does not compromise the project for others. For example, finish up that season's fieldwork and associated data collection before you depart. At the very least, when giving notice, allow a sufficient time interval for your position to be filled by someone else, so the project is not compromised. Situations where a student leaves abruptly can lead to bad feelings, and this can come back to haunt you later in your career. Professors will understand if the student truly cannot remain in a graduate program. Causing undue hardship or distress for the other members of the project, however, when staying on for an extra couple of weeks or a month is possible, will allow everyone involved to adapt to the upcoming change.

Conflicts between advisors and students can be very difficult for all concerned. Generally, most professors do not want to be at loggerheads with their students, but sometimes significant difficulties arise. We cannot tell you which resolution approaches will work every time, since many conflicts are situation specific. Inappropriate or threatening behavior should never be tolerated. If such interactions occur, consult with your department chair, the university's Office of Graduate Studies, or its Human Resources Department. Most conflicts, however, arise from differences of opinion or a lack of communication. In such instances, a third party may be able to help. For example, a department chair could serve as a mediator or participate in meetings until the situation is resolved. If a conflict still persists, then leaving the program may be the best solution. No matter what happens, always maintain your professionalism. Personal attacks are never appropriate.

PREPARING FOR THE NEXT STEP
Writing Your Thesis or Dissertation

A top-quality thesis or dissertation requires properly collected data and appropriate statistical analyses. Ideally, a statistician has participated in the design of an experiment, prior to any data collection. It is not necessary to have such a person on your graduate committee, although it may still be a good idea. The important point is that it is beneficial to consult with a statistician, both during the experimental design and in the data analyses. This will help in the review process for your project, which extends from your thesis or dissertation defense to publication.

After analyzing the data, your next step is to complete the tables and figures, which summarize your findings. This process will help you with the formal writing of your thesis or dissertation. If your project proposal was well crafted, then the introduction, literature review, objectives, hypotheses, study area, and methods sections should already be completed (although some minor revisions may be necessary), so all that's left is to write up the results and discuss them.

The results section should describe the significant findings from your data analyses. This should be relatively easy, if you describe the important facts from each of your tables and figures. Remember that the tables and the figures must be able to stand alone, without someone having to read the body of the text. The titles of these items should provide fully descriptive information about their contents. The main text should connect the results arising from each table and figure.

The discussion section should be a comparison of what is significant in your research with what others previously have or have not discovered. A discussion is not a repetition of what is in the results section. Your conclusions should be presented either in a separate section or wherever else is appropriate, depending on a journal's particular style or your university's guidelines. Here, briefly summarize your significant findings and state what can be learned from them. This section should be simple and straightforward, and perhaps can be bulleted, to make it easier to understand.

An important point when writing your thesis or dissertation is selecting a presentation style that is appropriate for the journal in which you hope to publish it. This will provide you with general guidelines and a specific format to follow. Each field has certain journals where most research within that discipline is published. For wildlife, these may be the *Journal of Wildlife Management* or the *Wildlife Society Bulletin*. Selecting an appropriate journal style beforehand will help facilitate the subsequent submission of your chapters for publication. In some cases, however, a university's Office of Graduate Studies (OGS) has a required thesis or dissertation format, which may differ from that for a journal. Many OGSs allow students to write individual chapters as though they are separate scientific articles. This approach will help when you prepare the chapters for publication. PhD students, however, are required to initially publish their original work in their dissertations. If you are a PhD student and want to publish an article that will later be part of your dissertation, you must first get permission to do so from your university's OGS.

Last, but far from least, thoroughly proofread your work. Make sure you have not misspelled words or used incorrect grammar. Too often students concentrate on the material they want to present and ignore proper grammar, punctuation, and spelling. Your thesis or dissertation represents you and the quality of your work. Let it be exemplary.

PREPARING FOR A THESIS OR DISSERTATION DEFENSE

The first step in preparing for a defense is to set a date for it. Make sure all your committee members have approved what you have written in your thesis or dissertation, and clear a convenient day and time with all of them before scheduling a particular date. The defense will involve explaining and supporting all aspects of your research: the initial question, and then your methods, results, and conclusions. While this may appear intimidating, it shouldn't be. You ought to know more about the topic than anyone else in the room, as long as you have been actively engaged in the study design, data collection, statistical analysis, and writing. If someone else collected your data, or if you consulted with a statistician on the analyses, then make sure you understand the hows and whys of what they did. Be well prepared to answer any questions related to your project and to general wildlife ecology.

Presentation is an important component of any defense. Normally, this involves a public session (where you give a general talk about your study to students and faculty) and a private, closed-door session (where you defend your research and your results in front of your committee). For the public component, you should prepare a 30- to 50-minute talk that discusses the various parts of your thesis or dissertation. The exact length and content of this presentation will vary, depending on your advisor and department. Therefore, be sure to meet with your major professor and review the appropriate guidelines prior to your defense. Remember, you have the luxury of providing minute details in what you have written, but your talk will have a limited time frame. Make good use of photographs, reduce the size of tables to show only the most relevant facts, and simplify the important figures. Be sure to finish your talk within the allotted time period, so there will be plenty of time for the formal defense. Although not required, it is courteous to bring light refreshments for your committee and the attendees at your talk. This helps make your defense a more pleasant experience for all.

Employment after Graduate School

Looking for employment should start before you defend your thesis or dissertation. To successfully find a job, you need to distinguish yourself from all the other graduate students who are applying for it. This may involve taking classes and obtaining training beyond your university's required curriculum. Additional courses in statistics, computing, and geographic information systems (GIS) may help bolster your resume and provide you with important skills currently sought by hiring managers.

Finding employment often is about whom you know, and who knows you. Work hard during graduate school to earn a strong recommendation from your major professor. Keep in mind that if you wish to work for a state or federal agency, they usually move personnel around internally before advertising whatever position remains open to outsiders. By that point, what is available is usually the least desirable position, or one in a less preferred location. You should expect this and not pass up on such opportunities. The important thing is to get your foot in the door. Once you are hired by an agency, you then will be able to apply for better positions or locations when they arise

As you begin searching for employment, you will notice that some positions (e.g., with consulting firms) pay more than others. If you are in the wildlife profession for the money, however, you are in the wrong field. To be good at your job, you must have passion for your work. If you do not, it will never satisfy you. If you are offered a position prior to completing your degree, you should weigh the options carefully and discuss this with your advisor before accepting the job. If you can complete your degree requirements and research project obligations while also performing well as a newly hired employee, then take the position and work after hours to complete your degree requirements. Such a scenario can be difficult, however, and students often struggle to complete their degree once they have accepted a job. At the very least, finishing a degree requires more time if you are simultaneously working at a job.

Developing as a Professional

Becoming a professional involves doing more than just what is required for your coursework and research. It means publishing your research, participating in one or more professional societies, and assisting with activities within your area of expertise. Publishing papers from your thesis or dissertation is a good first step toward this goal. Being a member of a professional organization, such as The Wildlife Society, at each stage of your career (e.g., in a student chapter, then from a state chapter to a regional section, and, finally, at the national level) also is a start, but belonging is not enough. Participation is important, too. Become involved. Present your research or volunteer for committee work. Attend as many professional conferences as you can afford during your undergraduate years and your graduate program, because this facilitates networking. The more conferences you attend, the more people you meet, who will also become acquainted with you. Presenting papers or posters at these meetings will be a good start toward your professional career. Participating in committees (which should begin when you are still a graduate student) is another important way to become involved. Initially become a committee member and then, as you gain experience (and only if your time as a graduate student allows), advance to a committee chair. Be an early volunteer for these positions, even though they may be time consuming or difficult, and do the best job you can. Remember, learning about employment opportunities, and sometimes getting a job, often is highly dependent on who you know.

You should develop this same drive and passion at your workplace. Be the first to volunteer for something that needs to be done. But be sure not to overpromise and then underdeliver. Find the right balance for you. Above all, respect your fellow workers. A favorite Native American expression fits this perfectly: "Oh Great Spirit, grant that I may not criticize my neighbor until I have walked a mile in his moccasins." If you follow this general advice, you will be in a much better position to find a fulfilling career, one where you can make a positive difference.

SUMMARY

We have presented a professor's perspective on several factors to consider when deciding whether graduate school is the correct choice for you, as well as suggestions on how to be successful should you enroll in graduate school. Based on our years of experience, the following highlight our primary pieces of advice:

- Enroll in graduate school only if you are sincerely interested in pursuing an advanced degree. Deciding to attend graduate school by default (e.g., because you have no immediate job opportunities) nearly always ensures failure.
- Display enthusiasm, motivation, creativity, adaptability, and perseverance. Be prepared to work long hours, both in the field and the classroom.

- Be professionally active in graduate school and after becoming employed. As a student, and then as a young professional, become involved by giving presentations at conferences and volunteering to serve on professional committees.

LITERATURE CITED

Krebs, C. J. 1998. Ecological methodology, 2nd edition. Addison Wesley Longman, Menlo Park, CA, USA.

Schwartz, M. A. 2008. The importance of stupidity in scientific research. Journal of Cell Science 121:1771.

Graduate School: A Student's Perspective

MONIKA BURCHETTE, LINDSEY PHILLIPS, HOLLEY KLINE, LIANNE KOCZUR, BLAISE KORZEKWA, SHAWN CLEVELAND, AND TERRA RENTZ

It is the last year of your undergraduate career. The past four years or more of your life are finally coming to an end. So what's next? What happens after everything you have worked so hard for comes to a close and you walk across that stage and are handed your diploma? Do you stay in the academic world and get an MS degree and possibly a PhD degree? Do you get a job and go straight into the workforce? Or do you work for a few years and then return to school to earn an MS degree? The decision is totally up to you!

In this chapter, we provide an idea of what graduate school is like. There are numerous opinions about graduate school, from a variety of sources, and our objective is to provide you with a student's perspective on what is actually involved in the process of graduate education. With any possible life-changing decision, there are always positive and negative aspects; we address both. By the time you finish reading this chapter, we hope you will have a clearer idea about the physical and mental demands of graduate school.

AS AN UNDERGRADUATE

Starting to prepare for graduate school really begins during your undergraduate years. You should start now to build up your resume and begin to network with professionals in the wildlife field, both of which are paramount to either a future job or graduate school. In most cases, people see it before they ever see you. Your resume gives future employers a first impression of you and your capabilities. This is why it is crucial to have an effective one (chapter 7).

NETWORKING TIPS: KNOWING THE RIGHT PLACE AND THE RIGHT TIME

In addition to having a strong resume, honing your networking skills will enhance your chances of success in the profession. The wildlife field is very small. Everyone is linked by one or two degrees of separation. This is not the time to be introverted. Get out and meet people. They are your colleagues within the profession. The following are some tips to help you network in all the right places (also see chapter 6):

- Professional society meetings (e.g., student chapter meetings at your university, state wildlife meetings, annual national meetings) are great places to meet and form connections with other professionals. Not only will you get to know your prospective employer or graduate school major professor, but you also will be able to interact with professionals and experts from all aspects of the wildlife field (chapter 5).
- Faculty members are well connected with others in all areas of the wildlife field. This can be beneficial when it comes to networking. Get to know your professors as colleagues outside the classroom. They can help connect you with the right people to advise you on your next professional step.
- Internships, seasonal positions, and volunteer work are also great ways to network. Supervisors and coworkers can be valuable references. They may have worked for an organization or at a location that appeals to you, or they may be able to help you expand your skill set through their personal connections.
- First impressions matter—so make a good one! Look and act like a professional when networking, so others will feel confident about putting you in touch with their

Networking at professional conferences is a great way to interact with other wildlife professionals, meet experts in their fields, and learn about potential job and graduate student positions (photo: Wini Kessler)

colleagues. A good impression goes beyond a simple handshake. Proper attire, posture, and mannerisms, in addition to direct eye contact and a firm handshake, demonstrate that you are someone worth taking seriously.

- Social media is a useful tool in projecting your professional image. With all of the outlets available today, it is extremely simple for someone to find a lot of information about you. Use caution, however, when posting on Facebook, Twitter, Instagram, and the like. Unprofessional posts could be detrimental to the image you are trying to convey.

HOW TO PICK A GRADUATE PROGRAM

You have decided to go to graduate school, but now what? How do you find graduate programs? What are you supposed to do to apply? What is graduate school like? What should you expect from your field seasons? What is required in a proposal or a thesis? Here we address some of these questions by shedding light on the graduate school process.

Finding a Program That Is Right for You

Searching for a graduate program can be a daunting task, but there are numerous ways to approach it. The easiest is to begin your search by looking at wildlife job boards online. Professors who have funding for research projects and need graduate students to work with them regularly post openings there. A few possible online sites include Texas A&M University Job Board, The Wildlife Society Career Center, the Ornithological Societies of North America, the Society for Conservation Biology, the Physiological Ecology Section of the Ecological Society of America (ESA), and the Society for Range Management (see appendix B). Each of these job boards contains information about available possibilities.

When you start your search, do not limit yourself to only a single wildlife species of interest. Spend time as an undergraduate working on various projects, to determine what it is that appeals to you and, specifically, what it is about those activities that excites you. When you apply for positions, no one is going to be impressed by an "I love wolves" statement. What *will* catch their attention is discussing predator/prey dynamics in which wolves can serve as a model species. Consider the following when reviewing job postings:

- Closely evaluate the location of the school and the site where the research project will take place to see if they are where you want to be and are relevant to what you want to do. The place, however, is less important than the question the project is based on and your interest in it. Graduate school is not a way of life—it is just a means to advance to the next level. Colleges that have amazing programs may not be in your ideal locality.
- Read the project description thoroughly. Consider how your experiences and interests relate to those of your potential major professor, as well as to the project itself.
- Check the list of qualifications for the project. Do you meet all of them? If not, don't count yourself out yet! Contact the sponsoring professor and see how strict these requirements are. You never know what may result from a conversation with that person.
- Note the personal qualities that are sought for the position and evaluate how well you fit them. For example, if a job posting says, "The successful applicant will work as part of a team of graduate students, research associates, and undergraduate assistants," then your application should clearly address how you have done this in the past, or how aspects of your personality would make you a good team member.
- Be sure to pay careful attention to the closing date for each position that interests you. Submit your application as soon as possible. If a research project fits within your long-term goals, don't let that opportunity slip away.

Some professors may only post new graduate student positions on their personal websites. If you have a school you would like to attend or a faculty member with whom you would like to work, make sure you reach out to them and determine what new projects are on the horizon. Even if no position is posted, email the professor directly, inquiring about any current or upcoming opportunities on their research team. Showing initiative, while maintaining professionalism, will help get your resume into that person's hands.

Once you have identified multiple programs and positions that suit your interests and skills, start the application process. (That's right—we used the plural! You should apply for positions in several programs.) Some postings will list certain required documents. Even if nothing is stated, a few standards still apply. Applications typically include a cover letter (or letter of interest), detailing how you found out about the position and why you think you are the best candidate; a copy

POSITION ANNOUNCEMENT
CAESAR KLEBERG WILDLIFE RESEARCH INSTITUTE
TEXAS A&M UNIVERSITY-KINGSVILLE
DICK AND MARY LEWIS KLEBERG COLLEGE OF AGRICULTURE, NATURAL RESOURCES, AND HUMAN SCIENCES

Position Title: Ph. D. and M. S. Graduate Assistantship: Aflatoxin effects on bobwhite quail

Position Description:
The M.S. graduate student selected will conduct research on the development of an easy-to-use qualitative method for the detection of aflatoxin in grains.

The Ph. D. candidate selected will investigate one or more of the following research components: 1) length of recovery time from acute aflatoxin poisoning, 2) effects of aflatoxin on bobwhite reproductive success, and/or 3) effect of metabolizing fat after aflatoxin exposure, plus a theoretical research component developed by the Ph. D. candidate.

Location: Texas A&M University-Kingsville, Kingsville, Texas (approx. 45 miles south of Corpus Christi, Texas, along the Gulf of Mexico coast).

Qualifications Required:
For M. S. candidate: B. S. in wildlife ecology, biology, or closely related field.
For Ph. D. candidate: M. S. in wildlife ecology, biology, or closely related field.
For both positions: A strong work ethic, good verbal and written communication skills, ability to work independently and as a productive member of a research team, and ability to work under adverse field conditions (hot humid south Texas environment) are essential. Student must have a minimum 3.0 GPA and competitive GRE scores.
Preferred: Background in wildlife disease; past necropsy experience.

Stipend/Salary:
M. S. candidate: $1,650/month plus benefits (medical package has a 90-day waiting period)
Nonresident tuition waived (resident tuition fees apply but tuition reimbursement endowment available).
Ph. D. candidate: $1,850/month plus benefits (medical package has a 90-day waiting period)
Nonresident tuition waived (resident tuition fees apply but tuition reimbursement endowment available).

Start Date: August 1, 2015

Application Deadline: Begin reviewing applications 15 June 2015 and will continue until a suitable candidate is selected. Applications will be reviewed as received.

To Apply: Send a cover letter stating interests and career goals, resume/CV, copies of transcripts (originals will be required if selected), GRE scores, and 3 letters of reference to:

Dr. Scott E. Henke, Regents Professor and Chair
Caesar Kleberg Wildlife Research Institute
Texas A&M University-Kingsville
1150 Engineering Avenue, MSC 228
Kingsville, Texas 78363-8202
Phone: 361/593-3689
Scott.henke@tamuk.edu

Texas law requires that males 18 through 25 show proof of compliance with Federal Selective Service law in order to be eligible for employment. Selected candidate must pass a pre-employment background investigation to be hired for this position.

EEO/AA/ADA

An example of a job announcement from the Texas A&M Job Board. Postings also can be viewed on the university's Job Board website.

of your resume, with names and contact information for at least three references; your college transcript (an unofficial one is fine at this stage, but you will eventually need official copies); and your Graduate Record Examination (GRE) scores. Universities use GRE scores and grade point averages (GPAs) to evaluate how well you have functioned in an academic setting. Both help narrow down the first wave of candidates. Some institutions or programs require a minimum GRE score for admittance. This information is generally stated in the job advertisement, but if not, contact the professor in charge of the project to determine what the minimum standards are. A GPA of more than 3.0 (based on a 4.0 scale) demonstrates a prior level of commitment to your education. Keep in mind, however, that scores are just one indicator of your qualifications and should not preclude you from trying for a position you strongly desire. After you have applied, wait a few weeks to contact the person in charge and ask about the status of the position and your application. If the post has been filled, be polite and thank that professor for considering you. You never know when he or she will have another project for which you can apply. Keep in mind that as an undergraduate, if you are accepted by a college or university, you become one of the hundreds to thousands of students admitted that year. In graduate school, however, a professor is selecting one student out of applicants from around the world. You want to be that person. If not, don't be disappointed. Not being chosen does not necessarily mean that you were not qualified, but rather that another person was better qualified. Use rejection as a learning experience. Find out what your deficiencies for that position were, so you can rectify those shortcomings in your next application.

The Next Steps: The Interview and Campus Visit
Applying for a graduate position means you are that interested in it, and being asked to interview for one shows that the individual doing the hiring is interested in you! At this point, do your research. Look at the sponsoring professor's website to obtain insight into who that person is. Where did he or she receive bachelor's, MS, and PhD degrees? What previous research projects did he or she work on? What is the focus of his or her current research? Familiarizing yourself with that faculty member and the project team will help in your decision-making process. Having the right advisor makes a world of difference during your graduate career—and beyond.

You should meet your prospective major professor before agreeing to join the research project and enter that school's graduate program. Visit the university and schedule a one-on-one meeting with the professor to see how well the two of you interact. This will also allow you to become acquainted with others on the faculty (who could be possible graduate committee members for you) and the professor's current students. Talking with other students on the project and, more broadly, in the relevant university program can give you an idea of what the graduate community is like and whether you want to become a part of it.

Potential major professors will sometimes offer to pay for these visits, if you ask. They often enjoy meeting prospective students! Even if you have to pay for the visit yourself, it is time and money well spent. If possible, find out about the status of the professor's previous students. Did most move onto successful careers in wildlife or natural resource areas, or did they completely change fields after finishing a degree? If many of them have done well, there's a good chance you will, too. If several have left the discipline, however, you may want to reevaluate your potential relationship with that professor.

Use the interview as a forum to discuss important issues that affect your research and your personal life. Inquire about the project's funding sources, as well as what will be expected of you. Some graduate positions are research centered, while others require you to be a teaching assistant (TA) for certain undergraduate classes. The latter can provide you with valuable experience, but it may take time away from your own studies. If you are required to become a TA as part of your graduate work, ask for the names of current TAs and find out from them which faculty members are better to work with. Try to meet those professors during your visit to the campus. Develop a relationship with them and express interest in being a TA for the courses they offer. You also should discuss specific details: the cost of tuition, what the assistantship requires, the dollar amount of the work stipend, the various aspects of data collection for the research project, the duration of the field season, the professor's expectations for his or her students, your expectations for that professor, and the length of the graduate program. It is important to get a clear picture of the steps required for completing an advanced degree, as well as of the overall flexibility of the proposed research questions. If you are returning to graduate school after pursuing a career, you may have your own ideas and research interests that do not fit well with those of your prospective major professor. If that is the case, you should explore other options.

Consider posing additional questions to your potential advisor and current students about the project and the university's overall wildlife program:

- What are the other graduate students on the project like?
- Will you be the first author on the papers you write and publish? What are the prospective professor's criteria for naming coauthors?
- Will you be expected to work on other students' projects as well as your own? Would additional publications be possible if you do this?
- Will the school provide insurance (e.g., health, life, dental, vision)?
- If you receive a stipend, will you be required to take credit-hour courses during that semester?
- Will you receive financial support to attend conferences? If so, how many will be paid for?
- How often is that faculty member on campus or otherwise available to his or her students? The professor and the students may have different perspectives. Make sure you get answers from both sides.

Accepting an Offer

If you are offered the position, there are additional questions you need to ask yourself:

- Did you get along with your potential advisor and his or her students?
- Was it intimidating to talk with them, or was it easy and natural?
- Are you comfortable with the expectations for the position and for the program?
- Can you be passionate about the project?
- What will this position mean for your family or pets?

Having a support system is critical, whether it is your immediate family, your extended family, fellow students, or the professional community. If you have a spouse or significant other, children, pets, or other dependents, it is important to consider their needs and the responsibilities of these relationships. Pursuing a graduate degree while raising (or starting) a family is possible, but several factors should be taken into account:

- First and foremost, are both your significant other and your children supportive of your plans to get a graduate degree? Do they understand the time commitment involved?
- Is moving your family an option? If not, can you be successful at graduate school while living apart from your family for extended periods of time? How far away is too far?
- Is there a viable employment opportunity or alternative for your significant other near the university?
- Will your potential major professor understand that you will need flexibility to address family concerns, especially if children are involved?
- Can you perform effectively during irregular work hours?
- Are your field seasons and locations restrictive? Or are they open (within reason), to allow participation by your significant other, children, or pets?
- Do the primary and secondary schools near the university meet your standards (if you have children)?
- Are the stipend (or salary) and the health care benefits associated with the position enough to meet your share of the financial obligations to your family?

Attending graduate school with a traditional or extended family is completely doable if you and your major professor are willing to be flexible. Nonetheless, you will be required to make tough choices about how to balance your personal life with your academic pursuits. You may have to adapt your habits and work patterns to accommodate the additional demands of a graduate degree.

Going to graduate school is a full-time job and should be viewed in the same light as any other exacting position. Here, and in any professional work environment, clear boundaries should be set. You and your major professor need to be aware of and in agreement on those limits. Find out from the start if you and your advisor can have transparent communications.

Is that professor willing to work with a graduate student who has family responsibilities? If not, then this may not be the right position for you. Be aware of the strain pursuing a graduate degree will have on your significant other and your family. Being supportive of them as they try to accommodate themselves to your new professional choice is as important as their support of you. Choosing to pursue an advanced degree can pay dividends for your career and your family. Although the choice to attend graduate school when you have family obligations is not as easy as it may have been when you were on your own, under the right conditions, the end result will be worth the temporary adjustment in your life.

GRADUATE SCHOOL ITSELF
Starting Graduate School

Now that you've selected a graduate program and have been accepted, you can begin the various steps required to complete your degree. The graduate school process is filled with many tasks that lead to the end goal of a thesis or dissertation, but getting there is half the fun. During this time, your life will be split among coursework; data collection; conference attendance; outreach; and, finally, thesis writing, thesis defense, and graduation. During all this, you need to make sure you balance your priorities. Maintaining a life outside of graduate school is critical in avoiding the danger of burnout.

Most programs have handbooks that outline timelines for important milestones during the graduate school process. Pay attention to them! One of your first responsibilities will be to select your graduate committee. This usually consists of three to five faculty and professional members, including your major professor or the professor who will serve as the direct point of contact for your research project. Your committee members are responsible for commenting on proposals, assisting with the design of the experiments for your project, reviewing your research, approving your coursework choices, and accepting your thesis or dissertation.

Once your graduate committee is selected, schedule at least one meeting with them to discuss both a research proposal and the coursework you plan to take. During this time, you may go over any side projects that are related to your main research focus. In some cases, depending on your research project, a committee may already have been selected for you. This typically is the case if several professionals were involved with the design and funding of that project. Update your committee regularly on your progress. Strive to maintain a positive relationship with committee members, as they can be paramount in making connections for future employment and research opportunities.

The Research Proposal

Writing a research proposal is one of the most important and tedious activities during graduate school. It is your responsibility to determine the current state of knowledge about your topic (usually in the form of an in-depth literature review) and then develop a research proposal to obtain new and relevant data. Before you begin writing a full-length proposal, you should submit an outline to your committee, both to organize your thoughts and to make sure that your idea has merit. Do not procrastinate in writing a proposal. Just reading the existing literature will take weeks. Maintaining thorough notes during meetings with your major professor and graduate committee to discuss your project will help when writing your proposal.

A research proposal will present your project topics, including their importance to wildlife management and conservation. It will also indicate the research location, describe how you plan to collect and analyze your data, and offer a proposed timeline for completing your research. The research proposal then becomes a contract involving you, your major professor, and the members of your committee. Think about additional elements this contract will require. For example, will you need to understand regression, multivariate statistics, and descriptive statistics? Many of these skills probably were not covered during your undergraduate career. Becoming familiar with data analysis techniques through coursework and independent research can improve the design of your project experiments and should streamline the future process of writing up your results. Once everyone has signed the agreement, it is your responsibility to uphold the standards in the contract.

Graduate school also includes coursework, not just a research aspect. You will be required to take a certain number of courses, which varies by institution. The amount of credit hours required per semester is less than that for undergraduates, but graduate courses have greater complexity and carry an increased workload. You often will have more flexibility in selecting the graduate courses you wish to take, but some universities have required core courses or specific types of classes you must enroll in. Good time management is extremely important. Completing this mandatory coursework must be balanced with your research and teaching loads. Graduate school is a full-time job, which often demands more than 40 hours of work a week.

Data Collection

At least some of your time will be spent in the field. Data collection routinely is a seasonal activity, which means you will not be out in the field year round, although some projects necessitate a balance of coursework, data collection, and teaching throughout the year. Many projects require students to spend long periods away from the university when conducting their research. The length of time varies among projects, but it is common to devote at least 60 hours per week to fieldwork, usually when living in remote locations. Depending on the type of research and the available funding, you may need to hire technicians to help you complete your field research in a timely matter. Not all projects have a budget for this type of assistance, however, so time management and hard work on your part will be critical.

Regardless of whether you are engaged in season-specific or year-round fieldwork, keep on top of the data. An efficient method is to perform an initial analysis as you are collecting the data and use it to better understand what you are uncovering. This will prepare you for the later phases of your project. It will also keep your major professor happy and allow you to take control during meetings with your committee, rather than have the discussions be led by your advisor.

Back up your data frequently! You will hear many horror stories about lost data, and you do not want to become one of these legends. Many universities will provide server space, which is backed up daily. This is good, but keeping your data in more than one location ensures that it will not be lost. Remember, servers crash, hard drives can become corrupt, and loose papers get lost. A good rule of thumb is to save data to the server daily, and back it up on an external hard drive weekly. If you are in the field, a similar rule applies. Stick drives offer cheap insurance. Save data to your hard drive and a stick drive every day, and then back it up on an external hard drive once a week. Every time you can, link to the university's server and back everything up once more. When you have finished collecting data, you can then proceed to the next steps: analyzing it, presenting your findings, and writing your thesis.

Outreach and Conferences

Now that you have gathered your data and understand how to answer your research question, it's time to make all that work come alive. There are several ways to do this. As part of the curriculum, many universities have poster sessions for students, where you present your data orally. Take advantage of every opportunity to display and discuss your data, rather than limiting yourself to only doing so at professional meetings. Many groups in the community, such as the Audubon Society, want to know what their local university is doing and like to be updated on current research. Presenting your data to such organizations is rewarding, increases your credibility, and may even result in long-term conservation measures. Also, the more you talk about your project with a broad variety of audiences, the more comfortable and knowledgeable you will become. Moreover, when you defend your thesis, you will have had experience presenting your data, which will reduce stress and hone your delivery style. By having already encountered tough questions in these early talks, you can use them to enhance your expertise. Working with community groups can also help you define and articulate the management implications of your research. Take the time to give such presentations. The intangible skills you develop will make you a better biologist, researcher, conservationist, educator, and champion for natural resources. This is well worth the extra effort!

Conferences offer opportunities to both network and present your data. Notice our emphasis here on networking, not on data presentation. You may ask yourself, "Should I go to a conference if I don't have data to present?" The simple answer to this is, "Yes!" Your major professor may not be able to fund your travel expenses if you are not delivering a talk or presenting a poster, but it is still worth the investment to find a way to get to conferences. You will be exposed to more job opportunities, new colleagues, previous acquaintances, collaborators, and, generally, fresh and exciting ideas. Take the time to go out during the evening and meet people in an informal setting. You never know where a late-night conversation might take you. If you are presenting your data at a conference, this is where you get to shine in front of your peers and potential employers. Be proud of the work you have done.

Thesis Writing

Your proposal, if done properly, is the first step toward completing your thesis. Once the proposal is approved, it is never too early to start writing. (Don't be surprised, however, if your final thesis looks entirely different from your proposal, due to changes in the techniques, the statistical analysis, funding, ideas, and interests among others.) By the time you have finished crafting your proposal, you will already have written several parts of your thesis: a well-developed introduction, clearly stated objectives, and a concise description of the methods you plan to follow. The results and discussion sections will be the last parts you need to complete for your thesis. The following are a few tips to maximize your approach to thesis writing:

- Update the methods section if changes occur during the field season, and notify your committee of any major shifts.
- Figure out the part of the day when you are most productive and try to write during those times. Keep away from emails, cellphone calls, text messages, and any other distractions. This is your best time to work, so really focus on your thesis. It is imperative to make progress during this stage.
- Capitalize on your team. If you are wrestling with a concept or an idea, it helps to work through it with your fellow students. You can often learn as much or more from your colleagues as you can from your major professor, so take the time to ask for their input. Also be willing to do the same for others.
- Be patient. You will revise your thesis many times with your major professor before it ever goes out to your graduate committee. Be prepared for this. Keep your previous versions and refer back to them. At times, you and your advisor (but mainly your advisor) will start adding and deleting the same ideas. This is a sign that you are getting close to the penultimate draft of your thesis, which is the one your committee members will review.

Once you have completed this next-to-last version and submitted it to your committee, you will have about two weeks to prepare for the formal thesis defense. In essence, you have been getting ready for this step from the moment you started on your project. Remember, you are now the expert on that research question! If you have put in the effort to get this far, you know your material.

At this point you also will be preparing for the public part of your defense, the oral presentation. If you gave talks about your data to others at various times along the way, most of the work is already done. In addition, you have already honed your presentation style. Put the finishing touches on a previous talk, practice it until you feel comfortable, deliver in front of others ahead of time, and enjoy the process. Don't worry if you did not have a previous opportunity to present your research; you have one now.

The same rules governing thesis writing also apply to presentations. Have your colleagues critique both what you say and how you deliver it. A word of advice—do not rely on notecards. Think about the best ones you have heard. Most (if not all) speakers did not depend on notecards. They were charismatic and sounded as if they were telling a story. To do the same, familiarize yourself with the order of your presentation. Minimize word use and maximize the use of images. Review your figures and tables, to be sure that they are legible and will be visible to those in the back of the room. Be conscientious about font size, color (avoid greens and reds), and layout, to make it easier on viewers. Remain professional, but also inject your personality and have fun with your presentation. If someone asks you a tough question and you do not have the answer, admit it and move on.

After your public presentation, you will meet privately with your graduate committee for the second portion of your defense. This will either occur immediately after your oral presentation or one to two days later. Try to schedule it right away, both to reduce stress and to move more quickly toward the completion of your degree. The closed-door side of your thesis defense can be intimidating, and it may last for several hours. Again, remember that you are the expert. Your committee wants to be sure that you thoroughly understand the work you have done, and they will ask some difficult questions. Remember, though, that your committee is not your enemy; they want you to succeed. It is okay to say "I don't know" in response to a question. No one expects you to know all of the answers. Keep in mind that you can ask questions, too. This is a great tool, as it lets you take a moment to breathe or gain insight on exactly what a certain committee member might be looking for in your response. You may be asked to go to the room's chalkboard or whiteboard and illustrate how aspects of your project fit together. If this makes you uncomfortable, practice beforehand with colleagues. Have them ask you questions where you must illustrate concepts on the board, including your statistical design and analysis.

After you have answered all of the committee's questions, you will be asked to leave the room so they can discuss the outcome. For the most part, if you did a good job, they will be covering whether you passed, revisions to your penultimate draft that should be incorporated in the final version, and timelines for publications. After what may seem like an hour (but generally is not), one of your committee members will walk out, shake your hand, and, if you have passed, offer you congratulations. You, your advisor, and the committee members will then discuss your next steps, including publications, a timeline for you to edit parts of your thesis or analyze data further, and whether committee members would like to see the thesis again before signifying their approval by completing the signature page. Get your edits done as quickly as possible, especially if you already have a job lined up. It becomes increasingly difficult to make time to finish this last step while working at a new, full-time job. Similarly, try and get your publications out as soon as you can. The longer you wait to publish your research results, the harder it is to make time to do so, and the further your thoughts will from the work you have just completed.

The process for finishing up a PhD is a bit different from that for an MS degree. Doctoral students have to successfully answer a written set of questions from each committee member, and then pass an oral exam. Both of these requirements are usually undertaken once your coursework is nearly completed. A student is considered to be a PhD candidate once these steps are accomplished. The last hurdle is to finish writing and then defend a dissertation, which typically is more theoretically based than a thesis.

Whether you are receiving an MS or a PhD degree, take a moment to reflect on your journey and be proud of what you have done. You've joined a new professional cohort and, during a few short years, acquired skills and abilities that are vital to our profession. The choices you made have set the stage for the next phase of your professional career.

Beyond Graduate School
You have finally finished your advanced degree! You survived! Now what? Your next move will depend entirely on the type of career you would like to pursue. If you want to work in a management position, an MS degree can get you there. On the other hand, if you want to continue conducting research or be a professor, then your next step is to find the right PhD program. For some, this may be a very simple decision. For others, it will require a lot of thought. Chapter 5 discusses the wide range of wildlife-related jobs in detail and should serve as a great resource to assist you in your pursuits. If you decide to end your graduate school career with an MS degree, you should start looking for jobs at least a few months beforehand. If you have an idea about where you want to work, or for which organization, inquire about possible openings by reaching out directly to the organization or to specific colleagues through emails. You can interview for a permanent position prior to completing your MS program. If you are the right person for the job, often the position will be held for you until you graduate. A graduate degree will make you a strong asset for a future employer.

THE WILDLIFE PROFESSION OUTSIDE OF GRADUATE SCHOOL

Graduate school is not for everyone, or at least not right away. Some people choose to complete their undergraduate education and head straight into a professional position. Choosing this path does not prohibit you from returning to graduate

school later to earn an MS or a PhD. Some people say that once you leave school you will never return, but this simply is not true. Many (if not most) wildlife professionals take a break between their undergraduate and MS degrees, and some even do so between their MS and PhD degrees. This is common, and the increased experience you gain is usually considered to be an advantage, both when applying to graduate school and in you future career.

There are numerous reasons to go straight into the workforce after completing your undergraduate degree, such as to gain experience, fulfill family commitments, reduce the amount of debt for student loans, or take a break from academics. Similarly, there are many arguments in favor of returning to school later to earn a graduate degree. An advanced degree is a way to increase your mobility, become eligible for higher paying jobs, and gain more experience. Remember, you can always return to school if you so desire (as long as you remained competitive during your undergraduate career). Each person is unique, and it is important to do what is best for your current situation and your future.

SUMMARY

This chapter addresses, from a student's perspective, numerous considerations in preparing for, selecting, attending, and completing graduate school. Contemplating whether to earn an advanced degree, and choosing when that might occur, are personal decisions. These will vary, depending on your own professional goals and the other factors in your personal life. The following are key points to consider:

- Prepare for graduate school while you are still an undergraduate.
- Network. This is a critical tool for increasing your success in the wildlife profession. Use it to identify and obtain a preferred graduate position.
- Find a graduate program that is right for you. This entails investigating a potential major professor, graduate program, and research topic, in addition to an introspective look at your own strengths, professional goals, personal and family needs, and timeline.
- Balance coursework, research, and writing once you are enrolled in a graduate program. There are at least five steps in successfully completing an advanced degree.

1. Develop a strong and transparent relationship with your major professor. This is the key to success.
2. Create a strong research proposal. This is the backbone of a graduate degree. Although this can be a tedious process, a well-developed proposal will lead to a more successful project and an eventual thesis or dissertation.
3. Practice good time management and organizational skills when engaging in field research and data collection. Familiarize yourself with your data early and often, and always back it up.

4. Strengthen what you are learning from your project by presenting the scientific information you've gathered through conferences and outreach efforts.
5. Start writing your thesis or dissertation early. Be patient, and rely on the strengths of your colleagues to get you through tough times. Remember, you are the expert when it comes your own research!

- Investigate the large variety of employment options offered in the wildlife field while in graduate school. This will guide you to the appropriate next steps. The wildlife field is vast and diverse. Remaining in academia is only one of several options.
- Celebrate your successes. Although completing a graduate degree is a great accomplishment, the real benefit is the journey. Enjoy it!

12

Professional Diversity

The Key to Conserving Wildlife Diversity

MICHEL T. KOHL, SERRA J. HOAGLAND, ASHLEY R. GRAMZA,
AND JESSICA A. HOMYACK

Our profession faces a challenge in managing increasingly complex environmental issues that require professionals with diverse perspectives and wide-ranging skills. Many have recognized that it is imperative for us to develop and retain a workforce reflective of changing demographics. By doing so, both the profession and wildlife in general will benefit substantially, because greater stakeholder participation will result in greater acceptance of conservation efforts and provide a more diversified funding base (Lopez and Brown 2011). Fortunately, governmental agencies, nongovernmental organizations, and professional societies recognize the power of a multifaceted workforce and are prioritizing human diversity. Some examples include the US Department of Agriculture's (2015) Cultural Transformation goals, the US Department of the Interior's (2012) Diversity and Inclusion Strategic Plan, and The Wildlife Society's (2015) Position Statement on Workforce Diversity. Such initiatives have made great strides in increasing racial, ethnic, and gender diversity within the wildlife profession by recruiting and retaining a rising number of candidates from minority and marginalized groups (Tuggle 2011). Much, however, still remains to be achieved.

Over the past hundred years, wildlife management has been transformed from a career path dominated by middle-class white males to one that is more reflective of societal demographics. The founders of the formal discipline of wildlife management and early members of The Wildlife Society, a professional organization for wildlife biologists, were representative of most natural resource professionals at that time. There was little inclusion of men from minority groups, and virtually no involvement by females. Today, a greater number of wildlife professionals have come from diverse backgrounds, various ethnicities, and nontraditional groups (e.g., nonhunters, people from urban areas). Recent trends in gender diversity are positive, with female membership in

TWS increasing from 1.5% to 22.2% in the years from 1937 to 2006 (Nicholson et al. 2008). Similarly, women's contributions to wildlife science—as authors of articles in TWS publications (e.g., the *Journal of Wildlife Management*, *Wildlife Monographs*, the *Wildlife Society Bulletin*) and through presentations given at the annual TWS conferences—have exhibited an upward trajectory since the 1990s. More women are starting to serve in TWS leadership roles, yet this organization has elected only two female presidents in its more than 75-year history.

Despite the increased involvement of women within the wildlife profession, racial and ethnic diversity across its membership and its leadership positions has lagged. For example, the current demographic estimate for the US population is 62.1% Caucasian, 17.4% Hispanic or Latino, 13.2% African American, 5.4% Asian, and 1.9% American Indian, Alaskan Native, Native Hawaiian or Pacific Islander (US Census Bureau 2015). In comparison, recent estimates of TWS membership demographics (for those who identified their ethnicity) are 94.4% Caucasian, 2.3% Hispanic or Latino, 0.4% African American, 1.8% Asian, and 1.1% American Indian, Alaskan Native, Native Hawaiian or Pacific Islander. Although TWS membership by underrepresented groups has increased slightly during the past few years and is more in line with the breakdown of ethnic diversity for those who enter college as wildlife majors, the numbers still fall short of those for the US population as a whole. Additionally, many important aspects of identity that shape a person's life experiences, values, and perspectives (e.g., cultural background, socioeconomic status, sexual orientation) are not yet addressed in recruitment and inclusion efforts.

These statistics demonstrate a significant deficit across our profession. Does human diversity really matter to wildlife conservation? We think the answer is a resounding "Yes!" The intricate relationships of a diverse biological community

The field of wildlife management has increased its diversity in recent years, with more females joining the profession (photo: Dustin Angell)

can help explain and emphasize the importance of inclusivity in a profession. Consider a forest comprised of conifer and hardwood trees of different heights and sizes. It also has many species of herbaceous plants growing near the ground, as well as standing dead trees and other deadwood distributed randomly throughout. Gaps in the canopy allow sunlight to reach the forest floor, providing ideal growing conditions for young trees. With plentiful resources, the community supports a wide array of invertebrates, mammals, reptiles, amphibians, birds, and even humans, due to the various environmental conditions available to each organism. The forest also helps produce clean water and provides many other ecosystem services. This complex community is resilient, resisting natural disturbances such as storms, lightning-ignited fires, and drought, thus providing suitable, long-term habitat types for an entire biotic community. In stark contrast, imagine a monoculture composed of aging trees. This community is devoid of most flora and fauna and is susceptible to disruptions, such as disease and insect infestation. The same lesson can be applied to the wildlife profession. A workforce lacking differences in identities, life experiences, and backgrounds limits

our ability to meet the multiple wildlife challenges of today and the future.

A diverse workforce also is needed to fill varied employment roles. The wildlife profession includes statisticians, spatial analysts, social scientists, economists, hydrologists, ecologists, and wildlife biologists. But life experiences and personal values are also important. We need people with dissimilar sociopolitical views, coming from circumstances that range from poor to affluent, and representing all walks of life. So how do we measure diversity? Most wildlife biologists have a strong background in scientific methods. As such, our primary approach to understanding diversity trends has been to categorize and count people in groups. Yet this practice oversimplifies human complexity. Individuals have both an observable identity, such as white male, and less obvious characteristics, such as low socioeconomic status or a disability. Unless we take the time to learn about each individual beyond those characteristics that can be relegated to simple checkboxes (e.g., sex, race), we cannot properly evaluate how a person's unique life experiences influence his or her knowledge, skills, and other abilities with regard to wildlife resources.

DIVERSITY: SEEN AND UNSEEN

While the concept is important as a whole, there are also different kinds of diversity that can and should be part of the wildlife profession. Rather than including an exhaustive list, we provide a broad overview of the types of human diversity that should be present in wildlife biology and related fields. We also discuss the unique knowledge and important perspectives that such a workforce can bring to the profession.

Primary diversity consists of external characteristics, which are visible and not easy to change. Examples include race and ethnicity. A high degree of cohesion is beneficial when individuals are comforted by familiar settings, such as being among groups of people with similar ages or ethnicities. Yet primary diversity commonly influences both conscious and unconscious bias and discrimination. In comparison, secondary diversity often cannot be seen and, if an individual so chooses, can be hidden from others. Examples include people's socioeconomic, educational, and veteran status, as well as the region where they grew up, their age, and their sexual orientation. It can be equally as important as primary diversity, since wildlife biologists must relate to a public rich in these secondary characteristics.

Primary diversity, such as ethnicity and gender, is essential if wildlifers hope to gain an understanding of new and unique wildlife management scenarios and the conflicts that can occur when cultures clash. A profession that includes both men and women is more effective, because they may sometimes work and manage situations differently.

Secondary diversity, such as the region where a person grew up, will play a larger and larger role in our various management decisions. Wildlife conservation will continue to shift toward urban and suburban settings, due to the increasing growth of populated areas and the subsequent adaptation of wildlife to these altered landscapes. As professionals, we are tasked not only with managing wildlife, but are also charged with increasing participation in wildlife-related recreational activities by people living in urban areas, as well as those of lower socioeconomic status, who may not have been involved in these types of activities historically. Veterans, nontraditional college students, and people from the lesbian, gay, bisexual, transgender, and queer (LGBTQ+) community are only a few of the groups who may approach wildlife management issues differently and have life experiences that are dissimilar to those of other students. A field rich in secondary diversity will increase our ability to implement innovative strategies that address wildlife management challenges.

Adding a high level of primary and secondary diversity to our profession may provide more avenues for communicating the benefits of wildlife and habitat conservation to a diverse public. For example, imagine a situation in which a state wildlife agency is hiring an urban wildlife biologist in a primarily African American district of Dallas, Texas. The Texas Parks & Wildlife Department might select two candidates for the position who have similar credentials but differ in the scope of their diversity. One is an African American male who grew up in the area; the second is a Caucasian male who was raised in rural Texas but has had more experience than the other in working with urban wildlife. In this scenario, the ability of the first candidate to relate to the public may outweigh that of the second potential employee, thus increasing the odds of success for the urban wildlife program. We do not mean to suggest that ethnicity should play a role in job selection, but rather to acknowledge the increased conservation potential provided by biologists who are able to relate both to the people they encounter and the wildlife they manage. The first candidate may share similar life experiences with the public and thus be able to guide the program in a direction that appeals to members of the community. This example, although simplistic, demonstrates that our profession needs job candidates who understand both the people and the landscapes in which they work, thus ensuring that human populations are aware and supportive of local wildlife issues.

CHALLENGES

Establishing an inclusive workforce is a complex undertaking. A critical mass of biologists from traditionally underrepresented categories will contribute to the creation of a more welcoming professional environment for those individuals. But underrepresented groups in natural resources face numerous difficulties that, individually and taken together, influence their entrance and retention in the profession. Unless all members of a crew, an organization, or a profession are actively heard, included, and empowered, the only thing we will have achieved is half-hearted symbolism. Shaping an inclusive professional culture will require each of us to recognize and hold in abeyance our own implicit biases and behaviors. Even the most talented and skilled individuals can be dissuaded and

handicapped by an unwelcoming environment. Fortunately, overt discrimination toward underrepresented groups is less tolerated than it once was, yet reports of sexual harassment and abuse of women in field settings indicate that negative experiences still occur (Clancy et al. 2014). Thoughtless "jokes," subtle suggestions about employees not being fit for work, and sometimes intentional bullying can drive out even the most dedicated would-be biologist.

Nor are challenges in our field limited to students and early-career professionals. Because the organizational structures of universities, wildlife agencies, and other employers were developed primarily by and for white males, lingering effects of these institutional systems are felt by all levels of wildlife professionals from other backgrounds (Acker 1992, Bird 2011). These unconscious biases can produce an uncomfortable atmosphere, as well as affect employment and promotion decisions. For example, research indicates that both men and women rate male applicants as being better qualified, and hiring managers of either sex are more likely to offer men a higher starting salary, even when resumes are otherwise identical (Moss-Racusin et al. 2012). People now recognize that microaggressions—subtle, pervasive, intentional or even unintentional slights or negative messages about people, based on race or membership in a socially marginalized group—further contribute to an unwelcoming atmosphere and decrease workforce talent. Examples of this behavior include comments arising from assumptions, such as that an Asian American is foreign born or that a woman is married and has children. Similarly, jokes or other remarks based on stereotypes, particularly when made by supervisors or people in positions of power, add to workplace issues that can be difficult to confront or harmful to internalize. For those belonging to more than one marginalized group, such as an African American woman, these challenges multiply. Negative biases affect professional success, and they may be one reason why women in research positions within the fields of science and engineering publish less overall, have fewer first-authored publications, and often feel overlooked in terms of opportunities for advancement (Angus 1995, Cameron et al. 2013). Understanding and identifying implicit biases are the first steps toward overcoming them.

We recognize that not all roles in the wildlife profession will be attractive to all individuals at all stages of their lives and careers. Hardships tend to hamper diversity, and retention can include 60- to 70-hour workweeks without overtime pay, few or no childcare facilities, and a lack of paid leave for ailing children or other family members. The logistics of fieldwork may impede some people from becoming wildlifers, because positions can require long and irregular hours (e.g., seven workdays on, followed by three days off, with early morning and late night schedules) and physically demanding work in remote and difficult environmental conditions. The instability and scarcity of funding for wildlife-related projects can lead to low salaries and a need for volunteer labor, further discouraging a diverse workforce. Fortunately, there are many types of positions in our profession that support wildlife conservation, including in education, management, research, and policy. By broadening our definition of *wildlife biologist*, we will reap maximum benefits from a more diverse workforce.

Other professional challenges are more likely to impede early-career members of underrepresented groups. Being the only person from a minority or marginalized group on a team or in an office creates additional daily stresses and energy demands. For example, individuals can feel isolated and may lack access to mentors who understand the unique challenges that underrepresented groups face. Some research suggests that institutional change does not occur until a critical mass equals roughly 15% of the workforce (Etzkowitz et al. 1994). Without diversity, minorities may feel excluded from or awkward participating in the formal and informal social networks that influence retention and career advancement. Further, when informal events center on traditional wildlife skills, such as hunting or skeet shooting, people from nonhunting or urban backgrounds may not be able to participate in such team-building opportunities, which enhance trust and help develop stronger relationships. Awareness and active leadership to create inclusivity are important skills to master.

Institutions can create formal opportunities to combat isolation through diversity networks that foster inclusion. It is vital, however, to ensure that they effectively reach their target audiences across all parts of the organization. For example, minorities in the wildlife profession living in rural and far-flung regions may benefit most from networking opportunities, but formal networks may unconsciously exclude them, due to the remoteness of their station areas. Mentors are important, because they can provide advice about wending your way along a successful career path, and lend a safe and empathetic ear. Sponsors are individuals in positions of power who can act to increase the visibility of and request promotions for individuals. For those in underrepresented groups, nurturing relationships with mentors and sponsors may lead to career advancement. Yet finding appropriate individuals who appreciate and understand the unique challenges that such groups face in wildlife careers is difficult. The situation should improve as human diversity in the profession increases.

Cultural conflict can prevent some individuals from entering the wildlife field. People from some minority communities might not be encouraged to become wildlife biologists, due to a fear of outdoor places and wildlife or to the social stigma attached to performing manual labor and working outside (Sexton et al. 2015). Families might also urge their children to pursue careers with higher pay or greater social status. In the United States, race, socioeconomic status, and the quality of a person's education are often linked, leading to potential difficulties for underrepresented groups to meet college admission standards or obtain financial support to attend a university. Further, it can be difficult to convince students from families with low socioeconomic and educational backgrounds to pursue a position as a minimum-wage wildlife technician, to go to graduate school, or to work 60 to 70 hours a week when they also have responsibilities to care for children or other family members.

A consistent theme among studies in the scientific disciplines is that real or perceived conflicts between careers and family are causing women and minorities to drop out of the career pipeline (O'Brien and Hapgood 2012). For many women, having and raising children coincides with the time when they start looking for their first professional job after undergraduate or graduate school. A gap in their work history, due to leaving to raise children, may put women at a permanent disadvantage in this competitive profession. Several studies suggest that women with children face a "motherhood penalty," because they are less likely to receive tenure or other career promotions, while men with children are viewed as being more committed than their childless counterparts (Goulden et al. 2009). Thus women may face difficult decisions regarding when to have children (and how many) versus career advancement, all while often taking on a disproportionate share of household chores and childcare. For wildlife professionals, additional challenges arise in finding adequate childcare and family-support systems in remote or rural locations, and conducting intense and lengthy fieldwork while pregnant or breastfeeding. Similarly, people from backgrounds where close physical connections to extended family is an important part of their culture may find it difficult to navigate through undergraduate and graduate degree programs, seasonal employment, and a job search across a wide geographic area while also meeting family demands.

It is common for natural resource professionals to be in personal relationships with other workers in this field, due to the isolating nature of the profession and their shared values about and appreciation for wildlife. Therefore, partners are faced with the difficult task of finding acceptable employment in the same locale for two people in a very competitive line of work. Dual careers are relevant when discussing greater diversity in the workplace, because a majority of women in the sciences indicate that their spouse's career limits their own employment, often leading to both professional and personal stresses (Primack and O'Leary 1993). When relationships begin during individuals' undergraduate degree programs, accommodating those partners during graduate school is an important but often overlooked part of increasing professional diversity. Unfortunately, the family-friendly institutional structures in place for faculty in academia (e.g., job placement for the spouse of a recent hire, help in finding childcare) are typically unavailable to undergraduate, graduate, or postdoctoral researchers at a university. Without access to these types of assistance, future biologists develop negative perceptions about becoming a faculty member early in their studies and often shift their careers away from an academic or research path.

Professional challenges for underrepresented groups in wildlife fields also are manifested in compensation gaps and fewer advancement opportunities in their jobs. Most research regarding disparities in pay compares men with women. These studies show that the pay gap begins immediately after graduation from college and, throughout a lifetime, can amount to women earning $1.2 million less than men (Women Are Getting Even 2015). Negotiating for fair and equal pay may narrow the compensation gap, but women and minorities working in organizations that do not make salary information public (e.g., private colleges, industry) face a disadvantage. It is difficult to bargain with an employer about your salary without information on what your peers are receiving. Similarly, women in the sciences are often dissatisfied with the smaller number of opportunities for promotion and the lesser recognition they receive for their work. These issues are contributors to a leaky pipeline, where only a small percentage of women who are trained as scientists remain employed in scientific research fields, particularly in academia. These factors also may compound feelings of job dissatisfaction and contribute to people from underrepresented groups leaving the wildlife field. There clearly are many opportunities to improve the practices, pathways, and institutions that nurture and employ natural resource professionals.

DIVERSITY AND INCLUSION
A Path Forward

The wildlife profession has made strides to recruit and retain students from minority and marginalized groups. From 2006 to 2012, the number of females in natural resource undergraduate programs in the United States has increased, from 34.6% to 39.9% (Sharik et al. 2013). Other underrepresented groups, however, continue to lag behind, both within wildlife programs at universities and in professional wildlife organizations (Davis et al. 2002). This lack of diversity is not due to a limited pool of potential and future employees, but rather to a dearth of creativity in identifying and engaging a more broadly representational group of people (Lopez and Brown 2011). Undergraduate students form the core target of recruitment strategies by professional organizations. For example, personnel in the US Fish and Wildlife Service's Southwest Region attend career fairs at colleges and universities that are historically associated with minority groups (Tuggle 2011). Nonetheless, the wildlife profession must diversify its recruitment strategies to engage a potential applicant pool prior to college and the choice of a career path.

Without a culture that fosters retention, recruitment efforts are wasted. Underrepresented groups often encounter intended or unconscious biases from peers, contributing to a movement away from the wildlife profession (Holland et al. 2012). It is important for each of us, regardless of our personal set of identities, to understand the value of a heterogeneous workforce and work toward providing a welcoming, culturally sensitive, respectful work environment. Although the burden of responsibility for ensuring a more diverse workforce is incumbent upon entrenched leadership and more privileged wildlifers, for those from underrepresented groups, Brown (2011) describes steps you can take to expand diversity in the wildlife profession. These include embracing who you are, remaining open to change, listening to others, building relationships, seeking guidance, and expressing thanks to those who help along the way. Prospective students or employees from underrepresented groups should research views about and institutional support of diversity prior to decisions regarding college attendance or job acceptance, in order to understand the culture of the prospective school or organization (table 12.1). All wildlife professionals must become skilled at recognizing and removing unconscious biases. We urge everyone to actively seek out knowledge- and skill-building opportunities, such as Exploring White Identity at Oregon State University or the Welcoming Diversity Workshop (see appendix B). These activities take each person into a discomfort zone, and the participants examine messages about other groups that they are not generally aware of (Harro 2000). Uncovering our implicit biases is difficult and can raise new and powerful emotions, but it is a valuable experience that enables each of us to better understand those around us.

A Student's Guide to Success
As a student, several of these challenges may seem like distant concerns, whereas others may confront you today. Nowadays, many government agencies, nonprofit organizations, and professional societies provide scholarships, employment, and networking opportunities that may interest you (table 12.2). Collectively, these groups assist students in the development of their own professional networks, including access to peers

Serra Hoagland (*right*) and Mike Dockry (*left*) were presenters at a To Bridge a Gap conference, where tribal entities and federal, state, and nonprofit organizations gather annually to discuss tribal relations issues (photo: Serra Hoagland)

Table 12.1. A starter guide for researching the diversity commitment of prospective universities, organizations, and employers

Component	Potential source of information
Diversity and inclusive climate	Prominent diversity and inclusion statement on web page
	Formal diversity networks or centers
	Diversity officer or coordinator on staff
	Diversity reports showing goals, progress, and prejudice-reduction learning opportunities
Partners with dual careers	Dual career statement, office, or officer
	Examples of spousal hires
Work/life balance	Delayed promotion reviews (e.g., tenure) for having/adopting a child or other family event
	Generous maternity/paternity leave
	Childcare/eldercare assistance
	Flexible working hours and work location
	Modified duties for illness/disability/pregnancy
	Lactation rooms
	Family housing
	Dependent and partner health care
Salary and promotion	Transparent system of salaries and promotions available
	Opportunities for lateral or vertical advancement
Mentoring programs	Formal mentoring of new employees
	Informal networking opportunities available to all

Table 12.2. Organizations dedicated to increasing diversity in science and to supporting the development of individuals from underrepresented groups

Organization	Website
American Indian Science and Engineering Society	www.aises.org
Association for Women in Science	www.awis.org
Ecological Society of America SEEDS Program	www.esa.org/seeds/
Ethnic and Gender Diversity Working Group of The Wildlife Society	http://wildlife.org/egdwg/
Minorities in Agriculture, Natural Resources and Related Sciences	www.manrrs.org
Minorities in Natural Resources Conservation	http://minrc.org
Native American Fish & Wildlife Society	www.nafws.org
Native Peoples' Wildlife Management Working Group of The Wildlife Society	http://wildlife.org/npwmwg/
Native Student Professional Development Program	http://wildlife.org/npwmwg/
Society for the Advancement of Chicanos/Hispanics and Native Americans in Science	www.sacnas.org

who encourage and support a diverse profession and cultural understanding. Seek out such organizations and begin building your professional network.

Although balancing educational demands with family and work obligations can feel overwhelming, we suggest researching diversity initiatives at your university (table 12.1), diversity organizations in natural resource fields (table 12.2), other student-body associations for minority groups (e.g., Native American Student Council, African American Student Council), or diversity centers that may be active on your campus. Participation in these groups may provide mentors, a welcoming atmosphere, and a sympathetic ear for students that otherwise feel isolated in their degree program.

First-generation college students from minority or low

socioeconomic backgrounds may encounter discouragement from their families about entering the wildlife profession, a historically low-paying field. These students may bypass an undergraduate wildlife degree and its initial financial hardships in search of equivalent salary for a job at a local business or company. If you are among this group, we hope you will use the careers portion of this book (chapter 5) to show family members that while entry-level positions in wildlife can be low paying, the potential for advancement in this profession, and thus a larger salary, is much greater than perceived.

Another obstacle for some minority students is the difficulty associated with leaving the region of the family unit. This may include either personal hardships or pressure applied from family members. We understand that such strong ties can make it especially uncomfortable to attend school or accept employment far from home. Technological advances, however, such as inexpensive long-distance phone services, video call services (such as Skype), and various other Internet applications can provide easy methods of communication, thus softening your transition away from your family unit.

For wildlife students from nontraditional backgrounds (e.g., urban residents) who lack experience in outdoor activities such as hunting or camping, it may feel comforting to acquire those skills. Organizations such as Conservation Leaders for Tomorrow (see appendix B) offer hands-on learning opportunities about the role of hunting in wildlife conservation and management. Student chapters of TWS and the Ecological Society of America provide ample occasions to expand your wildlife experiences. Through these groups, you can pursue leadership roles, undergraduate research projects, and jobs that will contribute to making you a more marketable wildlife biologist. Moreover, you will be able to display the value and importance of your unique skill sets to your peers.

Outreach efforts should engage diverse groups of elementary school–age children, to demonstrate the needs and benefits of a conservation-based career (Lopez and Brown 2011). We urge university wildlife students to become involved with educational programs that go to K–12 schools in the community and teach youth about the need for diversity within the wildlife profession. One of the most powerful tools is for young students from underrepresented groups to see others like themselves and follow in their footsteps.

If you are a student in the natural resource disciplines, you will soon enter the world of wildlife professionals. Once you have started on this career path, opportunities may open for you to move into supervisory and leadership roles. We ask you to return to this chapter then and remember that for our profession to reap the benefits of workforce diversity, we require leaders who are perceptive and cognizant of the need for multicultural, diverse perspectives and employees (Aghazadeh 2004). All of us must refocus our efforts from simple personal improvement and expand them to institutional transformation. As Bird (2010) explains, "if our only plan for advancing underrepresented groups is to continue enhancing personal skills for navigating the existing system, then the barriers will remain in place." We also need mentors and role models (Maughan et al. 2001), as well as a commitment to long- and short-term strategies that address challenges, such as the low enrollment of underrepresented groups in wildlife programs and the retention of candidates from all ways and walks of life in the workplace (Davis et al. 2002). Through these leadership decisions, we can guide our profession to a promising future.

SUMMARY

Today's society encompasses increasingly diverse backgrounds, values, and beliefs. Unless we develop and retain a professional workforce that understands and relates to this society, both our profession and the wildlife we manage are likely to suffer hardships across a wide range of conservation efforts, because not all stakeholders will feel that their views are shared by decision makers. Our profession has made strides to incorporate underrepresented groups, but we still face a significant diversity deficit, due to challenges like negative bias, microaggressions, socioeconomic barriers, and isolation. Together, these influence the recruitment and retention of such groups in what we do. Our hope is that this chapter will help you navigate these barriers through diversity-directed organizations and increase your marketability. We also challenge you to recognize and overcome your unconscious biases, so you can contribute to a welcoming atmosphere in our profession. With that, we leave you with a quote by Keith Basso (1996) about the Apaches, for whom wisdom "is achieved by relinquishing all thoughts of personal superiority and by eliminating aggressive feelings towards fellow human beings." Simply put, the wisest action our wildlife profession can engage in is an equitable representation across the wide spectrum of human diversity, to ensure that every creature on our planet has an opportunity to flourish.

LITERATURE CITED

Acker, J. 1992. From sex roles to gendered institutions. Contemporary Sociology 21:565–569.

Aghazadeh, S. M. 2004. Managing workforce diversity as an essential resource for improving organizational performance. International Journal of Productivity and Performance Management 53:521–531.

Angus, S. 1995. Women in natural resources: stimulating thinking about motivations and needs. Wildlife Society Bulletin 23:579–582.

Basso, K. 1996. Wisdom sits in places: notes on a western Apache landscape. Pp. 13–52 in S. Feld and K. Basso, eds. In senses of place. Santa Fe School of American Research Press, Santa Fe, NM, USA.

Bird, S. R. 2010. Moving the middle: unsettling systemic barriers to the advancement of women in academic STEM, www.advance.auburn.edu/powerpoints/SharonBird.ppt.

———. 2011. Unsettling universities' incongruous, gendered bureaucratic structures: a case-study approach. Gender, Work & Organization 18:202–230.

Brown, C. H. 2011. Lessons learned from life experience. Wildlife Professional 5:28–29.

Cameron, E. Z., M. E. Gray, and A. M. White. 2013. Is publication rate an equal opportunity metric? Trends in Ecology & Evolution 28:7–8.

Clancy, K. B. H., R. G. Nelson, J. N. Rutherford, and K. Hinde. 2014. Survey of academic field experiences (SAFE): trainees report harassment and assault. PLoS ONE 9:e102172.

Davis, R. D., Sr., S. Diswood, A. Dominguez, R. W. Engel-Wilson, K. Jefferson, A. K. Miles, E. F. Moore, R. Reidinger, S. Ruther, R. Valdez, K. Wilson, and M. A. Zablan. 2002. Increasing diversity in our profession. Wildlife Society Bulletin 30:628–633.

Eagly, A. H., and M. C. Johannesen-Schmidt. 2001. The leadership styles of women and men. Journal of Social Issues 57:781–797.

Etzkowitz, H., C. Kemelgor, M. Neuschatz, B. Uzzi, and J. Alonzo. 1994. The paradox of critical mass for women in science. Science 266:51–54.

Goulden, M., K. Frasch, and M. A. Mason. 2009. Staying competitive: patching America's leaky pipeline in the sciences. Berkely Center on Health, Economic & Family Security, Berkeley Law, University of California, Berkley, CA, USA.

Harris, C. R., M. Jenkins, and D. Glaser. 2006. Gender differences in risk assessment: why do women take fewer risks than men? Judgment and Decision Making 1:48–63.

Harro, B. 2000. The cycle of socialization. Pp. 15–21 *in* M. Adams, W. J. Blumenfeld, C. Castañeda, H. W. Hackman, M. L. Peters, and X. Zúñiga, eds. Readings for diversity and social justic. Routledge, New York, NY, USA.

Holland, J. M., D. A. Major, and K. A. Orvis. 2012. Understanding how peer mentoring and capitalization link STEM students to their majors. Career Development Quarterly 60:343–354.

Lopez, R., and C. H. Brown. 2011. Why diversity matters: broadening our reach will sustain natural resources. Wildlife Professional 5:20–27.

Maughan, O. E., D. L. Bounds, S. M. Morales, and S. V. Villegas. 2001. A successful educational program for minority students in natural resources. Wildlife Society Bulletin 29:917–928.

Moss-Racusin, C. A., J. F. Dovidio, V. L. Brescoll, M. J. Graham, and J. Handelsman. 2012. Science faculty's subtle gender biases favor male students. Proceedings of the National Academy of Sciences 109:16,474–16,479.

Nicholson, K. L., P. R. Krausman, and J. A. Merkle. 2008. Hypatia and the Leopold standard: women in the wildlife profession, 1937–2006. Wildlife Biology in Practice 4:57–72.

O'Brien, K. R., and K. P. Hapgood. 2012. The academic jungle: ecosystem modelling reveals why women are driven out of research. Oikos 121:999–1004.

Primack, R. B., and V. O'Leary. 1993. Cumulative disadvantages in the careers of women ecologists. BioScience 43:158–165.

Sexton, N. R., D. Ross-Winslow, M. Pradines, and A. M. Dietsch. 2015. The urban wildlife conservation program: building a broader conservation community. Cities and the Environment (CATE) 8:3.

Sharik, T. L., R. J. Lilieholm, and W. W. Richardson. 2013. Diversity trends in the U.S. natural resources workforce and undergraduate student population, www.naufrp.org/pdf/NRE%20Diversity%20Conference%202013%20Presentation%20on%20Enrollments%20and%20Workforce—Sharik%20V7.pdf.

The Wildlife Society. 2015. Standing position statement: workforce diversity within the wildlife profession, http://wildlife.org/wp-content/uploads/2015/04/SP_WorkforceDiversity1.pdf.

Tuggle, B. N. 2011. Making workforce diversity work. Wildlife Professional 5:6.

US Census Bureau. 2015. State and county quick facts, http://quickfacts.census.gov/qfd/states/00000.html.

US Department of Agriculture. 2015. Cultural transformation of USDA, www.dm.usda.gov/ct.htm.

US Department of the Interior. 2012. Diversity and inclusion strategic plan, www.doi.gov/pmb/eeo/whoweare/upload/Diversity-and-Inclusion-Strategic-Plan-Department-of-the-Interior-3-16-2012.pdf.

Westermann, O., J. Ashby, and J. Pretty. 2005. Gender and social capital: the importance of gender differences for the maturity and effectiveness of natural resource management groups. World Development 33:1783–1799.

Women Are Getting Even [WAGE]. 2015. What are the costs of the wage gap?, www.wageproject.org/files/costs.php.

APPENDIX A
Common and Scientific Names of Fauna and Flora Mentioned in the Text

Fauna

Mammals

African elephant (*Loxodonta africana*)
American black bear (*Ursus americanus*)
American marten (*Martes americana*)
Arabian oryx (*Oryx leucoryx*)
Armadillo (*Dasypus novemcinctus*)
Axis deer (*Axis axis*)
Beaver (*Castor canadensis*)
Blackbuck antelope (*Antilope cervicapra*)
Black-footed ferret (*Mustela nigripes*)
Bighorn sheep (*Ovis canadensis*)
Bison (*Bison bison*)
Bobcat (*Lynx rufus*)
Cheetah (*Acinonyx jubatus*)
Chipmunk (*Tamias* spp.)
Common mole (*Scalopus aquaticus*)
Coyote (*Canis latrans*)
Deer mouse (*Peromyscus maniculatus*)
Desert bighorn sheep (*Ovis canadensis nelsoni*)
Elk (*Cervus canadensis*)
Fallow deer (*Dama dama*)
Feral hog (*Sus scrofa*)
Fisher (*Pekania pennant*)
Flying squirrel (*Glaucomys* spp.)
Fox squirrel (*Sciurus niger*)
Golden lion tamarin (*Leontopithecus rosalia*)
Gray squirrel (*Sciurus carolinensis*)
Gray wolf (*Canis lupus*)
House mouse (*Mus musculus*)
Indiana bat (*Myotis sodalis*)
Moose (*Alces alces*)

Mountain lion (*Puma concolor*)
Mule deer (*Odocoileus hemionus*)
Opossum (*Didelphis virginiana*)
Pocket gopher (*Thomomys* spp.)
Pronghorn (*Antilocapra americana*)
Rabbit (*Sylvilagus* spp.)
Raccoon (*Procyon lotor*)
Red squirrel (*Sciurus vulgaris*)
Red wolf (*Canis rufus*)
River otter (*Lontra canadensis*)
Sika deer (*Cervus nippon*)
Striped skunk (*Mephitis mephitis*)
Stone's sheep (*Ovis dalli stonei*)
Thirteen-lined ground squirrel (*Ictidomys tridecemlineatus*)
Vole (*Microtus* spp.)
White-tailed deer (*Odocoileus virginianus*)
Wild burro (*Equus africanus*)
Wild horse (*Equus ferus*)
Woodchuck (*Marmota monax*)
Woodrat (*Neotoma* spp.)

Birds

Aplomado falcon (*Falco femoralis*)
Bald eagle (*Haliaeetus leucocephalus*)
Black-capped vireo (*Vireo atricapillus*)
Bobwhite quail (*Colinus virginianus*)
Boreal owls (*Aegolius funereus*)
California condor (*Gymnogyps californianus*)
Canada goose (*Branta canadensis*)
Ferruginous hawk (*Buteo regalis*)
Florida scrub-jay (*Aphelocoma coerulescens*)
Golden-cheeked warbler (*Dendroica chrysoparia*)

Golden eagle (*Aquila chrysaetos*)
Greater sage grouse (*Centrocercus urophasianus*)
Great horned owl (*Bubo virginianus*)
Harpy eagle (*Harpia harpyja*)
House sparrow (*Passer domesticus*)
Marbled murrelet (*Brachyramphus marmoratus*)
Mauritius kestrel (*Falco punctatus*)
Mexican spotted owl (*Strix occidentalis lucida*)
Mourning dove (*Zenaida macroura*)
Northern spotted owl (*Strix occidentalis caurina*)
Orange-breasted falcon (*Falco deiroleucus*)
Peregrine falcon (*Falco peregrinus*)
Pigeon (*Columba livia*)
Piping plover (*Charadrius melodus*)
Prairie falcon (*Falco mexicanus*)
Ridgway's hawk (*Buteo ridgwayi*)
Red-cockaded woodpecker (*Picoides borealis*)
Snow goose (*Chen caerulescens*)
Spectacled eider (*Somateria fischeri*)
Starling (*Sturnus vulgaris*)
Steller's eider (*Polysticta stelleri*)
Swainson's hawk (*Buteo swainsoni*)
Whooping crane (*Grus americana*)

Wild turkey (*Meleagris gallopavo*)
Wood duck (*Aix sponsa*)

Fish and Reptiles
Alligator (*Alligator mississippiensis*)
Blue iguana (*Cyclura lewisi*)
Brook trout (*Salvelinus fontinalis*)
Brown tree snake (*Boiga irregularis*)
Coho salmon (*Oncorhynchus kisutch*)
Desert tortoise (*Gopherus agassizii*)
Rainbow trout (*Oncorhynchus mykiss*)
Texas horned lizard (*Phrynosoma cornutum*)
Wyoming toad (*Bufo baxteri*)

Flora
Cheatgrass (*Bromus tectorum*)
Long-leaf pine (*Pinus palustris*)
Juniper (*Juniperus* spp.)

Insects
Coral Pink Sand Dunes tiger beetle (*Cicindela albissima*)
Karner blue butterfly (*Lycaeides melissa samuelis*)

APPENDIX B
List of Websites Mentioned within Chapters

Chapter 2

The Wildlife Society certification requirements   http://wildlife.org/learn/professional-development-certification/
The Wildlife Society Code of Ethics   http://wildlife.org/governance/code-of-ethics/
The Wildlife Society student chapters   http://wildlife.org/next-generation/student-chapters/

Chapter 4

The Wildlife Society certification program   http://wildlife.org/learn/professional-development-certification/certification-programs/

Chapter 5

Archbold field video   https://www.youtube.com/watch?v=st_0mgipTLo/
Archbold internships   www.archbold-station.org/html/research/internship/internship.html
Association of Fish and Wildlife Agencies   www.fishwildlife.org
Association of Zoos and Aquariums (AZA)   https://www.aza.org
AZA accredited zoos and aquariums   https://www.aza.org/current-accreditation-list/
AZA grants program   https://www.aza.org/conservation-funding-sources/
AZA strategic plan   https://www.aza.org/strategic-plan/
Blue Iguana Recovery Program   www.blueiguana.ky
CDC Public Health Fellowships   www.cdc.gov/fellowships/full-time/index.html
Cheetah Conservation Fund   www.cheetah.org
Continued service agreements   www.opm.gov/WIKI/training/Continued-Service-Agreements.ashx
Ecolog-L   https://listserv.umd.edu/archives/ecolog-l.html
Elephants for Africa   www.elephantsforafrica.org
Great Lakes Piping Plover Recovery Effort   https://glpipl.wordpress.com
Internet Center for Wildlife Damage Management   http://icwdm.org
Journal of Wildlife Rehabilitation   http://theiwrc.org/journal-of-wildlife-rehabilitation/
Marianas Avifauna Conservation Project   www.pacificbirdconservation.org/mariana-conservation-program-mac.html
The Marine Mammal Center   www.marinemammalcenter.org
National Association of Conservation Law Enforcement Chiefs   www.naclec.org
National Wildlife Control Operators Association   http://nwcoa.com
North American Game Warden Museum   www.gamewardenmuseum.org
North American Wildlife Enforcement Officers' Association   www.naweoa.org/j3/
Operation Migration   www.operationmigration.org

Organization of Biological Field Stations   www.obfs.org

Texas A&M University Job Board   http://wfscjobs.tamu.edu/job-board/

USA Jobs   https://www.usajobs.gov

USA Jobs Pathways program   https://www.usajobs.gov/StudentsAndGrads/

US Forest Service   www.fs.fed.us/working-with-us/jobs/

US Office of Personnel Management General Schedule (GS) pay scale   http://gogovernment.org/government_101/pay_and_the_general
    _schedule.php

US Small Business Administration   www.sba.gov

Western Association of Fish and Wildlife Agencies   www.wafwa.org

The Wildlife Society   www.wildlife.org

The Wildlife Society student chapters   http://wildlife.org/next-generation/student-chapters/

Chapter 10

Texas A&M University Job Board   http://wfscjobs.tamu.edu/job-board/

Chapter 11

Ornithological Societies of North America jobs   www.osnabirds.org/Jobs.aspx

Physiological Ecology Section, ESA jobs   www.esa-ecophys.org

Society for Conservation Biology jobs   www.careers.conbio.org

Society for Range Management jobs   http://rangelands.org/jobs/index.htm

Texas A&M University Job Board   http://wfscjobs.tamu.edu/job-board/

The Wildlife Society Career Center   www.careers.wildlife.org

Chapter 12

Conservation Leaders for Tomorrow   www.clft.org

Exploring White Identity at OSU   http://blogs.oregonstate.edu/liberalarts/2013/01/22/exploring-white-identity-at-osu/

Welcoming Diversity Workshop   http://ncbi.org/workshop-training-descriptions/welcoming-diversity-workshop/

Index